2nd International Conference on Machine Vision and Information Technology (CMVIT 2018)

AA006278

Journal of Physics: Conference Series Volume 1004

Hong Kong, China
23 – 25 February 2018

ISBN: 978-1-5108-6237-1
ISSN: 1742-6588

Printed from e-media with permission by:

Curran Associates, Inc.
57 Morehouse Lane
Red Hook, NY 12571

Some format issues inherent in the e-media version may also appear in this print version.

Printed by Curran Associates, Inc. (2018)

For permission requests, please contact the Institute of Physics
at the address below.

Institute of Physics
Dirac House, Temple Back
Bristol BS1 6BE UK

Phone: 44 1 17 929 7481
Fax: 44 1 17 920 0979

techtracking@iop.org

Additional copies of this publication are available from:

Curran Associates, Inc.
57 Morehouse Lane
Red Hook, NY 12571 USA
Phone: 845-758-0400
Fax: 845-758-2633
Email: curran@proceedings.com
Web: www.proceedings.com

Table of contents

Volume 1004

2nd International Conference on Machine Vision and Information Technology (CMVIT 2018)

23–25 February 2018, Hong Kong

Accepted papers received: 4 April 2018
Published online: 25 April 2018

Preface

Papers

IOP Conf. Series: Journal of Physics: Conf. Series **1004** (2018) 011001 doi:10.1088/1742-6596/1004/1/011001

PREFACE

International Conference on Machine Vision and Information Technology (CMVIT) is an annual conference which offers a place for participants to present and to discuss their innovative recent and ongoing research and theoretical work and/or their applications. The first CMVIT was held in Singapore in 2017. This year is the second CMVIT annual conference, which was held in Hong Kong, China, during 23-25 February 2018. CMVIT 2018 is organized by Asia Pacific Institute of Science and Engineering, sponsored by HKSME. The conference provides a useful and wide platform for researchers and engineers to present their latest research findings and exchange their ideas in a broad range of Machine Vision and Information Technology topics. The participants of the conference were from different regions of the world, with background of either academia or industry.

There were 126 submissions including 102 papers and 24 abstracts. After rigorous peer review, 43 papers and 9 abstracts were accepted. This proceeding covers a wide range of topics in machine vision and images, pattern recognition and basic technologies, and information technology and interaction, including object image classification, tracking, object detection, and deep convolutional networks.

We would like to thank all of those who supported CMVIT 2018. Particularly, we would like to thank the organizing committee for their invaluable advices in the organization and the program committee for the peer review of the papers.

We sincerely hope that CMVIT will be an excellent forum for discussions that will lead to fruitful research studies and other new scientific ideas and engineering products that will be beneficial to the society and the industry.

Associate Professor Sai-Keung Wong, National Chiao Tung University, Taiwan

IOP Conf. Series: Journal of Physics: Conf. Series **1004** (2018) 011002 doi:10.1088/1742-6596/1004/1/011002

COMMITTEES

General Chair

Prof. Javad Alirezaie, Ryerson University, Canada

Program Chair

Rhandley Cajote, University of the Philippines Diliman, Philippines

International Technical Committee

Fay Huang, National Ilan University,Taiwan

Liming ZHANG, University of Macau, Macao

Xianghua Xie, Swansea University , UK

Yuan Xiao Chen, Macau University of Science and Technology, Macao

Shih-Hao Chang, Tamkang University, Taiwan

Renoh Johnson Chalakkal, The University of Auckland, New Zealand

Sai-Keung WONG, National Chiao Tung University, Taiwan

Samad Roohi, Tabriz Islamic Art University, Iran

Nabil El Kadhi, University of Buraimi, Oman

Jungong Han, Lancaster University, UK

JIMIN XIAO , Xi'an Jiaotong-Liverpool University, China

Kun Hua, Lawrence Technological University, USA

IOP Conf. Series: Journal of Physics: Conf. Series **1004** (2018) 011003 doi:10.1088/1742-6596/1004/1/011003

Peer review statement

All papers published in this volume of *Journal of Physics: Conference Series* have been peer reviewed through processes administered by the proceedings Editors. Reviews were conducted by expert referees to the professional and scientific standards expected of a proceedings journal published by IOP Publishing.

IOP Conf. Series: Journal of Physics: Conf. Series **1004** (2018) 012001 doi:10.1088/1742-6596/1004/1/012001

Image Augmentation for Object Image Classification Based On Combination of Pre-Trained CNN and SVM

Yoshihiro Shima[1,*]

[1]School of Science and Engineering, Meisei University, Hino, Tokyo, 191-8506, Japan

*shima@ee.meisei-u.ac.jp

Abstract. Neural networks are a powerful means of classifying object images. The proposed image category classification method for object images combines convolutional neural networks (CNNs) and support vector machines (SVMs). A pre-trained CNN, called Alex-Net, is used as a pattern-feature extractor. Alex-Net is pre-trained for the large-scale object-image dataset ImageNet. Instead of training, Alex-Net, pre-trained for ImageNet is used. An SVM is used as trainable classifier. The feature vectors are passed to the SVM from Alex-Net. The STL-10 dataset are used as object images. The number of classes is ten. Training and test samples are clearly split. STL-10 object images are trained by the SVM with data augmentation. We use the pattern transformation method with the cosine function. We also apply some augmentation method such as rotation, skewing and elastic distortion. By using the cosine function, the original patterns were left-justified, right-justified, top-justified, or bottom-justified. Patterns were also center-justified and enlarged. Test error rate is decreased by 0.435 percentage points from 16.055% by augmentation with cosine transformation. Error rates are increased by other augmentation method such as rotation, skewing and elastic distortion, compared without augmentation . Number of augmented data is 30 times that of the original STL-10 5K training samples. Experimental test error rate for the test 8k STL-10 object images was 15.620%, which shows that image augmentation is effective for image category classification.

1. Introduction

A convolutional neural network (CNN) is a powerful machine-learning technique used in the field of deep learning. A simple neural network has been used as a machine-learning technique in the field of character recognition and image classification [1]. However, a huge amount of labeled data are needed for training such a neural network [2]-[6]. An easy way to leverage the power of CNNs, without investing time and effort in training, is to use a pre-trained CNN as a feature extractor. For generating complex decision surfaces, a support vector machine (SVM) is an extremely economical way of representing complex surfaces in high-dimensional spaces, including polynomials and other types of surface [7][8].

The first focus of this paper is a novel method that we propose that combines a CNN and SVM for classifying object images. The second is on image augmentation for expanding data sets. Pattern distortion is useful for improving accuracy. As a pre-trained CNN, Alex-Net [4], trained for large-scale object image datasets, is used for the extractor of character pattern features. An SVM is used for the trainable classifier. The performance of the proposed object image classification is experimentally evaluated by using the STL-10 database without and with distortion. In this paper, a CNN and a SVM are combined for image classification. As a pre-trained CNN, Alex-Net [4] pre-trained for the large-scale object-image dataset ImageNet is used as extractor of features. To avoid the consuming time for training, we use a pre-trained CNN. An SVM is used for trainable classifier. The performance of classification of object images is experimentally evaluated by using the STL-10 object-image dataset [9]. Instead of training, Alex-Net [10][11], pre-trained for ImageNet [4] is used.

In this paper, we make the following contributions. First, we show in experiments that data augmentation method such as cosine translation are useful for improving the accuracy of image category classification based on a pre-trained CNN and SVM. Second, we show in experiments that error rates are increased by other augmentation method such as rotation, skewing and elastic distortion.

2. Related work

2.1. Classification by deep learning

As showcased by the significantly improved state-of-the-art accuracy demonstrated from ILSVRC-2010 to ILSVRC-2014, massive progress has been made in large-scale object recognition [5]. At ImageNet LSVRC-2010, large, deep CNN was trained to classify 1.2-million high-resolution images into 1000 different classes [4]; on the test data, it achieved top-1 and top-5 error rates of respectively 37.5% and 17.0%. In contrast, at ILSVRC-2014, GoogleNet achieved top-5 error rate of 6.7% for image classification. It exploited an improved CNN architecture combining the multi-scale idea and increasing the depth and width of the network [5]. After that, the STL-10 dataset was used as a benchmark for developing unsupervised learning [9].

2.2. Augmentation by pattern distortion

To improve the learning of a machine, new training samples can be created by using prior knowledge on transformation-invariance properties. If the number of training samples is small and if the distribution to be learned has transformation-invariance properties, expanding the training set, that is, generating additional data using pattern distortion, may improve the classification performance. Yann LeCun et al. report that the results of classification on the MNIST database were improved by applying affine transformations [1]. Patrice Y. Simard et al. propose elastic distortion for expanding data sets [12]. Fabien Lauer et al. compare the recognition rate between elastic distortions and affine transformations [3].

3. Image category classification for object images

A scheme for classifying images in categories from object images is proposed as follows. It is based on a combination of a pre-trained CNN and a SVM. The CNN was pre-trained for object images and used as a feature extractor.

3.1. Target images of category classification

The STL-10 dataset are used as object images. The number of classes is ten as shown in figure 1. Training and test samples are clearly split. The STL-10 dataset is a benchmark for unsupervised learning of object images [9]. Images were acquired from labeled examples on ImageNet. The dataset contains 10 classes: "airplane," "bird," "car," "cat," "deer," "dog," "horse," "monkey," "ship," and "truck." Images are 96×96 pixels in color. Number of training images is 500 per class. Number of test images is 800 per class. The standardized testing protocol is recommended for unsupervised feature learning of the unlabeled dataset [9]. The standardized testing protocol was not used, since the pre-trained Alex-Net was used and trained by the labeled object images of ImageNet.

3.2. Recognition architecture combining pre-trained CNN and SVM

As shown in figure 2, the usual method of recognizing individual patterns essentially consists of two modules. The first module is a feature extractor; the second module is a trainable classifier [1][3]. The feature extractor transforms the inputted raw image into feature vectors. The classifier is trained by using a large amount of feature vectors and class categories from a dataset of raw images. To evaluate

IOP Conf. Series: Journal of Physics: Conf. Series **1004** (2018) 012001 doi:10.1088/1742-6596/1004/1/012001

classification accuracy, a feature vector is passed to the classifier, and the class category of the inputted raw image is outputted.

We use the combination of CNN and SVM. Combined structure for image category classification of object images is shown in figure 3. The first module is a pre-trained CNN used as a feature extractor. The second module is a SVM used as a trainable classifier. The CNN is pre-trained for a large-scale object-image dataset. Using features extracted by using a CNN for digit recognition was previously reported [3]. In our study, however, a CNN pre-trained on an object-image dataset is used for object image classification [10][11].

Figure 1. Examples of ten test object images from STL-10 dataset, which contains 10 classes: "airplane(s0)," "bird(s1)," "car(s2)," "cat(s3)," "deer(s4)," "dog(s5)," "horse(s6)," "monkey(s7)," "ship(s8)," and "truck(s9)."

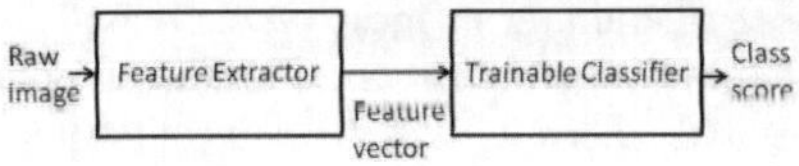

Figure 2. Usual structure for pattern recognition.

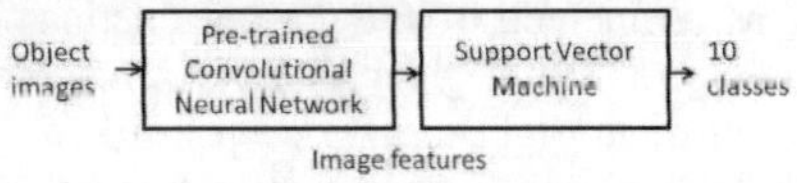

Figure 3. Combined structure for image category classification.

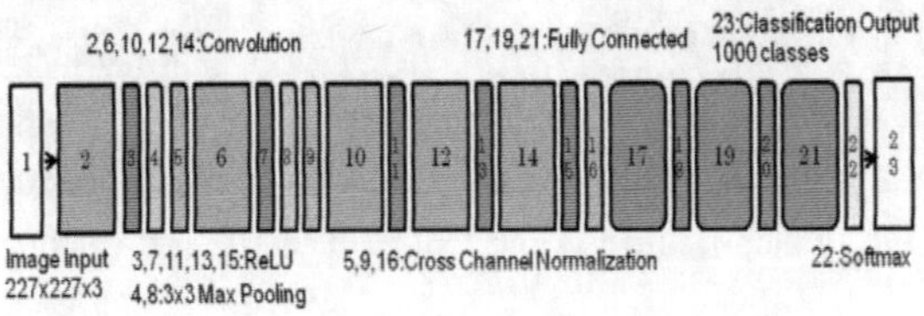

Figure 4. Layer architecture of CNN (Alex-Net).

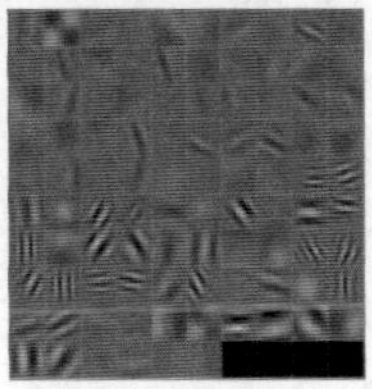

Figure 5. First convolution layer weights (96 $11\times11\times3$ convolutions in Alex-Net).

3.2.1. Pre-trained CNN as feature extractor. As a pre-trained CNN, Alex-Net was downloaded and used for an image-feature extractor [10][11]. Alex-Net has been trained on the ImageNet dataset, which has 1000 object categories and 1.2-million training images [4][5]. The layer architecture of Alex-Net is shown in figure 4.

The first layer defines the dimensions of an input image as 227×227×3. The layer weights were pre-trained for ImageNet. The first layer of the network learns filters for capturing blob and edge features. The intermediate layers are a series of five convolution layers and three fully connected layers, interspersed with rectified linear units (ReLU) and max-pooling layers. The final layer is the classification layer, which has 1000 classes. A pre-trained CNN for an object-image dataset is used as a feature extractor. First-convolution-layer weights are shown in figure 5 [10][11]. Instead of training object images such as those in the STL-10 database, a CNN pre-trained for object images, i.e., Alex-Net, was used as a feature extractor. The 17th layer of Alex-Net, named "fc7," is connected to the SVM classifier. Feature vectors obtained at the 17th layer are passed to the SVM classifier [10][11].

3.2.2. SVM as category classifier. A multiclass SVM classifies data by finding the best hyperplane that separates all data points of one class from those of the other classes. The SVM can represent complex surfaces including polynomials and a radial basis function. The best hyperplane is the one with the largest margin between the two classes. In other words, the margin is the maximal width of the slab parallel to the hyperplane that has no interior data points. The support vectors are the data points that are closest to the separating hyperplane [7][8].

The multiclass SVM is trained using CNN features. A stochastic gradient descent (SGD) solver is used to speed up the training when high-dimensional CNN feature vectors, which each have a length of 4096 layers, are used [10]. To measure the classification accuracy of the trained SVM classifier, test image features are extracted by the CNN and passed to the SVM classifier.

3.3. Augmentation by pattern distortion

As data augmentation methods, affine transformation and elastic distortion are well-known in character recognition [3][12][13]. They are used to generate new samples from original samples and to expand training sets. Simple distortions such as translations, rotations, scaling, and skewing can be generated by applying affine displacement fields to patterns. Elastic distortion is an image transformation for imitating the variations in handwriting styles [12].

We use the pattern transformation method with the cosine function [13]. We also apply some augmentation method such as rotation, skewing and elastic distortion [14]. Figure 6 shows examples of patterns augmented by the cosine function. The number of training images was increased by 31 times as large as the original 5k training samples. By using the cosine function, the original patterns were left-justified, right-justified, top-justified, or bottom-justified. Patterns were also center-justified and enlarged. Figure 7 shows examples of patterns augmented by rotation, skewing and elastic distortion.

(a)original (b)left-justified (c)right-justified (d)top-justified (e)bottom-justified (f)center-enlarged

Figure 6. Training-data augmentation of STL-10 object images. Original images are translated and justified to the left, right, top, or bottom by cosine function. Images are also center-justified and enlarged.

IOP Conf. Series: Journal of Physics: Conf. Series **1004** (2018) 012001 doi:10.1088/1742-6596/1004/1/012001

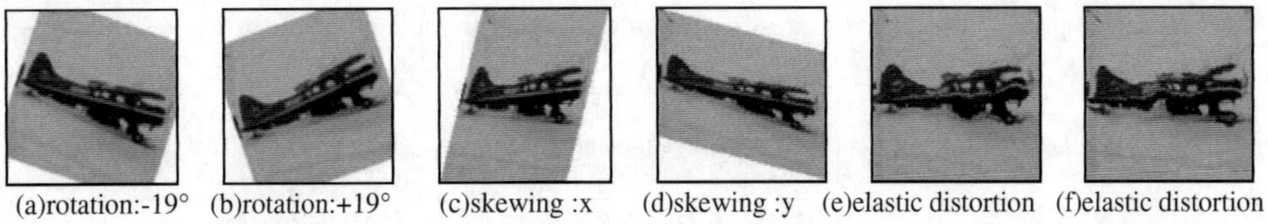

Figure 7. Other training-data augmented STL-10 object images. Original images are rotated, skewed to x- and y-direction, and elastic distorted by random numbers.

4. Experimental results concerning category classification

MATLAB [10][11] was used for experimentally classifying object images. As a benchmark, the STL-10 dataset were used for test and training samples. Error rates for the dataset were compared with and without augmentation.

4.1. Error rate with and without augmentation

As test samples, 8,000 full-test images from test STL-10 dataset were used. First, as training samples for the SVM classifier, 5,000 training images from the training 5k STL-10 dataset were used without augmentation [9]. The influence of the training-set size was measured by picking training images randomly from the training 5k STL-10 dataset. Picking ratio was varied as 0.1, 0.2, 0.5, 0.8, and 1.0. Number of training samples was changed to 500, 2500, 4000, and 5000, respectively. Test error rate of the proposed method is shown in figure 8. As the best score, test error rate is 14.95%. Average error rate of five tests is 16.055%. Table 1 shows the comparison of test error rates for training dataset without augmentation and with cosine transformation for expanding the training set. Test error rate is decreased by 0.435 percentage points from 16.055% to 15.620% by augmentation. Number of augmented data is 30 times that of the original STL-10 5K training samples. Test error rates are compared by augmentation method, which is shown in table 2. Error rates are increased by other augmentation method such as rotation, skewing and elastic distortion.

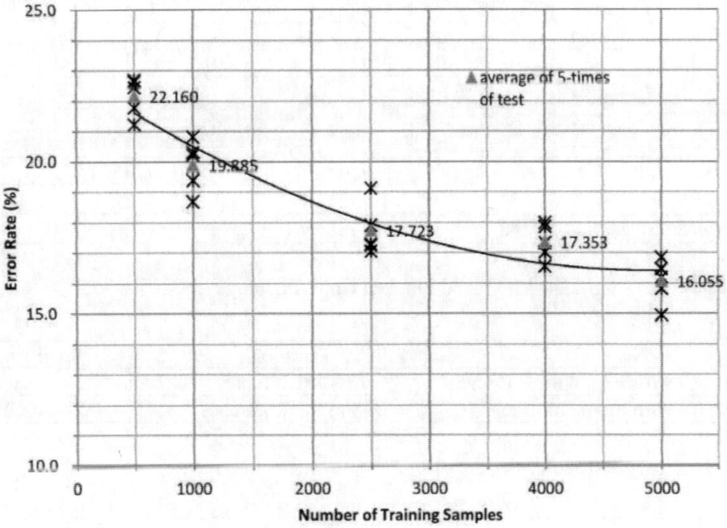

Figure 8. Classification test error rate for STL-10 object images with various sizes of training sets. Best test error rate is 14.95% at average of 16.055%.

IOP Conf. Series: Journal of Physics: Conf. Series **1004** (2018) 012001 doi:10.1088/1742-6596/1004/1/012001

Table 1. Comparison of test error rates for training dataset without augmentation and with cosine transformation for expanding the training set.

training dataset	STL-10 without augmentation; 5000(=10×500) images	STL-10 augmented by cosine function; 155000(=31×10×500) images
test dataset	STL-10 test dataset 8000 (=10×800)	STL-10 test dataset 8000 (=10×800)
average error rate for five tests	16.055%	15.620%
minimum error rate	14.95%	15.50%
maximum error rate	16.85%	15.78%
five test error rates	14.95, 15.81, 16.19, 16.48, 16.85%	15.50, 15.53, 15.60, 15.70, 15.78%

Table 2. Test error rate comparison of augmentation method.

augmentation method	**average error rate for five tests**	**number of training images**
rotation	18.645%	105000(=21×10×500)
skewing	16.625%	205000(=41×10×500)
elastic distortion(1)	17.005%	155000(=31×10×500)
elastic distortion(2)	16.655%	20000(=4×10×500)
cosine function	15.620%	155000(=31×10×500)

Table 3. Confusion matrix for STL-10 database's 8k test set with cosine transformation.

Predicted **True**	**class s0**	**s1**	**s2**	**s3**	**s4**	**s5**	**s6**	**s7**	**s8**	**s9**	**Total error**	**Error rate**
airplane :s0	739	8	3	2	1	2	1	2	23	19	61	7.63%
bird :s1	9	721	2	26	12	8	2	18	1	1	79	9.86
car :s2	9	3	724	6	4	1	0	1	11	41	76	9.50
cat :s3	4	35	0	625	54	33	16	31	1	1	175	21.88
deer :s4	3	20	0	37	707	1	20	10	1	1	93	11.63
dog :s5	4	31	1	83	38	517	66	56	3	1	283	35.38
horse:s6	2	26	5	38	53	24	624	18	2	8	176	22.00
monkey :s7	1	41	1	60	41	15	3	637	1	0	163	20.38
ship :s8	22	5	4	1	0	1	0	0	750	17	50	6.25
truck :s9	13	3	26	3	1	2	5	1	30	716	84	10.50
Total error	67	172	42	256	204	87	113	137	73	89	1240	15.50

Table 4. Measured elapsed time for image classification.

	Training	**Classification**
STL-10 without augmentation	120 s/5k samples	0.285 s/sample
STL-10 with augmentation	3480 s/155k samples	0.285 s/sample

Table 5. Specification of experimental system.

CPU	Intel Core i5-6400® (3.3 GHz/4 cores)
Main memory	16 GB PC3L-12800 (1600 MHz)
Graphic board	NVIDIA® GeForce® GT 730 (4 GB)

IOP Conf. Series: Journal of Physics: Conf. Series **1004** (2018) 012001 doi:10.1088/1742-6596/1004/1/012001

4.2. Misclassified samples and confusion matrix

Examples of misclassified test images are shown in figure 9. Number of test STL-10 dataset images is 8k. Ground truth (left) and incorrect answer (right) are displayed below each image. Some object images are genuinely ambiguous.

An example of a confusion matrix for the STL-10 database's 8k test set is shown in Table 3. It was made for the case where the error rate of 15.50% was achieved. The correct rate of images "airplane (s0)," "bird(s1)," "car(s2)" and "ship(s8)" was higher than that of images "cat(s3)," "dog(s5)," "horse(s6)" and "monkey(s7)."

4.3. Processing time

Elapsed time for the training and classification procedure of MATLAB [10][11] is listed in table 4. Clearly, training time increases with increasing number of training samples. For each test sample, classification time for the mixture dataset is similar to that for the discrete dataset. Specification of the experimental system is listed in table 5.

Figure 9. Examples of misclassified test object images in STL-10 dataset.

4.4. Discussion about classification accuracy

The state of the art of supervised learning on STL-10 dataset is the accuracy rate of 70.1 %, which was obtained by K. Swersky, et al. with 1k training set [15]. The state of the art of unsupervised learning is the accuracy of 73.15 %, which was obtained by S. Yang, et al. [16]. Testing protocol of our work is different from those of state-of-the-art learning methods. So it is not appropriate to compare with our model. Our supervised model archived 84.38 % test-set accuracy, exceeding the previous state-of-the-art. The pre-trained Alex-Net is used with 1000k training samples on ImageNet, by combining the SVM with 5k training set.

5. Conclusion

It was experimentally shown that using a pre-trained CNN as feature extractor for object images is a very promising approach. As a feature extractor, Alex-Net (pre-trained for the large-scale object-image dataset ImageNet) was used. As a trainable classifier, a support vector machine was used in combination with Alex-Net. Without augmentation, average test error rate was 16.055% for the test 8k STL-10 database and the training 5k STL-10 database. With augmentation of the training STL-10 datasets by cosine function, average test error rate is improved to 15.620%.

IOP Conf. Series: Journal of Physics: Conf. Series **1004** (2018) 012001 doi:10.1088/1742-6596/1004/1/012001

References

[1] Lecun Y, Bottou L, Bengio Y and Haffner P, "Gradient-based Learning Applied to Document Recognition," *Proceedings of the IEEE* (Volume: 86, Issue: 11), pp. 2278-2324, Nov. 1998.

[2] LeCun Y, Cortes C and Burges C J C, The MNIST DATABASE of Handwritten Digits, [Online]. Available: http://yann.lecun.com/exdb/mnist/, Feb. 2017.

[3] Lauer F, Suen C Y and Bloch G, "A Trainable Feature Extractor for Handwritten Digit Recognition," *Pattern Recognition*, 40(6), pp. 1816-1824, Elsevier, 2007.

[4] Krizhevsky A, Sutskever I and Hinton G E, "ImageNet Classification with Deep Convolutional Neural Networks," *Neural Information Processing Systems*, 2012.

[5] Russakovsky O, Deng J, Su H, Krause J, Satheesh S, Ma S, Huang Z, Karpathy A, Khosla A, Bernstein M, Berg A C and Fei-Fei L, "ImageNet Large Scale Visual Recognition Challenge," *arXiv pre-print*, arXiv: 1409.0575 v3[cs.CV], Jan. 2015.

[6] Zeiler M D and Fergus R, "Visualizing and Understanding Convolutional Networks," *ECCV 2014, Part 1*, LNCS 8689, pp. 818-833, 2014.

[7] Burges C J C and Schölkopf B, "Improving he Accuracy and Speed of Support Vector Machines," in *Advances in Neural Information Processing Systems* 9, Joprdan M, Mozer M and Petsche T, Eds. 1997, The MIT Press.

[8] Decoste D and Schölkopf B, "Training Invariant Support Vector Machines," *Machine Learning*, 46, pp. 161-190, 2002, Kluwer Academic Publishers.

[9] Coates A, Lee H and Ng A Y, "An Analysis of Single Layer Networks in Unsupervised Feature Learning," AISTATS, 2011, [Online]. Available: https://cs.stanford.edu/~acoates/stl10/, Feb. 2017.

[10] MatConvNet, [Online]. Available: http://www.vlfeat.org/ matconvnet/ pretrained/, accessed Feb. 2017.

[11] MATLAB, Computer Vision System Toolbox, R2016a, MathWorks, [Online]. Available: http://www.mathworks.com/help/vision/examples/image-category-classification-using-deep-learning.html, 2016.

[12] Simard P Y, Steinkraus D and Platt J C, "Best Practices for Convolutional Neural Networks Applied to Visual Document Analysis," *Seventh International Conference on Document Analysis and Recognition*, Aug. 2003.

[13] Guo J, Sun N, Nemoto Y, Kimura M, Echigo H and Sato R, "Recognition of Handwritten Characters Using Pattern Transformation Method with Cosine Function," *Trans. IEICE Japan*, J76-D2(4), pp. 835-842, 1993 (in Japanese).

[14] Shima Y, Nakashima Y and Yasuda M, "Pattern Augmentation for Handwritten Digit Classification based on Combination of Pre-trained CNN and SVM," *6th International Conference on Informatics, Electronics & Vision (ICIEV2017)*, Sept. 2017.

[15] Swersky K, Snoek J and Adams R P, "Multi-Task Bayesian Optimization," *Proceedings of the 26th International Conference on Neural Information Processing Systems (NIPS'13),* Volume 2, pp. 2004-2012, Dec. 2013.

[16] Yang S, Luo P, Loy C C, Shum K W and Tang X, "Deep Representation Learning with Target Coding," *Proceedings of Twenty-Ninth AAAI Conference on Artificial Intelligence (AAAI'15)*, pp. 3848-3854, Jan. 2015.

IOP Conf. Series: Journal of Physics: Conf. Series **1004** (2018) 012002 doi:10.1088/1742-6596/1004/1/012002

Unsupervised Feature Learning for Heart Sounds Classification Using Autoencoder

Wei Hu[1], Jiancheng Lv[1,*], Dongbo Liu[1] and Yao Chen[1]

[1]College of Computer, Sichuan University, Chengdu, 610065, China;

[*]lvjiancheng@scu.edu.cn; Tel.: +86-138-8050-3702

Abstract. Cardiovascular disease seriously threatens the health of many people. It is usually diagnosed during cardiac auscultation, which is a fast and efficient method of cardiovascular disease diagnosis. In recent years, deep learning approach using unsupervised learning has made significant breakthroughs in many fields. However, to our knowledge, deep learning has not yet been used for heart sound classification. In this paper, we first use the average Shannon energy to extract the envelope of the heart sounds, then find the highest point of S1 to extract the cardiac cycle. We convert the time-domain signals of the cardiac cycle into spectrograms and apply principal component analysis whitening to reduce the dimensionality of the spectrogram. Finally, we apply a two-layer autoencoder to extract the features of the spectrogram. The experimental results demonstrate that the features from the autoencoder are suitable for heart sound classification.

1. Introduction

According to the World Health Organization, cardiovascular disease (CVD) is the number one cause of death globally. Cardiac auscultation [1] is the most widely used method of hearing and interpreting heart sound (HS). However, it is usually influenced by a doctor's mood, experience, or other subjective factors. A doctor needs many years of clinical experience to make an accurate diagnosis. For these reasons, it is necessary to have an intelligent, automatic HS auscultation system. Therefore, the accuracy of cardiac auscultation system should be significantly improved.

HS is complex sound caused by the vibration of the heart valve opening and closing, systolic and diastolic tendon and muscle movement, and the impact of the blood flow on the blood wall, while is low frequencies signal. Normal heart rate is 72 beats per minute, so a cardiac cycle is equal to 1/72 minutes, or 833 ms per beat [2]. A cardiac cycle can be divided into four HSs, but usually only the first two can be heard. The first HS (S1) is caused by vibration during the systole, in which the mitral and tricuspid valves shut suddenly. The second HS (S2) is mainly due to valve vibration caused by ventricular diastole, in which the aortic and pulmonary valves close suddenly.

Usually, heart murmur is the first signal of a pathological change in the heart. Heart murmurs can be seen in healthy people, but more in patients with cardiovascular disease. The location of the heart murmur in the cardiac cycle has a special pathological significance for CVD diagnosis. The heart murmur can occur in any location in an HS signal. While it is easy to locate the murmur visually from signal graph, it is extraordinary difficult to locate it in an auscultation system.

To improve the accuracy of HS classification, it is important to find an appropriate data representation (the features) of the HS signal. These features must maintain a certain similarity with the original HS, which can show characteristics of heart sound. A HS has timing, frequency, and morphology characteristics [3], which increases the difficulty of extracting features. So far, many methods for extracting features have been proposed, and achieved good results [4]. F. Yaghouby and

A. Ayatollahi [5] present an effective arrhythmia classification algorithm using the heart rate variability (HRV) signal, basing on the Generalized Discriminant Analysis (GDA) feature reduction technique and the Multilayer Perceptron (MLP) neural network classifier. A radial wavelet neural network classifier is proposed for heart murmurs, which uses the extended Kalman filter as learning algorithm, extracting features from real cardiac cycles [6]. S-transform methods are commonly used to extract HS features. These methods are very suitable for the analysis of an unsteady signal, and their frequency features are superior to those of the Fourier transform method [3]. Four additional HS signal characteristics, namely activity, complexity, mobility, and peaks of power spectral density plots have been used as the input of a neural network [7] for classification. Higher-order cumulants are good at dealing with nonlinear signals [8] and perform HS classification very well. Moreover, [9] proposed an optimal multi-scale decomposition of a wavelet feature extraction method and used an Support Vector Machine classifier to recognize normal and abnormal HSs. Wu et al presents an approach for heart sounds identification, which ensured 95\% of accuracy using wavelet transform to extract the envelope of PCG signals [10]. A new heart sound classification technique was presented in 2014, which improves in performance of the diagnostic system, by combining Linear Predictive Coding coefficients, used for future extraction, with a classifier built upon combining Support Vector Machine and Modified Cuckoo Search algorithm [11].

In recent years, deep learning (deep neural network) has achieved a huge success in image recognition, voice recognition, natural language processing, and other research fields [12]. The artificial neural network [13] is a complex information-processing model simulating the human brain's nervous system, which has the ability to extract features using a nonlinear transformation [14]. Neural networks have been successfully used in many medical applications, such as decision-making, clinical diagnosis, and prediction [15-17]. The autoencoder (AE) [18] is a kind of neural network with a three-layer structure that is based on unsupervised learning using the back propagation (BP) algorithm [19]. It makes the output value equal to the input values, so that it can extract features which is conducive to classification [20]. In this paper, we first find the envelope of the original HSs signals by calculating the average Shannon energy to extract the cardiac cycle. We convert the time-domain signals of cardiac cycles into spectrograms and reduce their dimensionality using principal component analysis [21]. Finally, we apply two-layer AEs to extract the spectrogram features. The experimental results show that the AE features are effective for HS classification.

2. Methods

Figure 1 shows the flow of our algorithm. Because a HS changes cyclically, we apply the average Shannon energy to extract an envelope for dividing the signal into cardiac cycles. We then use the cardiac cycle as data input, and use frequency spectrum analysis to obtain its spectrogram. Because of the high dimensions of the spectrogram, we adopt PCA [21] to reduce its dimensionality. Finally, features are extracted by an unsupervised learning AE neural network. After the AE processing, we use soft max to classify the HS.

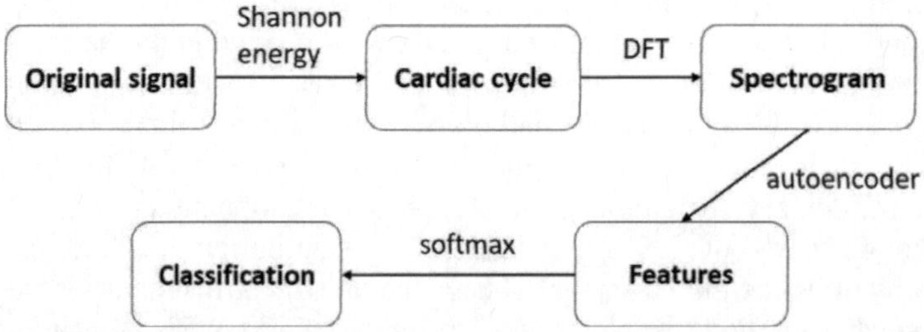

Figure 1. A flowchart of our algorithm. We first cut cardiac cycles from original HS signal. Then we convert each cardiac cycle into spectrogram. Spectrograms are feed to autoencoder neural network. Finally to classify.

2.1. extraction of cardiac cycle

IOP Conf. Series: Journal of Physics: Conf. Series **1004** (2018) 012002 doi:10.1088/1742-6596/1004/1/012002

The cardiac cycle is the basis of HS classification. The traditional method uses the HS essential elements(s1 and s2) for classification, but because a pathological HS contains murmurs, this classification was prone to error. In this paper, to avoid using the essential elements for HS classification, we use a segmentation method based on the extraction of cardiac cycles.

According to the characteristics of the HS signal, we adopt two adjacent peaks of the S1 to define a cardiac cycle, so the main task is to find the peaks. We use the average Shannon energy to extract the HS envelope, then it is simple to find the highest points. The two adjacent peaks of the envelope can be defined the cardiac cycle. Based on the Shannon energy, the normalized average Shannon energy can extract the HS envelope and is able to reduce low-amplitude noise so that the low-amplitude signal is easier to distinguish [22].

First, we use the following equation to normalize the original sound signal.

$$x_{norm}(k) = \frac{x(k)}{\max(|x(k)|)} \tag{1}$$

Shannon energy emphasizes a moderate-amplitude signal, and relative to a high-amplitude signal, the low-amplitude signal is very weak. Consequently, Shannon energy is good at reducing the difference in the envelopes of the high-amplitude and low-amplitude signals. This reduction enables a low-amplitude signal to be more likely to be found.

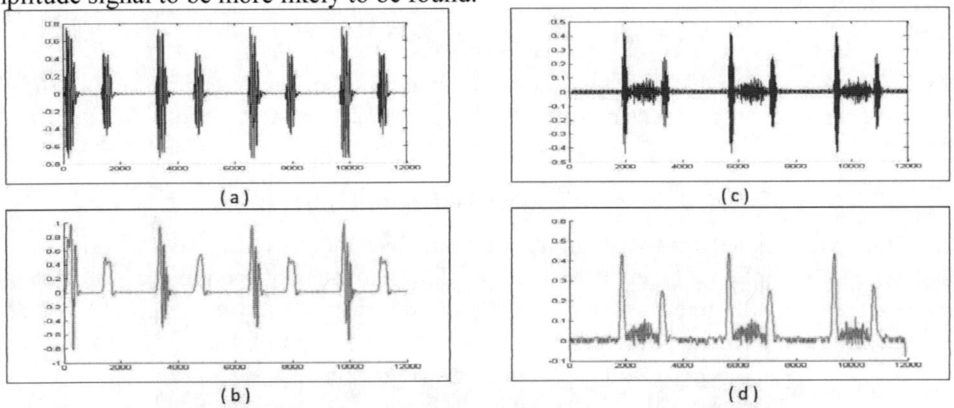

Figure 2. Examples of Heart Sound envelope. (a) and (c) are original Heart Sound signal.(b) and (d) are their envelope calculated by average Shannon energy.

We define the average Shannon energy as:

$$E_s = -\frac{1}{N}\sum_{i=1}^{N} x_{norm}^2(i) \log x_{norm}^2(i) \tag{2}$$

Where x_{norm} is the normalized signal from Equation (1) and N is the length of the signal window. In our experiment, average Shannon energy takes 0.02 s per window to calculate, and 0.01 s to step through the entire signal.

The energy envelope is calculated as follows:

$$P\{t\} = \frac{E_s(t) - M(E_s(t))}{S(E_s(t))} \tag{3}$$

Where $M(E_s(t))$ is the average of $E_s(t)$,and $S(E_s(t))$ is the standard deviation, $P\{t\}$ is the normalized Shannon Energy envelope that we need.

Figure 2 shows the envelopes obtained from our experiment. Figure 2(a) shows a normal HS, including S1 and S2. In Figure 2(b), the normal HS envelope has been extracted using average Shannon energy. Figure 2(c) is a HS signal with murmur, and Figure 2(d) is the envelope of Figure 2(c). This figure shows that the method can accurately extract the envelope of HS signal, which helps us to extract the cardiac cycle.

After obtaining the HS signal envelope, we calculate the highest point of the envelope, that is to say, we aim to obtain the highest points of S1. In our experiment, in order not to separate S1, we translate all the peaks by a distance, then extract the cardiac cycles from the envelope.

2.2. converting into Spectrogram

The original HS signals are obtained by sampling according to time, and every sample point presents the amplitude of the sound. If we directly extract features from the original signal, we can only obtain the time-domain characteristics. However, there are more HS features in the frequency-domain. Therefore, we need to find a data representation that can describe HS both in the time and frequency domains to extract the HS features accurately.

A two-dimensional map is obtained by a continuous spectral analysis of the cardiac cycle. In this map, the x-axis is time, the y-axis is frequency, and the value of each pixel reflects the energy density at the corresponding time and frequency. This map is called a sonogram or spectrogram [23]. This spectrum reflects the relationship of the HS signal over time and frequency, and is important in HS analysis.

HS is a continuous signal, and each single sample point of this signal only represents the amplitude of the HS. We put the cardiac cycle in a frame, and the frame length generally is a power of two, so a cardiac cycle is $x_n(\mathrm{m}), \mathrm{n} = 0,1,\ldots,\mathrm{N}-1$, where m is the frame number, n represents each sampling point in a frame, and N is the length of a frame (number of sampling points in a frame). There is an overlap between frames to help extract detailed frequency information. Discrete Fourier transform(DFT) of signal x(m) is:

$$\mathrm{X(n,k)} = \sum_{n=0}^{N-1} x_n(m) e^{-j\frac{2\pi kn}{N}} \quad (4)$$

Where $0 \leq \mathrm{k} \leq \mathrm{N}-1$, and $|X(n,k)|$ is short time amplitude spectrum estimation of x(n), then we can calculate energy spectrum density function(the power spectrum function) of time m, namely P(n,k):

$$\mathrm{P(n,k)} = |X(n,k)|^2 = \left(X(n,k)\right) \otimes \left(conj\left(X(n,k)\right)\right) \quad (5)$$

Where "$\otimes$" denotes element-wise multiplication, and "conj( )" is the function to compute conjugate value of complex number. P(n,k) is two-dimensional nonnegative real function. We can obtain the spectrogram with time n as the x-axis, k as the y-axis, and the value of P(n,k) as the value of the pixel.

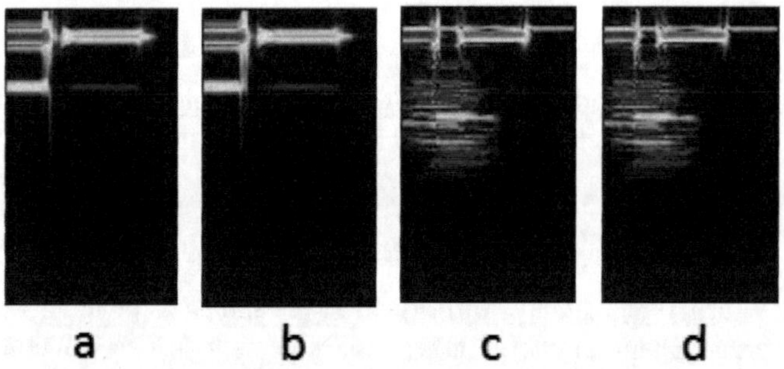

Figure 3. Spectrograms of cardiac cycles: (a) and (b) are spectrograms of 2 cardiac cycles in the same normal HS; (c) and (d) are from HS with murmur in our experiment. The more yellow the color, the greater the level of the frequency, and conversely, the more blue color, the smaller.

Figure 3 shows that spectrograms of different cardiac cycles in a HS are very similar, so spectrum analysis does not change the HS structure. Furthermore, the spectrograms show us that a normal HS is concentrated in the low frequency, the frequency of a murmur is higher than that of a normal HS, and most of the high frequencies are zero, which conforms to HS characteristics.

2.3. extracting feature with AE

The artificial neural network is a layer system that tries to simulate the way the brain thinks, where a node represents a neuron of brain, the connections between nodes correspond to neuron connections, and the value of the connection weight reflects the degree of connection. Unsupervised learning does not need to know the data label, as it directly models the whole data set to learn useful information

[24]. Applying unsupervised learning to HS feature extraction, we may extract features end-to-end without understanding signal processing and the characteristics of the HS signal.

AE is a typical artificial neural network structure for unsupervised learning, as shown in Figure 4. AE has three layers, the input layer, hidden layer, and output layer [18]. After training convergence, we obtain a representation of the input x, namely, the value of the hidden layer. The representation is not controlled manually, but learned from all training data via an neural network, where $w^{(1)}$ is the feature.

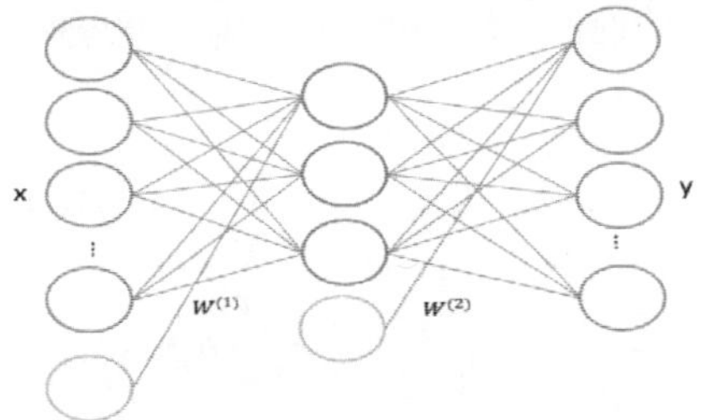

Figure 4. A autoencoder structure: x is input, y is output, and $w^{(1)}$ is weight of the first layer, $w^{(2)}$ is weight of the second layer. Each small circle represents a neuron, and the straight line represents the connection between the neurons. Neurons on the same layer are not connected.

The label of HS is its target output. So we use the BP algorithm to adjust its parameters and obtain the weight of each layer. To do so, we obtain a representation of input x (the value of the hidden layer), and $w^{(1)}$ is defined as the feature. AE is a kind of neural network that tries to reconstruct the original input signal, which is the same as the restricted Boltzmann machine [25]. That is to say, AE tries to learn the function $h_{w,b}(x) \approx x$.

Through forward propagation, we calculate the value of the hidden layer and output:

$$z^{(2)} = x * W^{(1)} + b^{(1)} \tag{6}$$

$$a^{(2)} = f\left(a^{(2)}\right) \tag{7}$$

$$z^{(3)} = a^{(2)} * W^{(2)} + b^{(2)} \tag{8}$$

$$y = f(z^{(3)}) \tag{9}$$

Where, $a^{(2)}$ is the value of hidden layer, $a^{(1)} = x$, $a^{(3)} = y$; $b^{(l)}$ denotes the bias of l layer; $z^{(l)}$ is the input of l layer; $f(x)$ denotes the activation function; y is the output of network. Bias is also the training parameter, which has the effect of regulating the network and enhancing the adaptability of the network. We use sigmoid function as our activation function:

$$f(x) = \frac{1}{1+\exp(-z)} \tag{10}$$

We compute all the activations of the nodes using the equations above that describe the forward propagation step. As stated above, AE tries to learn the function $h_{w,b}(x) \approx x$, that is to say, AE tries to make y equal to x, or reduce the reconstruction error. We define the cost function:

$$\mathrm{J}(x, \mathrm{y}) = \frac{1}{2}\|y - x\|^2 \tag{11}$$

Combining equations (8) and (9):

$$\mathrm{J}(\mathrm{W}, \mathrm{b};\, x) = \frac{1}{2}\left\|f\left(f\left(x * W^{(l)} + b^{(1)}\right) * W^{(2)} + b^{(2)}\right) - x\right\|^2 \tag{12}$$

In order to make J be smallest, we apply gradient descent to update the network parameters W and b:

$$W_{ij}^{(l)} = W_{ij}^{(l)} - \alpha \frac{\partial}{\partial W_{ij}^{(l)}} J(W, b; x) \tag{13}$$

$$b_i^{(l)} = b_i^{(l)} - \alpha \frac{\partial}{\partial b_i^{(l)}} J(W, b; x) \tag{14}$$

Where α is learning rate,which is usually a very small value. Because $\frac{\partial}{\partial W_{ij}^{(l)}} J(W,b;x)$ and $\frac{\partial}{\partial b_i^{(l)}} J(W,b;x)$ are very difficult to compute directly, we apply back propagation(BP) [26] to calculate them. Using the above equations and algorithm, the network can update W and b.

We impose some constraints on AE to help the network learn a better representation, such as the number of nodes in a hidden layer. In addition, we introduce the sparse constraint [27], which is used frequently in AE. Assume that when output of the node is close to 1, the node is thought of as activated, and when the node output is close to 0, we consider the node to be inactivated. The sparse constraint aims to inactivate most of hidden nodes.

Recall $a_j^{(2)}$ denotes the activation of hidden node j, then we compute the average activation of the hidden.

$$\hat{\rho}_j = \frac{1}{m} \sum_{i=1}^{m} a_j^{(2)} \tag{15}$$

Obviously, we would like to make $\hat{\rho}_j$ close to 0, namely $\hat{\rho}_j = \rho$, where ρ is a small value closing to $0(\rho = 0.05)$. To achieve the constraint, we add an penalty factor to J as follow:

$$\mathrm{KL}(\rho || \hat{\rho}_j) = \sum_{j=1}^{m} \left(\rho log \frac{\rho}{\hat{\rho}_j} + (1-\rho) log \frac{1-\rho}{1-\hat{\rho}_j} \right) \tag{16}$$

Where m is the number of hidden nodes. Now our cost function is:

$$J_{sparse}(W,b) = J(W,b) + \beta \sum_{j=1}^{m} KL(\rho || \hat{\rho}_j) \tag{17}$$

Multi-layer artificial neural networks have excellent ability to learn feature extraction, and the features learned have more essential description to the data, which facilitates visualization and classification [28]. The paper also proposes that the training difficulty of multi-layer neural network can be overcome by using layer-wise pre-training. Since AE is a simple network of three layers, we increase the number of layers to extract more descriptive features in our works.

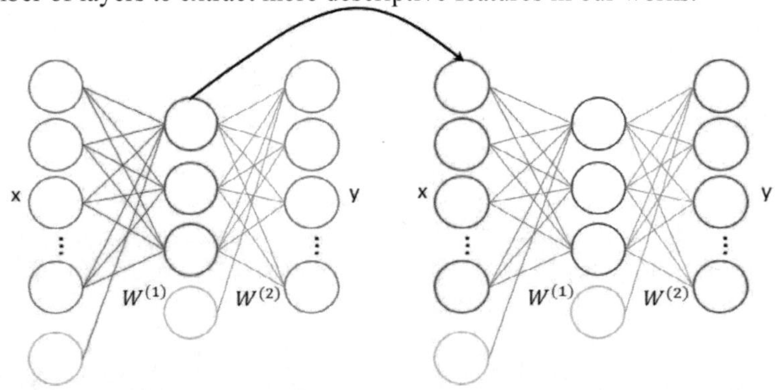

Figure 5. This is a two layer autoencoder neural network. The first AE hidden layer is put as the second AE input layer. Different AEs train their weights respectively.

As shown in Figure 5, we first trained an AE network, and after the network convergence, we used the AE hidden layer as the input layer of another AE. Training step-by-step reduces the training difficulty of a multi-layer network.

Fine-tuning [18] is a common tip in AE training. It can improve the accuracy of classification, and has been widely used in neural network. It can improve the classification accuracy and has been widely used in neural networks. It is likely that we can find a good initial point via fine-tuning, then globally train the neural network.

3. Results

3.1. Experiment preparation

Accuracy is a simple and effective experimental evaluation metric, which is the ratio of the number of classification correct and the number of all data.

As far as we know, there is no publically available standard HS dataset so far. In our experiment, most HS signals were collected by ourselves through Bluetooth stethoscopes. In addition, data from the Internet is also included. We integrated these data and divided them into two categories, namely, normal HS and abnormal HS. This is a simple and effective classification, and using this classification, we can quickly diagnose whether the HS is normal or abnormal. In our dataset, there are positive class 1126 and negative class 1525. Of these, thirty percent serve for test sets, seventy percent for training sets.

Furthermore, according to the location of the murmur in a cardiac cycle, we mark an HS signal into seven classes (normal, early systolic murmur, mid-systolic murmur, late systolic murmur, whole systolic murmur, diastolic murmur, and systolic click). The classification method pinpoints the location of murmur, which is more useful for CVD diagnosis. Of course, it greatly increased the difficulty of classification. Similarly, thirty percent serve for test sets, seventy percent for training sets.

3.2. Experiment results

The network structure we adopt is described above. In the first experiment, The first AE (1-AE) had 196 nodes for the input layer and 100 nodes for the hidden layer nodes. After AE training convergence, we gain the feature(the hidden layer). Then we used softmax for classification, which has fine-tuning structure. Table 1 shows that the accuracy has reached 99.13%.

Further, we then design the second AE with 100 nodes for the input layer and 64 nodes for the hidden layer nodes. Through layer-wise pre-training, we obtained a better representation of the data. In the experiment, the accuracy reached 99.48%, which is superior to only one-layer AE. The experiment results show that the feature of HSs extracted by multilayer network has more essential expression to HS, and is more conducive to classification.

Table 1. Different models classification rate on two-class

method	accuracy in train set(%)	accuracy in test set(%)
1-AE	99.5	99.13
2-AE	100	**99.48**
RBF	68.6	72.63
MLP	99.93	98.83
SVM	100	97.83

Table 2. Different models classification rate on seven-class

method	accuracy in train set(%)	accuracy in test set(%)
1-AE	99.73	97.11
2-AE	99.97	**98.55**
RBF	73.60	81.10
MLP	98.60	95.04

Table 2 also presents some representative experimental results of other methods. This table shows that the radial basis function (RBF) neural network performs very poorly, indicating that the RBF has difficulty learning an effective representation from the lower-dimensional data reduced using PCA. The SVM performs very well on the training set, but the test set results are poorer than those of the neural network method. A support vector machine constructs a hyperplane, or in a high or infinite dimensional space, which can be used for classification and regression. In addition, the performance of MLP is good (98.83%), but the performance of AE is better still. It is proved that the accuracy of Heart Sound classification can be improved by using the feature extracted by AE in the same network structure.

In our second experiment, we recognized the seven-class data. Apparently, The seven classification is more difficult than the two classification. If our method performs well on the seven category, it is sufficient to show that our approach is successful. Again, we adopted a two-layer AE for feature extraction and classification.

Table 2 shows that the accuracy of seven classification is indeed lower than that of the two classification several. And we can see that the feature extraction of AE based on unsupervised learning for HS classification is better than that of other methods(the accuracy is the highest). Above all, the hight accuracy(98.55%) proves that we can find the exact location of the heart murmur by our method.

4. Discussion

Traditional methods first extract HS features, then process and classify them. A neural network provides an end-to-end model that combines feature extraction and classification. Each layer of the model represents a kind of feature. Further, if the layers are deeper, the features are more abstract. We can interpret each layer of the neural network as a data space, and transform the data into a different space. The aim of the neural network is to convert data into a space suitable for classification. Our experimental results also show that the neural network classification of HS has achieved excellent results. Enhancing the accuracy of HS classification is a precondition for establishing an automatic auscultation system. Using the increasingly popular deep neural network method, the accuracy is already close to 100%. Therefore, it is possible that we can accurately locate the location of the heart murmur, which can help the patient find his heart problems at an early date.

To obtain a good classification model, deep neural network requires a large amount of data for training. Now that big data is becoming more available, data are not the problem. A portable automatic auscultation system with high accuracy could soon appear in everyone's home. At that point, patients will be able to listen to their own HS signals anytime and anywhere, and as early as possible be able to protect their own hearts.

5. Conclusions

In this paper, we proposed a set of HS classification methods based on autoencoder. First, cardiac cycles are extracted by extracting the HS envelope to avoid the need to identify the HS composition. A spectrogram that consists of the time-domain and frequency-domain characteristics of cardiac cycles is fed into an AE. We use an unsupervised learning multi-autoencoder to extract high-level features. In the first experiment, the proposed method recognized whether the HS was normal or abnormal quickly and accurately. In the second experiment, the proposed method was able to recognize the location of the murmur, which is important for CVD diagnosis. These experimental results show that the proposed method performs better than other methods in accuracy. So, the intelligent and automatic HS auscultation system is realized in the future, and it could help doctors or patients better identify heart problems.

Acknowledgments

This work was supported by the National Science Foundation of China (Grant No. 61375065 and 61625204), partially supported by the State Key Program of National Science Foundation of China (Grant No. 61432012 and 61432014).

References

[1] BENDER J R. Yale university school of meddicine heart book. New Yourk:Willian Morrow and Company 1992

[2] I. Khurana. Textbook of Medical Physiology. Elsevier, Noida, UP, 2006, pp. 1204-1209.

[3] G.Amit, N.Gavriely, J.Lessicick, N. Intrator. Acoustic Indices of Cardiac Functionality. International Conference on Bio-inspired System and Signal Processing(BIOSINGALS) 2008, vol 2, pp 77-83

[4] P. Sengottuvelan, R. Arulmurugan. Object classification using substance based neural network. Math. Prob. Eng 2014

[5] F. Yaghouby, A. Ayatollahi, R. Soleimani. Classification of Cardiac Abnormalities Using Reduced Features of Heart Rate Variability Signal. World Applied Sciences Journal 6 (11), 2009, pp. 1547-1554

[6] J.E. Guillermo, L.J.R. Castellanos, E.N. Sanchez, A.Y. Alanis. Detection of heart murmurs based on radial wavelet neural network with kalman learning. Neurocomputing 2015, 164, pp. 307-317
[7] Ibrahim Abdel-Motaleb, Rohit Akula. Artificial Intelligence Algorithm for Heart Disease Diagnosis using Phonocardiogram Signals. IEEE transaction on biomedical engineering 2012, Vol. 51, NO. 7, pp. 1196-1202
[8] C. Ahlstrom, P. Hult, P. Rask, J.E. Karlsson et al. Feature extraction for systolic heart murmur classification. Annals of Biomedical Engineering 2006, 34 (November (11)), pp. 1666-1677
[9] Wang Y, Li W.Z, Zhou J.L, Li X.H, Pu Y.F. Identification of the normal and abnormal heart sounds using wavelet-time entropy features based on OMS-WPD. Futur. Gener. Comput. Syst 2014, 37, 488-495
[10] Wu JB, Zhou S, Wu Z, Wu XM. Research on the method of characteristic extraction and classification of phonocardiogram. In: Systems and Informatics (ICSAI), 2012 International Conference on. pp. 1732–1735
[11] G. Redlarski, D. Gradolewski, A. Palkowski, "A system for heart sounds classification", PloS one 2014, pp. e112673
[12] LeCun Y, Bengio Y, Hinton G. Deep learning. Nature 2015, 521, 436-444
[13] J. C. Lv, K. K. Tan, Z. Yi, S. Huang. A family of fuzzy learning algorithms for robust principal component analysis neural networks. IEEE Trans. Fuzzy Syst. 2010, vol. 18, no. 1, pp. 217-226
[14] X. Yang, J. Zheng. Artificial neural networks. Handbook of Research on Geoinformatics 2009, p. 122
[15] Alkan. A, Koklukaya. E, Subasi. A. Automatic seizure detection in EGG using logistic regression and artificial neural network. Neurosci
[16] Sawa. T, Ohno-Machado. L. A neural network-based similarity index for clustering DNA microarray data. Comput. Biol. Med 2003. 3, 1-15
[17] Tan. A-H, Pan. H. Predictive neural network for gene expression data analysis. Neural Networks 2005, 18, 297-306
[18] P. Vincent, H. Larochelle, Y. Bengio, P.-A. Manzagol. Extracting and composing robust features with denoising autoencoders. Proceedings of the Twenty-fifth International Conference on Machine Learning (ICML'08) 2008, (W. W. Cohen, A. McCallum, and S. T. Roweis, eds.), pp. 1096-1103, ACM
[19] LeCun. Y, Boser. B, Denker. J.S et al. Handwritten digit recognition with a back-propagation network. Advances in Neural Information Processing Systems 1990, 2, 396-404, Morgan Kaufman
[20] A. Ng. Sparse autoencoder. CS294A Lecture notes 2011, Stanford University, pp. 72.
[21] Jian Cheng Lv, Zhang Yi, Y. Li. Non-divergence of stochastic discrete time algorithms for PCA Neural Network. IEEE Transactions on Neural Networks and Learning Systems 2015, vol. 26, no. 2, pp. 394-399
[22] H Liang, S Lukkarinen, I. Hartimo. Heart Sound Segmentation Algorithm Based on Heart Sound Envelogram. Computers in Cardiology 1997 24: 104
[23] Greenberg. S., Kingsbury. B. The modulation spectrogram: In pursuit of an invariant representation of speech. Proc. ICASSP 1997, pp.1647-1650
[24] B. Fritzke. Growing cell structures -a selforganizing network for unsupervised and supervised learning. Neural Networks 1994, 7(9):1441-1460
[25] G. E. Hinton. A practical guide to training restricted Boltzmann machines. Tech. Rep. 2010. UTML TR 2010-003
[26] Rumelhart D E and McClelland J L. Parallel Distributed Processing: Explorations in the Microstructure of Cognition. Cambridge, MA: MIT 1986,1
[27] A. Ng. Sparse autoencoder. CS294A Lecture notes 2011, Stanford University, pp. 72
[28] G. E. Hinton, R. Salakhutdinov. Reducing the dimensionality of data with neural networks. Science 2006, vol. 313, no. 5786, pp. 504-507

IOP Conf. Series: Journal of Physics: Conf. Series **1004** (2018) 012003 doi:10.1088/1742-6596/1004/1/012003

A Real-Time Method to Estimate Speed of Object Based on Object Detection and Optical Flow Calculation

Kaizhan Liu[1], Yunming Ye[1,*], Xutao Li[1] and Yan Li[2]

[1]Department of Computer Science, Shenzhen Graduate School, Harbin Institute of Technology, Shenzhen, China
[2]Shenzhen Polytechnic, Shenzhen, China

*yeyunming@hit.edu.cn

Abstract. In recent years Convolutional Neural Network (CNN) has been widely used in computer vision field and makes great progress in lots of contents like object detection and classification. Even so, combining Convolutional Neural Network, which means making multiple CNN frameworks working synchronously and sharing their output information, could figure out useful message that each of them cannot provide singly. Here we introduce a method to real-time estimate speed of object by combining two CNN: YOLOv2 and FlowNet. In every frame, YOLOv2 provides object size; object location and object type while FlowNet providing the optical flow of whole image. On one hand, object size and object location help to select out the object part of optical flow image thus calculating out the average optical flow of every object. On the other hand, object type and object size help to figure out the relationship between optical flow and true speed by means of optics theory and priori knowledge. Therefore, with these two key information, speed of object can be estimated. This method manages to estimate multiple objects at real-time speed by only using a normal camera even in moving status, whose error is acceptable in most application fields like manless driving or robot vision.

1. Introduction

Speed, as a physical quantity to describe one object's motion referring to another one, is objective and quantizable. We human cannot directly measure the value of speed by our sensors, so normally we can just estimate it through other indirect methods. Speed estimation plays an important role in many application fields, such as vehicle speed detection in transportation industry, object motion analysis in robot field, pedestrian behavior analysis in manless driving field. To practically apply speed estimation method to above application fields, three basic conditions are needed to be reached: 1) the running speed of method should be or be closed to real-time, 2) the ability to simultaneously measure multiple objects' speed, 3) keeping valid even the observer itself is moving.

On the other hand, thanks to the flourish of deep learning on computer vision field, some traditional difficult tasks are provided with new solutions. In object detection field, a series of CNN-based algorithm such as RCNN [1], fast RCNN [2], SSD [3], YOLO [4] were proposed, where SSD and YOLO is able to reach real-time speed. These algorithms are able to output bounding boxes of multiple objects from each frame image. In optical flow calculation field, Philipp Fischer *et al.* [5] designed out a CNN to calculate optical flow, which shows approximate performance compared to the traditional algorithms and also reaches the real-time speed.

Though above CNN-based algorithms are in specific niche which has indirect relation with speed estimation, neither of them is able to achieve speed estimation singly. With the idea of sharing output information of multiple networks so as to figure out useful message that each of them cannot provide singly, this paper proposes a general speed estimation method based on combination of object detection network and optical flow calculation network (ODFSE). It uses the bounding box from object detection network to select out optical flow of object from the whole flow image provided by FlowNet, and uses the label and image size of object, by means of optics theory, priori knowledge and camera calibration. Shown by experiment result, this method is able to estimate speed of different type of object in real-time speed even the camera itself is in moving status.

2. Related Work

Normally we estimate the speed of object by two main types of method: physical method (e.g. Radar and Laser) or image processing method [6]. These methods have respectively advantages but also have respectively disadvantages.

About the physical method, the principle is: an ejector of radar or laser ejects the wave or light signal and a receiver receives the returned signal blocked by object. Since the speed of object will change the signal's frequency, by comparing the frequency of received signal with ejecting signal we can calculate out the speed. This method has been widely used in market. Company like VST, Bushnell, Lai Lai and so on develop a number of embedded devices of vehicle speed estimation, running at real-time speed, 1 km/h error and measuring range is 18-300km/m. Apart from the benefits, this kind of method shows some drawbacks: 1) can only measure speed of vehicle and ball, 2) can only measure one object at the same time, 3) measuring distance is limited, 4) device itself must be still.

About the image processing method, there are multiple solutions. Yu *et al.* [7] uses camera calibration method to get the relationship between camera and real world, then using image matching method to calculate the change of location of vehicle to estimate the speed of vehicle. Wang [8] used priori knowledge of vehicle's size to achieve object tracking, then use frame difference to calculate out the speed of vehicle. Yan *et al.* [9] used the information of fixed point in camera to calculate the speed of camera itself. A.L. Harvey *et al.* [10] considers the correlation of two neighbouring frames to estimate the distance vehicle moves each frame thus transforms it to its speed. Similarly, S. Pumrin *et al.* [11] uses frame difference method and camera calibration method to achieve speed estimation. The methods mentioned above are all used to measure the speed of vehicle. As for other objects, Huei-Yung Lin and Chia-Hong Chang [12] found that the motion blur's parameter has relationship with speed of spherical object and use it to calculate out the speed of it, this method is only suitable for spherical object. Above methods are well-behaved in some specific scenes but have a number of drawbacks: 1) most of them can only be used to measure speed of vehicle, some might be used to spherical object, 2) the precision and robustness are limited by the precision and robustness of image processing algorithm itself 3) most of these methods require camera to be still because most of them use frame difference method to detect object, which needs the background to be unchanged.

3. Speed estimation method ODFSE

The process of ODFSE is: Firstly, using the bounding boxes from object detection network to select out the object parts of the whole optical flow image from optical flow calculation networks. Secondly, figuring out the transformation relationship between camera and real world thanks to the information of object type and object width and height in pixels provided by object detection network, where the former one helps to get the real size of object with prior knowledge and the latter one helps to get the image size of object. Lastly, transform the optical flow to real speed with above information together with optics theory and camera calibration. The framework of the method is shown in Figure 1.

IOP Conf. Series: Journal of Physics: Conf. Series **1004** (2018) 012003 doi:10.1088/1742-6596/1004/1/012003

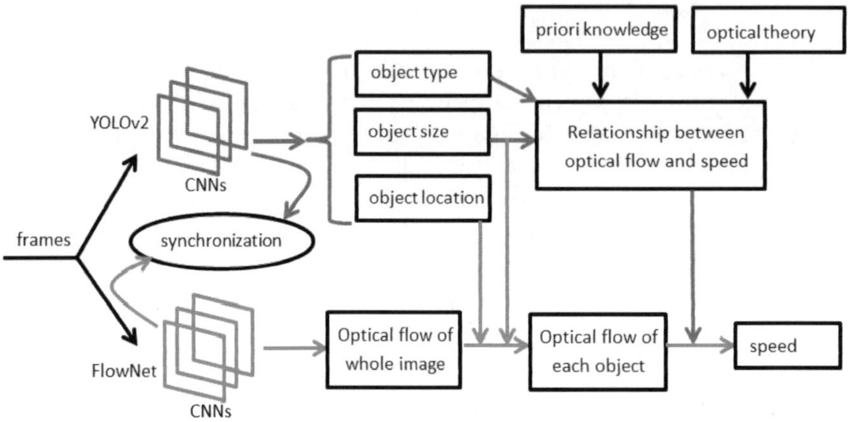

Figure 1. The framework of ODFSE.

3.1. Object detection

To achieve object detection in real-time, we choose YOLOv2 [13] as the basic network. The reasons are: 1) the same as YOLOv1 [4], YOLOv2 is an end-to-end CNN which takes object detection problem as a regression problem, makes it possible to run in real-time speed, 2) YOLOv2 use 98 anchor boxes with pre-selected size to represent object, whose width and height are approximately equal to the width and height of object, 3) YOLOv2's detecting object relies on the whole image information instead of partial information where CNN like SSD and Faster-RCNN rely, so it might perform better in the situation when objects are moving, which is the case in our work.

We at the beginning used the original network trained with COCO dataset which contains 80 types of object to detect a horizontal moving smart car in camera. Basically network is able to catch the smart car in still or slow-moving state. But as the speed of smart car increased, network began to lose the detection because the image of object became blurring and network did not get used to its features. To solve this problem, we make a tiny dataset where objects move in serious of level of speed to retrain the network. The dataset is divided in 10 groups, whose speed ranges from 50 cm/s to 150 cm/s with a 10 cm/s step. Each of them contains 5 videos representing 5 different types of object (e.g. bottle, motor car and so on). The learning rate is set to 0.001, epoch is 50. A compare of retrained network with original network is shown in figure 2, where (a) is the output of original network and (b) is the output of retrained network. We can see the retrained one is more capable to catch object with high speed than the original one.

(a) (b)

Figure 2(a). (a) Output of original network. It is not able to catch object in fast speed. (b) Output of retrained network. It is able to catch object in fast speed.

From the network we can get three useful information: object type, object size and object location, where object size and location are contained in object bounding box. The object type is used to get the prior knowledge of object real size, that is: if the object type is "Person", we then consider the object size is the size of an ordinary grown-up man with height of 170cm and weight of 30cm. About the object size we use the ratio of pixels of object and pixels of whole image to represent. The object location does not take part in calculation of speed but it helps the FlowNet select out the object part, then we are able to get the optical flow of each object.

3.2. Optical flow calculation

FlowNet [5] is a network to calculate optical flow. It is an end-to-end CNN which takes the front and back frame as input then output optical flow value of whole image. The principle of the network is using convolutional layer to extract features of the difference between the front one and back one, then using upconvolutional layer to put the features back to image size, thus the optical flow of whole image is calculated out. As the information is for the whole image, it actually cannot give out the respective optical flow value of each object. To achieve this, as described in 3.1, the object bounding box helps to select out object part as shown in figure 2(a), and the selected part is shown in figure 3(b).

(a) (b)

Figure 3. (a) Using bound box to select out the object part of optical flow. (b) The image contains both optical flow of object and background.

To calculate the optical flow value of object in figure 2(b), we need to classify the object with background. Here we use k-means algorithm to do so. As most pixels of the bounding box belong to object, the cluster containing the most elements should belong to the object, and the rest should belong to background. By taking the average value of both of them we manage to get optical flow of object and background. By the way, the ability to calculate out optical flow of background enables the method working even in a moving status.

3.3. Calculation of true speed

To calculate the real speed of object from above data, we need to find out the correspondence between camera optics parameter and real world.

The model of this correspondence is: assuming one point m_0 owns real speed v_0 (to testfiy in convinience here we assume the camera is still, and the speed direction is parallel to camera plane), its projection on image m_0' then owns speed v_0'. In unit time dt, the journey of them are $v_0 dt$ and $v_0' dt$. Following the theory of camera projection, there should be:

$$\frac{v_0 * dt}{z} = \frac{v_0' * dt}{f} \quad (1)$$

Where f stands for focal length of camera, z stands for distance between camera and object. Besides, the relationship of image speed v_0' and optical flow speed v_{opt} should be:

$$\mathrm{v_{opt}} = \mathrm{v_0'} * \mathrm{k_0} \quad (2)$$

IOP Conf. Series: Journal of Physics: Conf. Series **1004** (2018) 012003 doi:10.1088/1742-6596/1004/1/012003

Where $k_0$stands for a constant value.
From (1), (2) we got:

$$v_0 : v_{opt} = z : (f * k0) = z : k \quad (3)$$

Where $k = f * k_0$ is specific value as f is specific with specific camera.
Thus we find out the correspondence of v_0 and v_{opt}. Figure 4 shows the vector diagram.

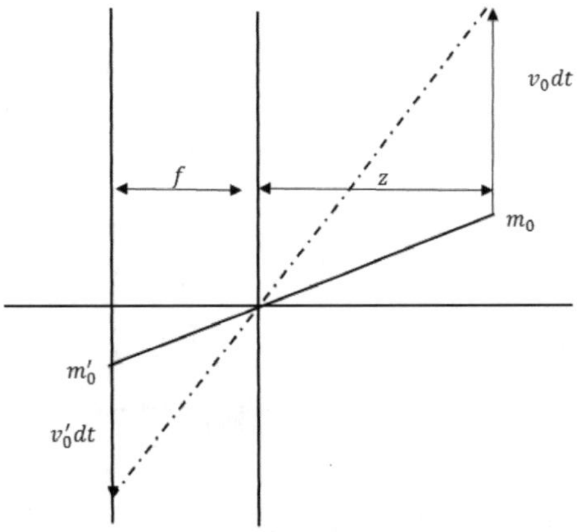

Figure 4. The vector diagram of camera focusing principle.

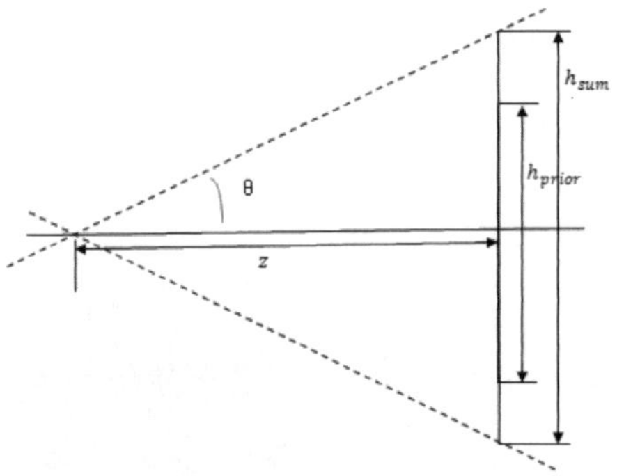

Figure 5. The vector diagram to show the relationship among z, θ, h_{sum} and h_{prior}.

The remaining problem is to find the specific value k and the formulation to get value z. As a specific value, we use calibaration to get value k. The process is: make an object moving in real speed v_1 in distance of $z_1$with camera, get the optical flow speed v_{opt1} from FlowNet, then we get:

$$k = v_{opt1} * z_1 * {v_1}^{-1} \quad (4)$$

About the value z, it is a physical quantity which can be measured by manpower or depth sensor, but as mentioned in chapter 3.1, we can use prior knowledge of object size to figure out the value even without above conditions. Figure 5 shows the principle.

h_{prior} is the prior height of object mentioned in chapter 3.1, h_{sum} is the height that camera can at most shoot in distance of z, θ is angle of view. So we can get the relationship:

$$z = \frac{h_{sum}}{\tan\theta} = \frac{h_{prior}}{r_{yolo} * \tan\theta} \quad (5)$$

Where $r_{yolo}$is the ratio of $h_{prior}$and h_{sum} that YOLO can offer.

$\tan\theta$ is the specific value of camera, which can also be accessed by calibration, the process is: take an object with known height h_2 (e.g. a ruler), adjust its distance with camera until the object can filling the screen of camera, record the distance z_2, then:

$$\tan\theta = h_2 * {z_2}^{-1} \quad (6)$$

From (4), (5) and (6):

$$v_0 = \frac{h_{prior} * z_2}{r_{yolo} * h_2} * \frac{v_1}{v_{opt1} * z_1} * v_{opt} \quad (7)$$

Since z_2, h_2, v_1, v_{opt1}, z_1 are known parameters, h_{prior} is the prior value, r_{yolo} are output of YOLO network, v_{opt} is output of FlowNet, we can calculate out the true speed v_0. Note that v_0 and v_{opt} are vector type that is we can not only calculate out the value of speed but also the direction of it.

Besides, for convinence we assume the camera is still, but as described in chapter 3.2, the optical flow can still be calculate correctly even the camera is moving.

4. Experiment

4.1. Measurement Platform

Since there is no public dataset about speed, we need to make a dataset with accurate and actual ground truth of speed. Here we set up a smart car which can load some tiny objects and move together with a constant speed. Also, two hall elements are installed beside its tyres so the actual speed of car and object can be captured in real-time. Thus, we are able to simulate the moving scene of an object and do the experiment to evaluate our method.

4.2. Serious of experiments

To prove the accuracy, robustness and stability of this method, we set up 3 experiments referring to different topics: 1) the accuracy and stability of our method; 2) speed value influence on method's effectiveness; 3) camera movement's influence on method's effectiveness. Here are some relative variables: 1) estimated object value v_o; 2) ground truth of object speed value v_g; 3) distance between object and camera z; 4) camera speed value v_c; 5) object type class T_o.

4.2.1. Accuracy and stability of estimated speed. Here we test the accuracy and stability of our estimated method and calculate the error and variance with ground truth of speed. We respectively use a bottle and a telephone as experiment object, and also set the camera speed v_c to 0 m/s, ground truth speed v_g to *approximate* 0.3cm/s, table 1 shows the experiment result in 10 groups, where the first 5 ones belongs to object 'Bottle' and the last 5 ones belong to object 'Telephone'.

Table 1. Experiment on accuracy of estimated speed.

	T_o	$v_c(\mathrm{ms}^{-1})$	$v_g(\mathrm{ms}^{-1})$	$v_0(\mathrm{ms}^{-1})$
1	Bottle	0.000	0.305	0.295
2	Bottle	0.000	0.278	0.256
3	Bottle	0.000	0.310	0.278
4	Bottle	0.000	0.331	0.315
5	Bottle	0.000	0.341	0.385
6	Telephone	0.000	0.357	0.328
7	Telephone	0000	0.383	0.351
8	Telephone	0.000	0.362	0.308
9	Telephone	0.000	0.352	0.332
10	Telephone	0.000	0.337	0.317

The average error between v_0 and v_g is: 8% and the variance is 0.115. The error is mostly from: 1) the quarter view of camera causes the precision loss in distant view, 2) error of bounding box, 3) error of measurement quantities. And the variance is mostly from: 1) the float up or down of size of bounding box when object is moving, 2) the error between neighbour frames caused by the sample interval of camera.

4.2.2. Speed of object influence on method's effectiveness Here we test if the method's effectiveness will be disturbed by the value of speed of object or not. We take a bottle as experiment object, and set the camera to be still, and then set the ground truth v_g ranges from 0.0m/s to 2.000m/s, table 2 shows the experiment result.

Table 2. Experiment on speed's influence on method's effectiveness.

	T_o	$v_c(\text{ms}^{-1})$	$v_g(\text{ms}^{-1})$	$v_0(\text{ms}^{-1})$
1	Bottle	0.000	0.000	0.000
2	Bottle	0.000	0.115	0.102
3	Bottle	0.000	0.363	0.382
4	Bottle	0.000	0.583	0.630
5	Bottle	0.000	0.703	0.615
6	Bottle	0.000	0.897	0.831
7	Bottle	0.000	1.512	-
8	Bottle	0.000	2.103	-

The estimated speed of first 6 groups basically match the ground truth, this shows the effectiveness of method is kept in slow moving status. But the last 2 groups cannot get the estimated speed, this is because the blur of image caused by high moving speed disable theYOLOv2 to catch object. Figure 6 shows the rendering of experiment where a smart car is moving in a slash.

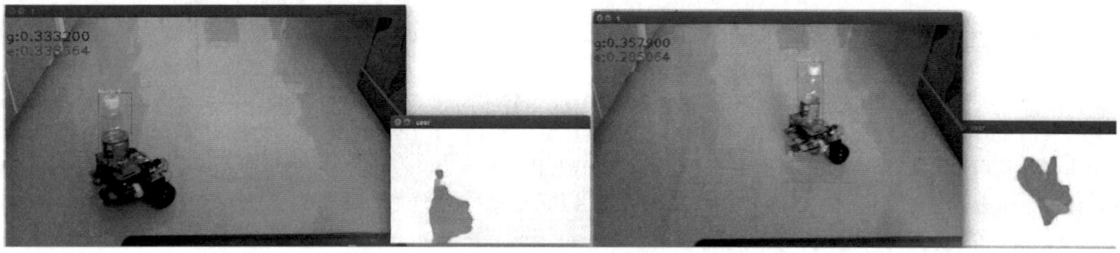

Figure 6. Rendering of experiment

4.2.3. Camera movement's influence on method's effectiveness. Here we test if the method's effectiveness will be disturbed by the movement of camera or not. When the smart car is moving, we also move the camera with a horizontal direction, and we compare the estimated speed and ground truth speed that labeled in image of figure 6.

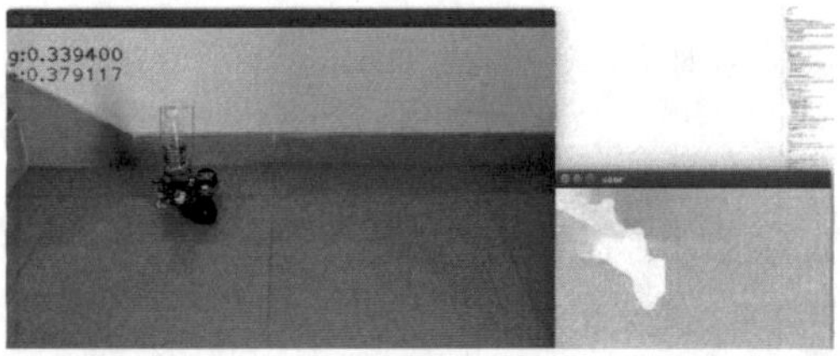

(a) State1, smart car is moving, camera does not move yet.

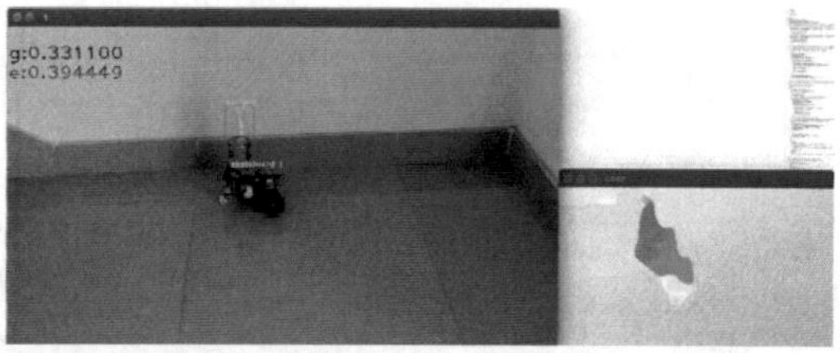

(b) State2, both smart car and camera are moving.

(c) State 3, camera is still moving, smart car stops.

Figure 7. Three states of experiment with camera moving.

From figure 7 we know that the estimated speed of all three states basically match the ground truth speed, which illustrates the method is still valid when camera is also moving.

5. Conclusion

Based on above analysis and experiments, the method mentioned in this paper is able to estimate the true speed of different types of object with a normal camera or video at a real-time speed, even when the observer itself is moving. This method can be used in fields like road accident analysis, players' motion analysis in count and so on.

Acknowledgments

The research was supported in part by Shenzhen Science and Technology Program under Grant Nos. JCYJ20160330163900579 and JCYJ20160413163534712.

References

[1] Girshick, Ross, et al. Rich feature hierarchies for accurate object detection and semantic segmentation[C]// IEEE conference on computer vision and pattern recognition, Columbus, OH, USA, 2014: 580-587.

[2] Girshick R. Fast r-cnn[C]// IEEE international conference on computer vision, Santiago, Chile, 2015: 1440-1448.

[3] Liu W, Anguelov D, Erhan D, et al. Ssd: Single shot multibox detector[C]// European conference on computer vision. Springer, Cham, 2016: 21-37.

[4] Redmon J, Divvala S, Girshick R, et al. You only look once: Unified, real-time object detection[C]// IEEE Conference on Computer Vision and Pattern Recognition, Las Vegas, NV, USA, 2016: 779-788.

[5] Dosovitskiy A, Fischery P, Ilg E, et al. FlowNet: Learning Optical Flow with Convolutional Networks[J]. 2015:2758-2766.

[6] Lin H Y and Chang C H. Automatic speed measurements of spherical objects using an off-the-shelf digital camera[C]// IEEE International Conference on Mechatronics, Beijing, China, 2005:66-71.

[7] Yu Y Y, Li X Y and Zhao Y T. A method to calculate the macroscopic optical flow speed of vehicle[C]// IGTA, Beijing, China, 2009:20-23.

[8] Wang J. A method to recognize vehicle and to measure its speed based on video[D]. Northern nationalities university, 2009.

[9] Yan L P, Sun C X, Wang X L, et al. A method to real-time measure speed based on vehicle-mounted camera. CN, CN103686083A[P]. 2014.

[10] Harvey L and Cohen A. Vehicle speed measurement using an imaging method[C]// International Conference on Industrial Electronics, Control and Instrumentation, Kobe, Japan, 1991: 1730-1733.

[11] Pumrin S and Dailey D J. Roadside camera motion detection for automated speed measurement[C]// IEEE Conference on Intelligent Transportation Systems, Singapore, 2002: 147-151.

[12] Lin H Y and Chang C H. Automatic speed measurements of spherical objects using an off-the-shelf digital camera[C]// IEEE International Conference on Mechatronics, IEEE, 2005:66-71.

[13] Redmon J, Farhadi A. YOLO9000: better, faster, stronger[M]. ArXiv preprint ArXiv:2016.612.08242.

IOP Conf. Series: Journal of Physics: Conf. Series **1004** (2018) 012004 doi:10.1088/1742-6596/1004/1/012004

Component Pin Recognition Using Algorithms Based on Machine Learning

Yang Xiao[1], Hong Hu[1,*], Ze Liu[1] and Jiangchang Xu[1]

[1]Harbin Institute of Technology Shenzhen Graduate School
Shenzhen, Guangdong 518055, China

*Email: honghu@hit.edu.cn

Abstract. The purpose of machine vision for a plug-in machine is to improve the machine's stability and accuracy, and recognition of the component pin is an important part of the vision. This paper focuses on component pin recognition using three different techniques. The first technique involves traditional image processing using the core algorithm for binary large object (BLOB) analysis. The second technique uses the histogram of oriented gradients (HOG), to experimentally compare the effect of the support vector machine (SVM) and the adaptive boosting machine (AdaBoost) learning meta-algorithm classifiers. The third technique is the use of an in-depth learning method known as convolution neural network (CNN), which involves identifying the pin by comparing a sample to its training. The main purpose of the research presented in this paper is to increase the knowledge of learning methods used in the plug-in machine industry in order to achieve better results.

1. Introduction

With the development of manufacturing, the demand for industrial automation has increased exponentially. The electronic manufacturing industry is a large-scale production industry that continuously produces automated equipment that is in high demand.

As an important part of the electronics manufacturing industry's automated production line, the plug-in machine is a number of regular electronic components automatically and standardly inserted in the printed circuit board of an item of mechanical equipment.

The basic control process of the plug-in machine is as follows: shoot a mark point into the printed circuit board, prepare the components, qualify with camera detection, and position the plug-in. The role of machine vision in the plug-in machine is mainly to improve the accuracy of the plug-in [1]. Machine vision can be divided into two parts: the first part is getting the mark coordinates and the second part is waiting for the device to be qualified to determine the location of the components and angle of deviation. This paper focuses on the second part of the machine vision work.

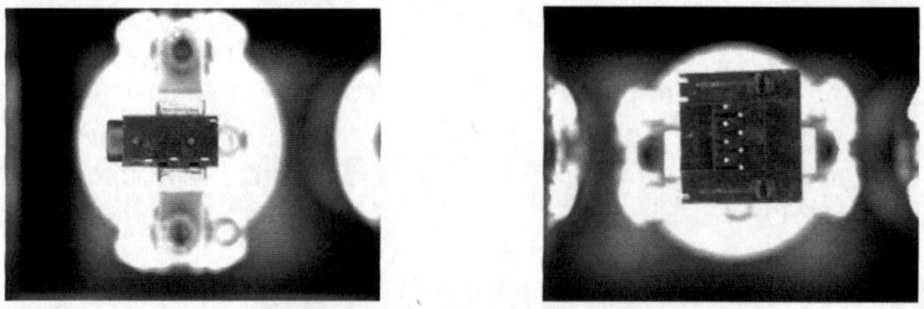

Figure 1. Component renderings

Figure 1 shows the effect of the device to be inserted. The main purpose of the algorithm is to identify the component pins. The core algorithm of traditional image processing is binary large object (BLOB) analysis, which has a high recognition rate and a good effect. However, when light conditions are not ideal, the effect will decline. In addition, when dealing with male/female connector components, a template-matching algorithm is needed to realize the effect. That is, the algorithm needs to be specifically for the corresponding component.

In the emergence of artificial intelligence, machine learning is undoubtedly the most amazing aspect. In this paper, we first extract the histogram of oriented gradient (HOG) characteristics of the samples, and then use the support vector machine (SVM) and the adaptive boosting machine (AdaBoost) learning meta-algorithms for classification learning. However, the merits of the machine learning effect depends heavily on the choice of features, such that if the feature selection is wrong it is likely that the test error will be too large. To some extent, convolution neural network (CNN) can overcome this traditional machine learning shortcoming, but CNN construction requires a lot of experience and getting good test results is not easy.

2. Image processing

For recognition of pin components, the traditional image processing is divided into three steps. First, the region of interest (ROI) is extracted from the acquired image. Then, the binarization of the image is performed. Finally, BLOB analysis is performed to obtain the pin area and the center of gravity position.

2.1 Image binarization

Figure 2 shows the extracted ROI area.

Figure 2. ROI area image

There are basically two binarization methods: the global threshold method or the local threshold method [2]. Because the analysis is highly dependent on binarization, it is a very important step when using tools.

Because of the high contrast between the pin image and the background, the global threshold method can be used. In this paper, the maximum interclass variance method (OTSU) is used. This method divides the image into two parts by the principle of minimizing the rate of misclassification, and then determining the segmentation threshold for the purpose of optimizing the weighted sum of the variance in the class. Figure 3 shows the effect of binarization.

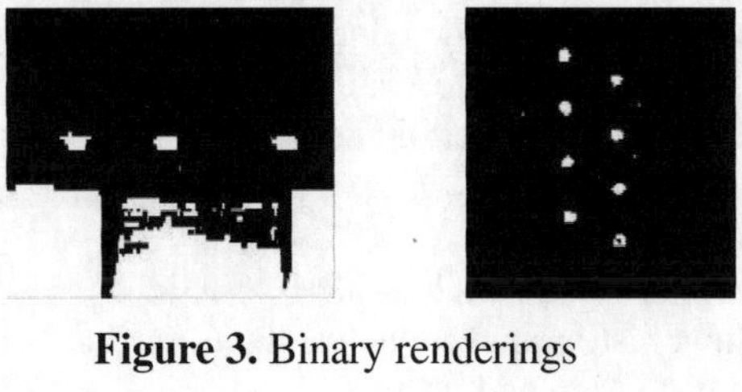

Figure 3. Binary renderings

2.2 BLOB analysis

BLOB analysis refers to blocks that have similar image features such as texture and color, and spatially connected pixels. The BLOB analysis tool separates the foreground and background of the image, and then gives morphological parameters of any foreground shape; it can also detect round objects. In this process, the BLOB does not parse a single pixel but rather analyzes the rows of BLOB

IOP Conf. Series: Journal of Physics: Conf. Series **1004** (2018) 012004 doi:10.1088/1742-6596/1004/1/012004

blocks. This approach can be faster than the pixel-based algorithm. It provides the number, location, area, perimeter, shape and orientation of the spots in the image for the machine vision application [3].

An important step in the BLOB analysis is determination of the connected area. The different analyzes according to the processing method are basically divided into two types: the pixel scanning-based processing method, and the run-length coding method. This article uses a process based on run-length coding whereby the connected area can be extracted to obtain the chain code table and the linear table for each connected area. The chain table can be used to calculate the length and perimeter of the boundary of the region. The linear table can calculate the parameters such as area and center of gravity. Figure 4 shows the results for the BLOB analysis.

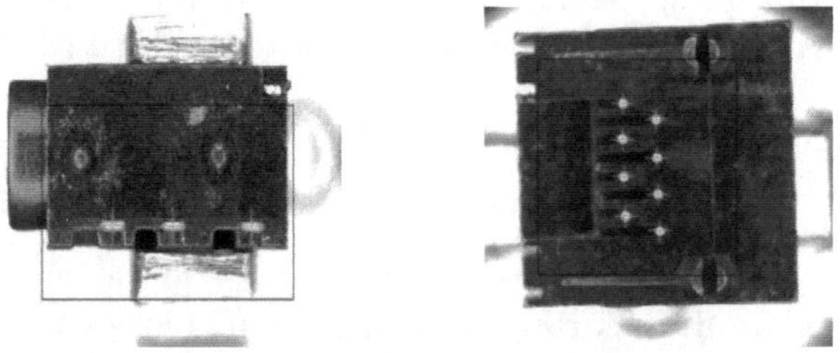

Figure 4. BLOB analysis results

3. Machine learning method

The traditional image processing method is largely determined by image quality problems such as lighting. This article therefore considers using a relatively intelligent algorithm to deal with the problems. Feature extraction of the sample is the most important aspect of machine learning, and so this paper uses the HOG feature with the SVM and AdaBoost algorithm classification. Figure 5 (a) and Figure 5 (b) are positive samples while Figure 5 (c) and Figure 5 (d) are negative samples. To maintain a positive and negative sample balance, 5,800 positive samples and 6,200 negative samples were collected.

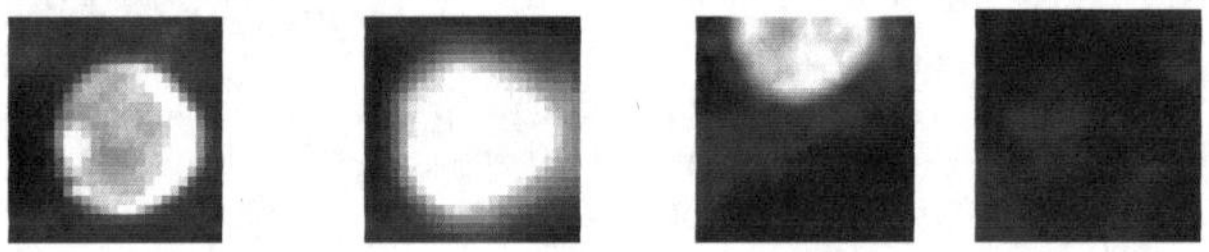

(a) (b) (c) (d)

Figure 5. Positive and negative samples

3.1 Extraction of sample HOG characteristics

The gradient direction histogram is simply referred to as the HOG feature and is attributed to Navneet Dalal and Bill Triggs [4]. The HOG feature is also an image feature that is used in image recognition and machine vision, with the most successful application being for pedestrian detection. The HOG feature is concerned with the change of the edge region and the edge shape of the target in the image. The gradient histogram is constructed according to pixels on the edge of the target, and then the histogram features are combined in a certain way for target detection. The specific implementation steps are as follows:

1) Image normalization.
2) Calculate the pixel gradient for each point.
3) Calculate the gradient histogram for each cell.
4) Block the gradient histogram and normalize it.
5) Combine all block gradient histograms to form HOG characteristics.

The HOG characteristics of the positive and negative images extracted from the pair of samples are shown in Figure 6.

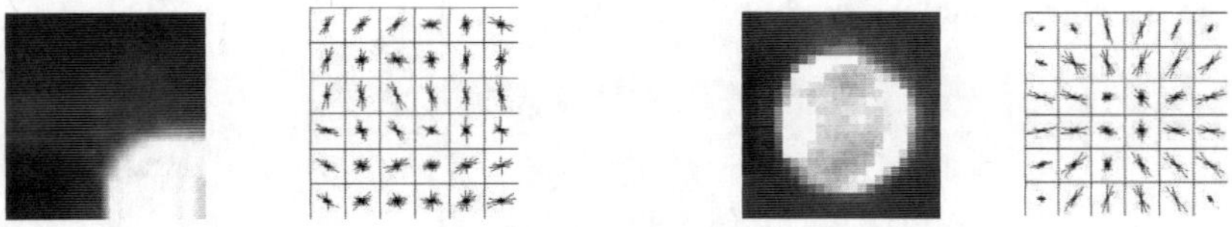

Figure 6. HOG charcteristic extraction

3.2 Support vector machines

The support vector machine (SVM) is a two-class model proposed by Corinna Cortes and Vladimir Vapnik [5]. The basic model is defined as the linear classifier with the largest interval in the feature space. The learning strategy is interval maximization, which can be transformed into a convex quadratic programming problem.

The SVM clasifies target and non-target objects to establish a separation between them. In general, when solving two kinds of target classification and target linear separability in two-dimensional space, the segmentation criterion is to construct an optimal classification line between the target and the non-target so that one side of the line is all target and the other side is all non-target. However, in reality there is often more than two types of target and there are often linear separabliliy problems. The solution to this lies in the choice of multi-classification and spatial dimension transformation.

The classification process is discussed here under two categories. Assume that the general linear equation is:

$$g(x) = w \bullet x + b \tag{1}$$

The hyperplane equation is:

$$w \bullet x + b = 0 \tag{2}$$

The optimal equation for the optimal hyperplane discrimination is:

$$y_i[(\omega \bullet x + b)] \geq 1, i = 1,2,\ldots,n \tag{3}$$

Satisfying the minimum of $\| \omega \|^2$, the optimal hyperplane can be obtained. In this way, it is transformed into a convex quadratic programming problem, and the Lagrangian duality is transformed into the optimization problem of the dual variable. Then the minimal optimization SMO algorithm is used to solve the dual problem.

By training the parameters w and b, the test set can be used to calculate the accuracy rate.

3.3 AdaBoost algorithm

Freund and Schapire when seeking a solution to specific application problems, devised a self-regulating boosting algorithm, which later became known as the Adaboost algorithm [6]. The main principle is that before the training of the model, the target image is given an initial proportion, which is the degree of contribution of the target to the classification accuracy of the classifier. During the training process, each target image of the corresponding weights change with the degree of training according to certain rules. Finally, each target image in the trained model also corresponds to a specific proportion.

The following is the basic process of the Adaboost algorithm:

(1) Initialize all training samples with weights of 1 / N, where N is the total number of samples.

(2) Train weak classifiers and get the corresponding weight α, set the number of weak classifiers to m.

1. Train weak classifier ym(x) and make the weight error function εm minimum. The εm expression is as follows:

$$\varepsilon_m = \sum_{n=1}^{N} \omega_n^{(m)} I \qquad \left(y_m(X_n) \neq y_i \right) \tag{4}$$

2. Calculate the specific gravity α, an expression for the weak classifiers is as follows:

$$\alpha_m = \ln \left\{ \frac{1 - \varepsilon_m}{\varepsilon_m} \right\} \tag{5}$$

3. Update the corresponding weight of the sample.

$$\begin{cases} \omega_{m+1,i} = \dfrac{\omega_{mi}}{Z_m} \exp\left(-\alpha_m y_i y_m(x_i)\right), \quad i = 1,2,\cdots,N \\ Z_m = \sum_{i=1}^{N} \omega_{mi} \exp\left(-\alpha_m y_i y_m(x_i)\right) \end{cases} \tag{6}$$

Among them, the role of Zm is a normalization factor whose role is to ensure that all of the ω is 1.

(3) The final output of the classifier is:

$$Y_M(x) = sign\left(\sum_{m=1}^{M} \alpha_m y_m(x)\right) \tag{7}$$

The AdaBoost algorithm is trained using the positive and negative samples of the classified component pins. Finally, the accuracy is calculated from the test set.

4. Deep learning method

The problem with the above machine learning is whether the input feature provides a good distinction between positive and negative samples. For many machine learning problems, feature extraction is a difficult task. On some complex issues, an effective set of features is designed by artificial means but this needs a lot of time and effort.

Since the artificial approach cannot extract the characteristics of the entity, we can use the more efficient method of in-depth learning. One of the core issues of in-depth learning is the automatic combination of simple features into more complex features. In-depth learning can be a layer of simple features gradually transformed into more complex features [7-11].

This paper classifies the images using the convolution neural network (CNN), which consists of two layers. One layer is the feature extraction layer where the input of each neuron is connected with the local acceptance domain of the previous layer, and the local feature is extracted. Once the local feature is extracted, its positional relationship with other features is also determined. The second layer is the feature mapping layer where each calculation layer of the network is composed of multiple feature maps and each feature map is a plane; the weights of all neurons on the plane are equal. The feature mapping structure uses the sigmoid function, which affects the function nucleus, as the activation function of the convolution network so that the feature map has the displacement invariance. In addition, since the neurons share weights on a map surface, the number of free parameters of the network is reduced. The classical convolution neural network model includes the Le-Net5 model [12] and the Inception-v3 model [13].

The convolution neural network used in this paper is similar to the Le-Net5 model.

First layer: convolution layer

The input is the original image pixel, the input layer size is $24 \times 24 \times 1$. The first convolution layer filter has a size of 5×5, a depth of 6, and no full 0 padding. The output matrix is $20 \times 20 \times 6$.

Second layer: pool layer

The input of this layer is the output of the first layer, which is the $20 \times 20 \times 6$ node matrix. This layer uses a filter size of 2×2 and the output matrix size of the layer is $10 \times 10 \times 6$.

Third layer: convolution layer

The input matrix for this layer is $10 \times 10 \times 6$, using a filter size of 3×3, a depth of 16, and no full 0 padding. The output matrix is $8 \times 8 \times 16$.

Fourth layer: pool layer

The input matrix of this layer is $8 \times 8 \times 16$, the filter size is 2×2, and the output matrix size of this layer is $4 \times 4 \times 16$.

Fifth layer: whole connection layer

The upper output $4 \times 4 \times 16$ matrix straightened into a vector. The output node of this layer is 80.

Sixth layer: whole connection layer

The input node of this layer is 80, and the output is 32 nodes.

Seventh layer: whole connection layer

The input node of this layer is 32, and the output is 2 nodes.

The above is the establishment of the CNN model used in this paper. The loss function is expressed by the cross entropy, which represents the proximity of the output vector and the expected vector. Given two probabilities p and q, the expression is:

$$H(p,q) = -\sum_{x} p(x) \log q(x) \tag{8}$$

The parameters in the neural network are adjusted by the back propagation algorithm [14] and the gradient descent algorithm. Finally, test the trained network with the test set to calculate the accuracy rate.

5. Analysis of results

5.1 Analysis of machine learning methods

The test accuracy rate of a small training sample set of 50 with a negative sample of 50 for training classification produced the following results. As table 1 shows, this is the test set accuracy.

Table 1. Accuracy rate of small sample training set

Algorithm	Accuracy rate%
HOG+SVM	95.8%
HOG+AdaBoost	96.2%

As the training effect is not ideal, the sample is increased to 500 with the number of negative samples also 500. As table 2 shows, this is the test set accuracy.

Table 2. Accuracy rate of medium sample training set

Algorithm	Accuracy rate%
HOG+SVM	96.1%
HOG+AdaBoost	96.3%

Although the effect of increasing the sample set is effective, it is not significant. Using all the training set of 5800 samples with 6200 negative samples, the test set accuracy is as the table 3.

Table 3. The accuracy of all recognition results

Algorithm	Accuracy rate%
HOG+SVM	96.2%
HOG+AdaBoost	96.5%

It can be seen that the difference in effect of using different classification methods is not disparate. The reason for this is probably due to inaccuracies in the sample itself, as well as to the HOG operator not having extracted the classification characteristics well.

5.2 Analysis of in-depth learning methods

Although the positive sample of 5800 and negative sample of 6200 is basically satisfied for the depth of the neural network. The accuracy of the test set is as the table 4.

Table 4. Accuracy of CNN method

Algorithm	Accuracy rate%
CNN	98.9%

The results of machine learning indicate that the HOG operator is not the best feature of classification.

6. Summary

Rather than a proposal to replace traditional methods of image processing with machine learning and in-depth learning, this paper simply presents the possibility of applying both as complementary methods. For example, since the greatest difficulty with the traditional machine learning is finding the best classification of the eigenvector, in-depth learning can be used to extract relatively good characteristics and thus achieve a better classification result.

Acknowlegement

The authors gratefully acknowledge the support provided by the National Nature Science Foundation of China under NSFC Grant 51375118, as well as the support from Shenzhen Government Fund JCYJ20160330161701343 and JCY20170413105740689.

References

[1] Zhang W Y, Zhao Q S, Wang D Y. Actualities and Developing Trend Of Machine Vision[J]. Journal of Zhongyuan University of Technology, 2008.

[2] Kassim A A, Zhou H, Ranganath S. Automatic IC orientation checks[J]. Machine Vision & Applications, 2000, 12(3): 107-112.

[3] Kim H T, Song C S, Yang H J. Algorithm for automatic alignment in 2D space by object transformation[J]. Microelectronics Reliability, 2006, 46(1): 100-108.

[4] Dalal N, Triggs B. Histograms of oriented gradients for human detection[C]// IEEE Computer Society Conference on Computer Vision & Pattern Recognition. IEEE Computer Society, 2005: 886-893.

[5] Cortes C, Vapnik V. Support-Vector Networks.[J]. Machine Learning, 1995, 20(3): 273-297.

[6] Freund Y, Schapire R E. Experiments with a new boosting algorithm[C]// Thirteenth International Conference on International Conference on Machine Learning. Morgan Kaufmann Publishers Inc. 1996: 148-156.

[7] Hinton G E, Osindero S, Teh Y W. A Fast Learning Algorithm for Deep Belief Nets[J]. Neural Computation, 2014, 18(7): 1527-1554.

[8] Lecun Y, Bengio Y, Hinton G. Deep learning[J]. Nature, 2015, 521(7553): 436.

[9] Hinton G, Deng L, Yu D, et al. Deep Neural Networks for Acoustic Modeling in Speech Recognition: The Shared Views of Four Research Groups[J]. IEEE Signal Processing Magazine, 2012, 29(6): 82-97.

[10] Schmidhuber J. Deep learning in neural networks: an overview[J]. Neural Networks, 2014, 61: 85-117.

[11] Ian Goodfellow, Yoshua Bengio, Aaron Courville. Deep learning[M]. MIT Press, 2016.

[12] Lécun Y, Bottou L, Bengio Y, et al. Gradient-based learning applied to document recognition[J]. Proceedings of the IEEE, 1998, 86(11): 2278-2324.

[13] Szegedy C, Vanhoucke V, Ioffe S, et al. Rethinking the Inception Architecture for Computer Vision[J]. 2015: 2818-2826.

[14] Rumelhart D E, Hinton G E, Williams R J. Learning representations by back-propagating errors[J]. Nature, 1986, 323(6088): 533-536.

IOP Conf. Series: Journal of Physics: Conf. Series **1004** (2018) 012005 doi:10.1088/1742-6596/1004/1/012005

Marker Registration Technique for Handwritten Text Marker in Augmented Reality Applications

N Thanaborvornwiwat[1, 2] and K Patanukhom[1]

[1]Department of Computer Engineering, Chiang Mai University, Chiang Mai, Thailand.
[2]Graduate School, Chiang Mai University, Chiang Mai, Thailand.

*karn@eng.cmu.ac.th

Abstract. Marker registration is a fundamental process to estimate camera poses in marker-based Augmented Reality (AR) systems. We developed AR system that creates correspondence virtual objects on handwritten text markers. This paper presents a new method for registration that is robust for low-content text markers, variation of camera poses, and variation of handwritten styles. The proposed method uses Maximally Stable Extremal Regions (MSER) and polygon simplification for a feature point extraction. The experiment shows that we need to extract only five feature points per image which can provide the best registration results. An exhaustive search is used to find the best matching pattern of the feature points in two images. We also compared performance of the proposed method to some existing registration methods and found that the proposed method can provide better accuracy and time efficiency.

1. Introduction

Augmented Reality (AR) is a technology that combines vision of real-world and virtual world together by generating virtual images that are accurately overlaid with real-world scene. AR systems allow user to interact with virtual objects imposed on real environment. The AR systems have been applied in many fields such as entertainment, education, medical application, industrial design, visual art, etc. Tracking and registration of real environment's points are significant processes to correctly overlay the virtual objects on the real scenes. The registration techniques are typically based on feature point extraction and matching procedures. Many feature points are extracted from the real scenes. The feature points in different scenes that are projected from the same 3D real-world coordinate are matched. Then camera pose in each scene can be estimated based on matching results and some additional information from depth camera [1], device orientation [2], or specific marker [3]–[5].

In this work, we develop a marker based mobile AR application that allows children to write a word in a paper to create its corresponding virtual object imposed on the paper. Fig. 1 illustrates examples of the proposed application where virtual 3D object or translated word can be created on scene images. Children can create their desired virtual objects and arrange scene composition by writing words and placing papers. The application starts from localizing texts in a scene. Then characters are recognized from those text images and are grouped into words. Each synthetic marker of the words that have virtual objects to overlay in the scene is created as a binary image of a printed word in a normal frontal view. The feature points are extracted from the markers and the texts in the scenes and are matched. Homography matrices between the marker and the scene are estimated from

IOP Conf. Series: Journal of Physics: Conf. Series **1004** (2018) 012005 doi:10.1088/1742-6596/1004/1/012005

the matching results. Based on estimated homography matrices, the virtual objects can be rendered into the scene image. An overview of the proposed AR application is given in Fig. 2.

Figure 1. Example of the proposed AR application.

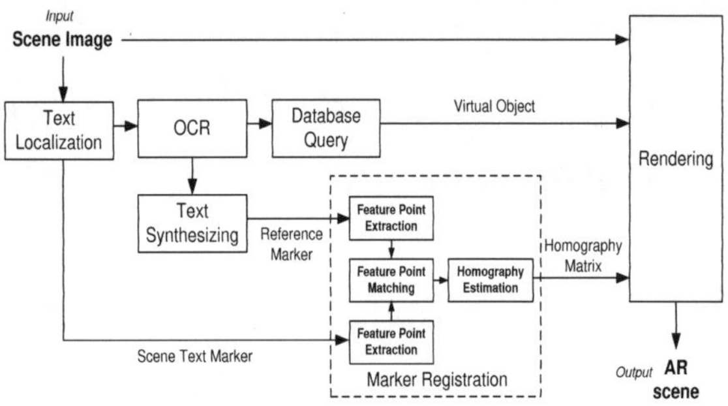

Figure 2. Overview of the proposed AR application.

There are many approaches developed to detect text regions in the natural scene image [6]–[8]. In this work, we use a Maximally Stable Extremal Regions (MSER) [9] which is one of regions-based techniques implemented in many researches [7], [8]. One of advantages of the MSER based approaches is that it is invariance to affine transformation which is suitable for our application.

After the scene texts have been detection, those handwritten texts can be recognized by several methods such as using some specific features [10] or using deep-learning frameworks [11].

Some existing marker-based AR applications [3]–[5] use special pattern of markers such as four corners of a square marker [3], five shades of grayscale marker field [4] and circular marker [5] to estimate the homography. However, they cannot be used for our applications because of variation of handwritten text markers.

For registration of the printed text marker and the handwritten scene text, typical marker registration that based on general local feature matching such as SIFT [12], SURF [13], MSER [9] cannot perform well in this kind of application. The printed text marker typically consists of few characters as shown in Fig. 1 therefore it will produce less feature points and more similar descriptor for each feature point.

There are some registration approaches that focus on content specific contours such as conic shape [14] or specific characters [15]. These algorithms required specific shapes appears in target images as clues for registration which is not flexible for our application that cannot guarantee the appearance of any specific shape in the short text marker.

On the other hand, geometric rectification approaches using page layout [16] which are widely used in rectification of documents before OCR require multiple text lines or line segments in documents for analyzing vanishing points. This kind of methods is not suitable for the text markers that consist of few words or only one text line.

As a result, the registration and the homography estimation of the handwritten text markers are still a challenging problem. In this work, we develop novel approach for finding correspondences of text marker in the synthetic text marker and the real-world scene marker that is more robust to camera pose and handwritten style than the conventional methods. Most of contents presented in this paper do not focus on whole AR application but mainly focus on our proposed marker registration technique. The MSER-based text localization process applied in this work is briefly summarized in Section 2. The proposed registration method is presented in detail in Section 3. Section 4 demonstrates an experiment on the proposed method in comparison with some existing methods [7], [11], [13]. Finally, there are conclusions of this paper and direction of future works given in Section 5.

2. MSER Based Text Localization

Localization of the text in the scene image is the first process in this text-based AR application. In this work, we modified the text localization process from [8]. The MSERs can be considered as regions with homogenous intensities that are different with their boundary intensities. The MSER can be extracted by binarizing a target image with various intensity thresholds. MSERs are some blobs at some thresholds which their areas are the most stable to the threshold variation in comparison with themselves in other thresholds. When MSERs are extracted from the image, they include text regions and non-text regions. The non-text region can be filtered out by essential shape parameters such as area, MSER intensity threshold, solidity, major axis length, minor axis length. Then, the text regions are clustered by their similarities in positions, sizes or colours. An example of MSER extraction is given in Fig. 3(a)-(c). In this work, after text regions are obtained, the result will be verified by observing surrounding areas outside the text region along its major axis to avoid missing of first or last character in the text. If there are any blobs that have equivalent properties to the characters in the text region then they will be included in that text region.

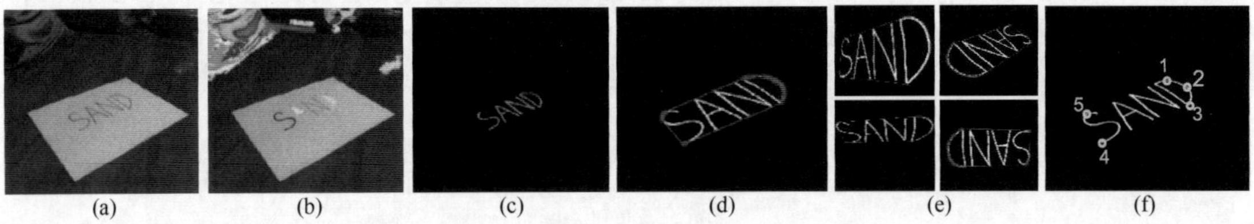

Figure 3. Text localization and feature point extraction process. (a) Scene image. (b) MSERs. (c) Text mask. (d) Convex hull. (e) Examples from other camera poses. (f) Polygon simplification.

3. Proposed Registration Method

The proposed text marker registration for homography estimation and connected direct view of real-world are presented in the section. In the registration process, we assume that a text is already recognized, and a reference marker is already synthesized. The registration process consisted of three sub-modules that are feature point extraction, feature point matching and homography estimation as shown in Fig. 2. The proposed feature point extraction is based on contour points of a convex area of the text region and polygon simplification which provide more robustness over handwritten styles or camera poses. Centroid distances, centroid angles, and histogram of the oriented gradient (HOG) [17] are used as descriptors of the feature points. For the feature point matching process, we apply an exhaustive searching for all possible matching to find the best four matched pairs. Since there are only four to six feature points extracted in the proposed scheme, the exhaustive searching process can still perform very fast. An order consistency of the contour points is used as a constraint to avoid wrong matching results and reduce size of searching space. Details of feature point extraction and matching processes are given in the following sections.

3.1. Feature Point Extraction

In order to locate the feature points, convex hulls of text regions in both reference and scene images are extracted. The convex hull is defined by a convex set of points that can visualized as the shape enclosed by a rubber band stretched around text region as shown in Fig. 3(d) where red dots are vertex point of outer contour of the convex hull. In general, we can expect that the contour's vertex points of convex hull obtained from the same text are robust to the camera pose as demonstrated in Fig. 3(e) therefore those vertices can be used as the feature points. In order to reduce a computational cost in matching process, we reduce the number of feature points by applying a polygon simplification [18]. Unimportant points that have large angle or small segment length are repeatedly removed until the desired number of points is reached. The remaining points from reduced polygon are feature points of the text marker. An example of the final feature points is given in Fig. 3(f) and Fig. 4. Based on experiments, reducing the number of feature points not only improve the time efficiency but can also improve matching accuracy.

To estimate orientation and scale of feature points, major axes are extracted from the convex hull of the scene text and all characters that have feature points located inside. We assume that the orientations of all feature points in the same text are approximately the same. Therefore, the orientations of all feature points are assigned as the orientation of the major axis of the convex hull of the scene text Φ. However, the orientation of the major axis is unsigned, the orientation of the feature points can be Φ or $-\Phi$. The scale σ of each feature point is assigned to be a length of the major axis of the host characters that it belongs to.

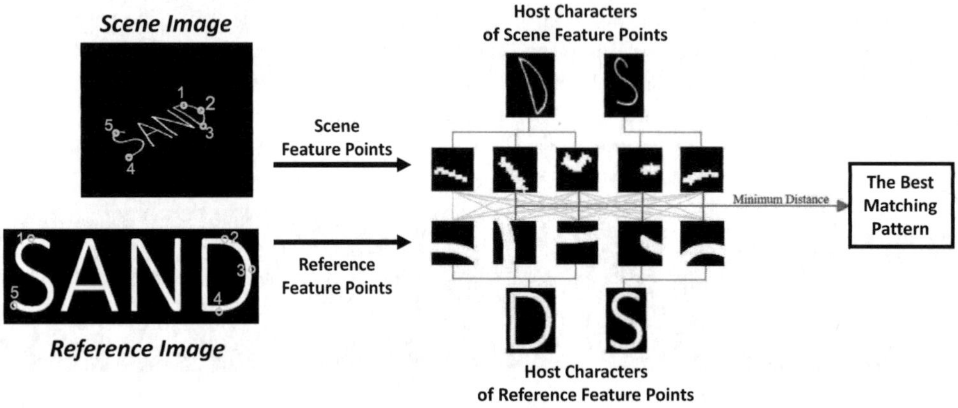

Figure 4. Feature point matching.

3.2. Feature Points Matching
In order to estimate the homography matrix that has eight unknowns, at least four corresponding points between two images are need. Let n and m are the number of feature points extracted from the scene image and the reference image, respectively. The parameters n and m are controllable in the polygon simplification step. Since the feature points are contour points which are ordering as circular sequence, an order consistency is used as a constraint for matching the feature points. As a result, there are $4\times\frac{n!}{(n-4)!4!}\times\frac{m!}{(m-4)!4!}$ patterns of choosing four corresponding feature points. Fig. 4 illustrates an example of the proposed feature point matching procedure. Let $P=\{(p_i,q_i)\}_{i=1}^{4}$ is any possible matching pattern where p_i is a feature point in the scene image and q_i is a corresponding point to p_i in the reference image. To find the best matching pattern, we divide the process into two stages.

The first stage is to find good candidates by using simple contour-based features that are distance and direction to a centroid of the text. Differences between the distance and the direction to a centroid of the text between two images are calculated by

$$E_d(P)=\sum_{i=1}^{4}\left|d(p_i)-d(q_i)\right| \text{ and } E_\theta(P)=\min_{\pm\Phi}\left(\sum_{i=1}^{4}\left|\theta(p_i)-\theta(q_i)\pm\Phi\right|\right), \quad (1)$$

Where $d(\cdot)$ is a distance from the corresponding feature point to the centroid of the text which is normalized by a length of major axis of the text and $\theta(\cdot)$ is a direction from the corresponding feature point to the centroid of the text. By combining E_d and E_θ, a dissimilarity function for the first stage is

$$E(P)=\frac{E_d(P)}{\max_P\left(E_d(P)\right)}+\frac{E_\theta(P)}{\max_P\left(E_\theta(P)\right)}. \quad (2)$$

All possible matching patterns are ranked by $E(P)$. The top K matching patterns that provide less $E(P)$ are selected as candidates. We also consider the distribution of the matched feature points. In order to reduce an error in the homography estimation, the matched feature points should be distributed in all four quadrants divided by the major and the minor axes in both scene marker and reference marker. The candidates that has high and their feature points does not distribute in all quadrants are removed. This filtering process significantly reduces the computational cost.

The second stage implements the HOG [17] to extract descriptors of feature points. HOG descriptors of windows surrounding every feature points and HOG descriptors of every host character that have feature points are extracted. An example of all windows and host characters of the feature points is illustrated in Fig. 4. Sizes of the windows are relative to scale σ of each feature point. In this paper, we use the windows size of 0.25σ in the experiment. While orientations of the windows and the host characters are selected from $\pm\Phi$ that provide the minimum value in the calculation of $E_\theta(P)$. Final descriptor of each feature point is obtained by concatenating the HOG descriptor from its local windows and the HOG descriptors of its host character. The best matching pattern is defined as

$$P_{\text{opt}} = \arg\min_{P}\left(\sum_{i=1}^{4}\left|h(p_i) - h(q_i)\right|\right) \tag{3}$$

Where $h(\cdot)$ represents the final descriptor of the corresponding feature point.

4. Experimental results

In this section, we conduct an experiment to evaluate performance of the proposed registration scheme. The experiment is conducted on 1,560 scene images of multiple views of 120 handwritten markers from 12 English words with 10 writing styles. Examples of test images are given in Fig. 5. The resolution of the scene images is 750×1000 pixels. 120 reference text images are generated by using Calibri font in normal font style. The homography matrices are estimated for every test images. Accuracy of homography estimation and registration time and homography estimation time consumption (excluding text localization) are measured. Accuracy of homography estimation is defined as a percentage of the scene markers that are correctly rectified into their frontal view. We compare performance of the proposed methods with three conventional approaches that are MSER [7], SURF [13], and contour correspondences [15]. In the baseline MSER approach, the MSERs are extracted and HOG descriptor are used for matching the MSERs in both images. Centroid positions of MSERs use as positions of the feature points for find homography. Random Sample Consensus (RANSAC) [19] is applied for finding the best homography matrix in all three conventional approaches.

Figure 5. Examples of testing images.

Table 1. Performance comparison

Methods	Accuracy [%]	Time consumption [a]
Proposed ($n=m=5$)	97.37	1.0000
MSER + HOG [7]	60.96	1.4389
SURF [13]	34.81	2.4584
Contour Correspondence [15]	49.74	1.7972

[a] Normalized by the proposed method.

Table 2. Accuracy of the proposed method when the number of feature points is varying

The number of feature points	Accuracy [%]
4	91.99
5	97.37
6	92.63
7	89.42
8	86.03

The performances of four methods are demonstrated in Table 1. In order to ignore the hardware specification, time consumptions are presented as ratios of average time usage in each method comparing with the proposed method. In the proposed method, we choose the number of feature points as five because it provides the best accuracy comparing with the other number of feature points as shown in Table 2. Examples of the results from four methods are given in Table 3 that include visualization of feature point matching results, rectified scene markers by the estimated homography (expected to be the frontal view), and the reference markers rendered to the real scene. According to the results, the proposed methods can provide the highest accuracy and the least time consumption. The proposed method extracts only few feature points which mostly appear in both images. The proposed matching scheme uses the order of the circular sequence as matching constraint to avoid miss-matching and keeps the matched feature points over four quadrants of the marker to reduce the error of homography estimation. The simple contour-based features used for rejecting bad choices also helps to improve time efficiency. The SURF method obtains the lowest accuracy and time efficiency in this dataset because SURF method extracts many feature points which have no correspondence in the other image. In addition, the binary text markers that consists of only simple lines, curves and corners produce descriptors that are non-unique in many feature points resulting in incorrect matching as shown in Table 3.

On the other hand, the conventional MSER and the contour correspondence methods can provide more robust feature points and faster than SURF. For conventional MSER method, the number of feature points is typically equal to the number of character in the text marker therefore it cannot be used for the texts that consist of less than four characters. A problem also occurs when the text markers consist of the same character more than one. The homography estimation using the MSER's centroids produces more error than the proposed method which can be observed from the rectified scene texts in Table 3 because the character centroids are almost aligning in a straight line that may result in nearly ill-condition in the homography estimation.

For the contour correspondence method, the feature points distribute better than the conventional MSER method. However, there are still many wrong correspondences when using feature based on cross-ratio of areas.

5. Conclusion

In this paper, we presented the marker registration technique for AR application that is designed for dealing with the simple and low-texture handwritten text markers. The MSER and the contour simplification are used for extracting the feature points. The new feature point matching procedure is developed. Based on the experiment from 1,560 scene images with handwritten text markers, the proposed method provides higher accuracy, more robustness to handwritten styles and camera poses and requires less computational cost than the baseline methods.

Table 3. Examples of the results from four methods.

	Feature point matching result	**Rectified scene texts**	**Reference marker rendered to the scene**
Proposed			
MSER+HOG			

IOP Conf. Series: Journal of Physics: Conf. Series **1004** (2018) 012005 doi:10.1088/1742-6596/1004/1/012005

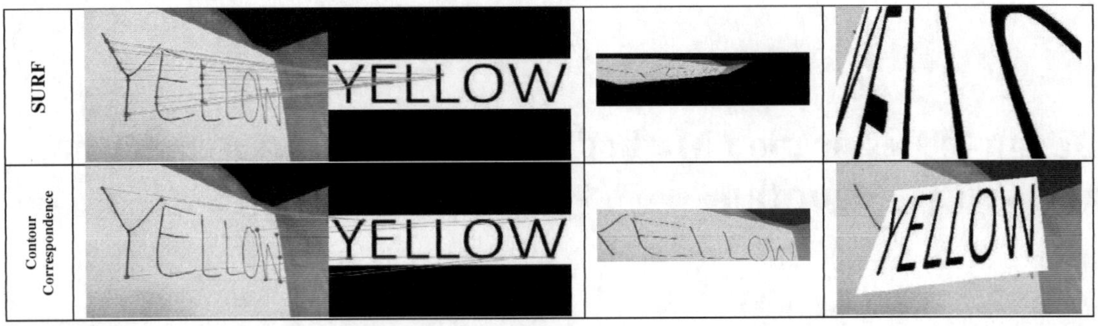

References

[1] Kim B S, Xu S, and Savarese S 2013 Accurate localization of 3D objects from RGB-D data using segmentation hypotheses Proc. IEEE Comput. Soc. Conf. Comput. Vis PatternRecognit. pp 3182–89.

[2] Kurz D and Benhimane S 2011 Gravity-aware handheld augmented reality 2011 10th IEEE nt. Symp. Mix. Augment. Reality, ISMAR 2011 pp 111–20.

[3] Park J, Jiang B, and Neumann U 1999 Vision-based pose computation: robust and accurateaugmented reality tracking Proc. 2nd IEEE ACM Int. Work. Augment. Reality pp 3–12.

[4] Herout A, Szentandrasi I, Zacharia M, Dubska M, and Kajan R 2013 Five shades of grey for fast and reliable camera pose estimation Proc. of the IEEE Computer Society Conference on omputer Vision and Pattern Recognition pp 1384–90.

[5] Pagani A, Koehle J, and Stricker D 2011 Circular Markers for camera pose estimation 12th Int. Work. Image Anal. Multimed. Interact. Serv. - WIAMIS 2011.

[6] Zhang H, Zhao K, Song Y Z, and Guo J 2013 Text extraction from natural scene image: A survey Neurocomputing 122 pp 310–23.

[7] Turki H, Halima M B, and Alimi A M 2016 Scene text detection images with pyramid image and MSER enhanced Int. Conf. Intell. Syst. Des. Appl. ISDA pp 301–6.

[8] Intaratat S and Patanukhom K 2017 Self-learning Structure for Text Localization 2017 15thIAPR International on Machine Vision Applications 2 pp 370–73.

[9] Matas J, Chum O, Urban M, and Pajdla T 2002 Robust Wide Baseline Stereo from Maximally Stable Extremal Regions Br. Mach. Vis. Conf. pp 384–93.

[10] Mingqiang Y, Kidiyo K, and Joseph R 2008 A Survey of Shape Feature Extraction Techniques," Pattern Recognit. Tech. Technol. Appl.

[11] Cireşan D C, Meier U, Gambardella L M, and Schmidhuber J 2011 Convolutional neural network committees for handwritten character classification Proc. Int. Conf. Doc. Anal. Recognition, ICDAR 10 pp 1135–39.

[12] Lowe D G 2004 Distinctive image features from scale-invariant keypoints Int. J. Comput. Vis.60 pp 91–110.

[13] Xu A and Namit G 2008 SURF : Speeded - Up Robust Features Eur. Conf. Comput. Vis. pp 1–30.

[14] Kumar P, Jawahar C V, and Narayanan P J 2004 Geometric Structure Computation from Conics Proceeding Indian Conf. Comput. Vis. Grraphics Image Process. pp 1–6.

[15] Kumar M P, Jawahar C V, and Narayanan P J 2004 Building blocks for autonomous navigation using contour correspondences Image Process. 2004. ICIP '04. 2004 Int. Conf. 2 pp 1381–84.

[16] Clark P and Mirmehdi M 2001 Estimating the Orientation and Recovery of Text Planes in a Single Image Proc. Br. Mach. Vis. Conf. 2001 pp 44.1-.10.

[17] Dalal N and Triggs B 2005 Histograms of oriented gradients for human detection Proc. 2005 IEEE Comput. Soc. Conf. Comput. Vis. Pattern Recognition, CVPR 2005 I pp 886–93.

[18] Veltkamp R C 2003 On the Implementation of Polygonal Approximation Algorithms OvidiuGrigore.

[19] Fischler M and Bolles R C 1981 Random Sample Consensus: A Paradigm for Model Fitting with Commun. ACM 24 pp 381–95.

IOP Conf. Series: Journal of Physics: Conf. Series **1004** (2018) 012006 doi:10.1088/1742-6596/1004/1/012006

Design of Restoration Method Based on Compressed Sensing and TwIST Algorithm

Fei Zhang[1] and Yan Piao[1]

[1]Changchun University of Science and Technology, Changchun of JiLin, China

*E-mail: piaoyan@cust.edu.cn

Abstract. In order to improve the subjective and objective quality of degraded images at low sampling rates effectively,save storage space and reduce computational complexity at the same time, this paper proposes a joint restoration algorithm of compressed sensing and two step iterative threshold shrinkage (TwIST). The algorithm applies the TwIST algorithm which used in image restoration to the compressed sensing theory. Then, a small amount of sparse high-frequency information is obtained in frequency domain. The TwIST algorithm based on compressed sensing theory is used to accurately reconstruct the high frequency image. The experimental results show that the proposed algorithm achieves better subjective visual effects and objective quality of degraded images while accurately restoring degraded images.

1. Introduction

The famous mathematicians D. Donoho and E. Candes put forward the compressed sensing theory in 2016 [1-2].Compressed sensing technology has made breakthrough progress in image reconstruction and encryption .However, the application of compressed sensing to the field of image restoration [3-4] is still in its infancy. Both signal reconstruction part of compressed sensing and image restoration need to solve the inverse problem [5].In addition, some methods in the field of image restoration [6-9] are also the reconstruction methods of compressed sensing.Therefore, it is possible to apply compressed sensing technology to the field of image restoration. Ma Jianwei proposed a method based on Poisson Singular Intergral (PSI) and Curvelet iterative hard threshold to reconstruct degraded images.The algorithm improves the subjective and objective effect of degraded image by using a small amount of observation data, but the algorithm does not consider the existence of noise.

The algorithm proposed in this paper obtains the sparse high-frequency signal and the non-sparse low-frequency signal by the wavelet transform of the degraded image.The algorithm uses compressed sensing to reconstruct sparse high-frequency signals and reconstruct the low-frequency signals by using the traditional restoration algorithm.Finally, the reconstructed high and low frequency signals are inverse transformed to obtain a restored image.Though the algorithm only extracted a smallamount of high-frequency observations,it can not only improves the subjective and objective effects ofdegraded images but also reduces the storage space of data and reduces the computational cost.

2. Compressed sensing

Compressed sensing theory mainly includes three parts: sparse signal transformation, signal measurement and signal reconstruction. Figure 1 shows the theoretical framework of compressed sensing.

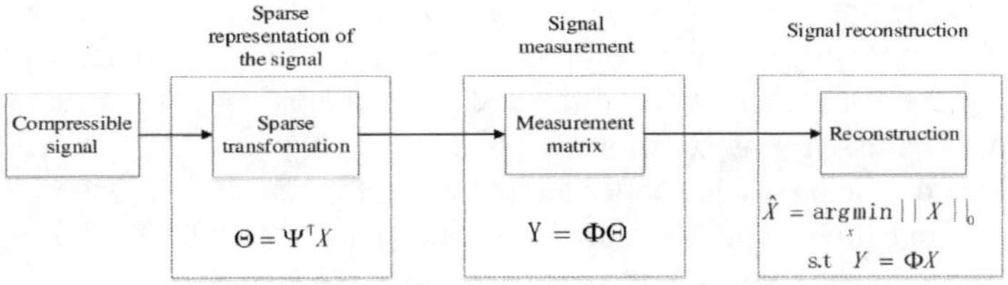

Figure 1 Theory frame of compressied sensing

Suppose $X = [X_1, X_2, ..., X_N]^T$ is an one-dimensional discrete signal,then any set of N-dimensional vectors can be represented by an orthogonal vector basis $\{\Psi_i\}_{i=1}^N$ of $N \times 1$ dimensions. Any signal can be expressed as:

$$X = \Psi\Theta \tag{1}$$

Where Θ is the transform coefficient.If there are only $K(K << N)$ nonzero coefficients (or the value is large) in Θ and all other coefficients are all zero (or the value is small).So the signal X is said to be K-sparse and Ψ is the sparse basis of the signal.Then, the transform coefficients can be linearly transformed with an observation base Φ ($M \times N$ dimension, $M << N$) to obtain the observation set Y ($M \times 1$ dimension), which is:

$$Y = \Phi\Theta = \Phi\Psi^T X \tag{2}$$

Sparse signal transformation and signal measurement can also be expressed by the matrix A^{CS}, then, the signal is observed as:

$$Y = A^{CS} X \ (A^{CS} = \Phi\Psi^T) \tag{3}$$

When A^{CS} satisfies the RIP criterion, that is to say, the observation matrix is not related to signal sparse basis.Then the Reconstruction of the original signal X from the observation set Y transforms into an optimization problem:

$$\min \|\Psi^T X\|_0 \ s.t. A^{cs} = \Phi\Psi^T X = Y \tag{4}$$

3. Restoration method based on compressed sensing and TwIST algorithm

3.1. Raditional recovery methods

The key to image restoration is to establish a model of image degradation and restoration.Assuming that the original image is $f(x, y)$, the degraded image is $g(x, y)$, the restored image is $\hat{f}(x, y)$, the random noise is $n(x, y)$ and H[·] is a function that combines all the degradation factors.Then, the image degradation and restortion model shown in Figure 2.

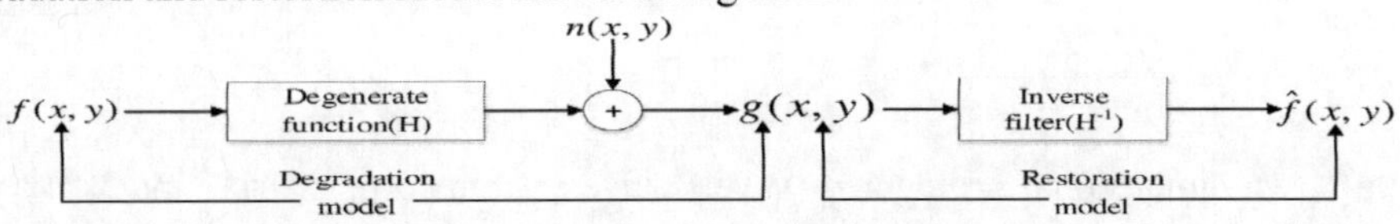

Figure 2 Model of image degradation and Restoration

As can be seen from the figure above, the degraded image can be expressed as:

$$g = Hf + n \tag{5}$$

The process of image restoration is the inverse filtering process. Then, the traditional image restoration model is expressed as:

$$\hat{f} = H^{-1}[g] = f + H^{-1}[n] \tag{6}$$

3.2. Image recovery based on Compressed Sensing

The study found that in the compressed sensing sparse transform, the frequency domain transform can decompose the image into high frequency part and low frequency part.However, the low-frequency part contains most of the information of the image, and the overall coefficient is larger and less sparse.Therefore, it is not suitable for compressed sensing sampling.The high-frequency image contains the details of the image, and most of the coefficients are 0 or approximately 0.So it is Sparse enough to apply to compressed sensing samples.

Therefore, for degraded images, this paper adopts wavelet transform to decompose the image into high-frequency signal and low-frequency signal.This paper uses the GPSR_BB algorithm to recover the low-frequency imag and uses compressed sensing method to recover the high-frequency image.The Gaussian random measurement matrix is used to observe the high-frequency signals and use zero to fill the unobserved part.Then using the TwIST recovery algorithm based on compressed sensing to recover the degraded image.Specific algorithm flow shown in Figure 3.

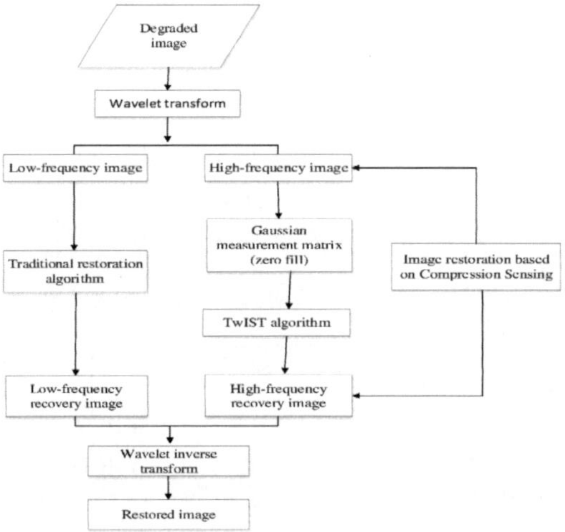

Figure 3 Image Restoration Based on Compressed Sensing

3.2.1. Image restoration based on Compressed Sensing. The objective function of traditional image restoration is expressed as:

$$J(x) = \frac{1}{2} \| g - Hx \|^2 + \lambda \varphi(x) \tag{7}$$

Where $\varphi(x)$ is the normalization factor and λ is a regularization parameter, which is used to adjust the proportion of the two parts of the objective function.Through Bayes estimation, the optimization problem of J in (7) can be transformed into the problem of maximum a posteriori estimation[11].It can be expressed as:

$$\hat{x} = \arg\min_x \frac{1}{2} \| Hx - g \|_2^2 + \lambda \| x \|_1 \tag{8}$$

Let W be a two-dimensional orthogonal wavelet inverse transform matrix, and $W^T W = I$.Then it performs a wavelet transform on x to get the coefficient vector θ.That is $\theta = W^T x$.The recovery model of sparse coefficient is obtained by Bayesian-MAP estimation.

$$\hat{\theta}_{\mathrm{MAP}} = \arg\min_\theta [\frac{1}{2} \| g - HW\theta]_2^2 + \lambda \| \theta \|_1 \tag{9}$$

The restored image formula is expressed as:

$$\hat{x} = W\hat{\theta}_{MAP} \tag{10}$$

Under the compressed sensing theory, the optimal solution obtained by l_1 is expressed as follows

$$\hat{x} = \arg\min_x \frac{1}{2} \| \Phi x - g \|_2^2 + \lambda \| \Psi^T x \|_1 \tag{11}$$

In the above equation, Φ is the observation matrix, Ψ^{T} is the transformation matrix.

IOP Conf. Series: Journal of Physics: Conf. Series **1004** (2018) 012006 doi:10.1088/1742-6596/1004/1/012006

Transform the formula(10) into the wavelet domain, the wavelet coefficient is θ and connect to the maximal posterior estimation of it, the wavelet domain image reconstruction, based on the compressed sensing theory, is shown as follows:

$$\hat{\theta} = \arg\min_{\theta}[\frac{1}{2}\| g - \Phi HW\theta \|_2^2 + \lambda \| \theta \|_1] \quad (12)$$

In the above equation, Φ is the observation matrix, H is the degenerate function, W is the orthogonal wavelet transform matrix, and θ is the wavelet coefficient.

From the analysis above, we can see that (12) is the equivalent formula of signal reconstruction algorithm and image restoration algorithm under the compressed sensing theory.

3.2.2 TwIST algorithm. In solving the problem of (12),The TwIST algorithm mainly uses the two previous estimated values to update the current value.The update process is given by (13) and *(14)*.

$$x_1 = \Gamma_\lambda(x_0) \quad (13)$$

$$x_{t+1} = (1-\alpha)\cdot x_{t-1} + (\alpha - \beta)\cdot x_t + \beta \cdot \Gamma_\lambda(x_t) \quad (14)$$

In the formula, x_0 is the initial value, Γ_λ is a noise reduction processing function. α, β are the parameters for adjusting each iteration estimation value, x_{t+1}, x_t, x_{t-1} are the values when estimate t+1,t,t-1times, x_{t+1} is jointly decided by x_t and x_{t-1} .This is the so-called two-step iterative contraction.

Noise reduction processing function $\Gamma_\lambda(x_t)$ is defined as:

$$\Gamma_\lambda(x) = \Psi_\lambda(x + \Phi^T(g - \Phi x)) \quad (15)$$

$\Psi_\lambda(x)$ is defined as:

$$\Psi_\lambda(x) = \sum_i T(\langle x, \theta_i \rangle)\cdot \theta_i \quad (16)$$

Where, $T_\lambda^T(x)$ represents the threshold processing operator,It is represented by (17) and (18):

Soft threshold:

$$T(x) = sign(x)\cdot \max(|x| - th, 0) \quad (17)$$

Hard Threshold:

$$T(x) = x \cdot \max(|x| - th, 0) \quad (18)$$

Where, th represents the threshold.

Literature [7] made a detailed study on the convergence of TwIST algorithm.The convergence speed of the algorithm is decided by the parameters α and β. α, β is expressed as follows:

$$\alpha = \rho^2 + 1,\ \beta = \frac{2\alpha}{\xi_m + \xi_1} \quad (19)$$

Where, $\rho = \frac{1-\sqrt{k}}{1+\sqrt{k}} < 1, k = \frac{\xi_1}{\xi_m}$ .Here $\xi_m = 1$, under different circumstances, ξ_1 take a different value.

4. Experimental results and analysis

The experiment uses 256*256 size image, including Figure (A) Changchun University of Science and Technology's Family Floor and Figure (B) Cameraman.

(A) Family floor

(B) Cameraman

IOP Conf. Series: Journal of Physics: Conf. Series **1004** (2018) 012006 doi:10.1088/1742-6596/1004/1/012006

Figure 4 Test images

The test images (A) and (B) were degenerated by the uniform degenerate function ($L = 4$). A Gaussian white noise with a variance of 2 was added. After extracting 60% of the high frequency components of the degraded image in the wavelet domain, the proposed algorithm is compared with the proposed algorithm [7] and [9]. The simulation results are shown in Figure 5-6 and Table 4.1

(a) Blurred image (b) document [7] (c) document [9] (d) Our method

Figure 5 Restorated map(A) by different approachs

(a) Blurred image (b) document [7] (c) document [9] (d) Our method

Figure 6 Restorated map(B) by different approachs

Observing Figure 5-6, at the subjective aspects, all the three restoration algorithms have improved the quality of the degraded image effectively. However, the algorithm proposed in this paper is obviously superior than the one proposed in [7] and [9] in color contrast. Only the extraction of 60% of the amount of information cases of high-frequency part, there are some white spot appears in some detail part of the restoration algorithm in the literature [7] and [9]. While the more accurate algorithm in this paper has obtained of the restored image detail part. In order to show that the proposed algorithm is superior than the traditional one, we analyze the enlargement of Figure(A) and Figure (B) , they are shown in Figure7 and Figure 8.

(a) document [7] (b) document [9] (c)our method

Figure 7 Different restoration algorithm for map (b) restoration of the details of the map

(a) document [7] (b) document [9] (c) our method

Figure 8 Different restoration algorithm for map (B) restoration of the details of the map

It can be seen from the analysis of Figure 7 and Figure 8 that in the case of obtaining 60% information amount of the high-frequency information, it can be seen from the result of the restoration shown in image (A) that the contour of the restored house and the texture of the tile have the approximation degree to the original image. It is obviously better than the literature [7], [9] restoration algorithm. It can be seen from the result of the image (B) restoration that the proposed algorithm outperforms the traditional algorithm in the restoration of the contour of the face, the clothes and the structure of the camera. In the case of extracting only 60% of the amount of information, the details of the original image are reconstructed precisely without any loss of detail.

The objective of this algorithm is to analyze the quality of the restoration effect from the peak signal-to-noise ratio (PSNR) and the feature similarity (FSIM).

Table 4.1 Recovery performance comparison with different algorithm

	Figure A		Figure B	
	PSNR (dB)	FSIM	PSNR (dB)	FSIM
document [7]	25.01	0.853	25.62	0.842
document [9]	25.12	0.861	25.21	0.824
Our method	27.55	0.895	27.26	0.876

It can be concluded from Table 4.1 that under the same conditions, the image reconstructed by our algorithm is higher than the algorithms of [7] and [9].The PSNR of this algorithm is 1-2dB higher than that of the literature [7] and literature [9].In the aspect of FSIM, the proposed algorithm is 0.02-0.05 higher than that of the original algorithm [7] and [9], which shows that the structural similarity between the restored image and the original image is higher than the other two algorithms, and Image distortion is small.

5. Conclusion

In this paper, we propose a new TwIST image restoration algorithm based on compressed sensing in the traditional image restoration algorithm. The compressed sensing technology is applied to the field of image restoration, breaking the constraints of signal sampling by the Nyquist sampling law. Through the sparse decomposition and observation of degraded images, it achieves the subjective and objective effects that in a large extent to change the degradation of images can be observed with a small amount of data. At the same time it saved storage space and reduced the amount of computation. In this paper, compressed sensing technology is applied to the field of image restoration, which provides a new feasible idea and an effective solution with the use value for the future image restoration.

References

[1] Donoho D. Compressed Sensing[J]. IEEE Transactions on Information Theory, 2006, (4): 1289-1306.

[2] Candès E,Romberg J,Tao T.Robust uncertaintyprinciples: Ex-act signal reconstruction from highly incomplete frequency in-formation[J]. IEEE Trans. Inform.Theory,2006,52（2）:489-509.

[3] Xue ming. A Study on Application of Image Restoration based on Compressive Sensing and Sparse Decomposition[D].Xidian University,2009

[4] Cao Li-hong, Xu Zhi-jing.Research of fast compressed sensing image recovery algroithm based on multi-wave transform [J].EL Electronic Design Engineering,2015,23(03):176-178.

[5] Xiao Ting-yan, Yu Shen-gen, Wang Yan-fei. A numerical method for inverse problems.Bei Jing: Science Press,2003.

[6] Daubechies I,Defrise M,De Mol C.An iterative thresholding algorithm for linear inverse problems with a sparsity constraint[J].Communications on Pure & Applied Mathematics,2003, 57(11):1413-1457.

IOP Conf. Series: Journal of Physics: Conf. Series **1004** (2018) 012006 doi:10.1088/1742-6596/1004/1/012006

[7] Bioucas-Dias J M, Figueiredo MAT. A New TwIST:Two-Step Iterative Shrinkage/Thresholding Algorithms for Image Restoration[M]. IEEE Press, 2007.
[8] Bioucas-Dias JM, Figueiredo MAT.Two-Step Algorithms for Linear Inverse Problems with Non-Quadratic Regularization[C]// IEEE International Conference on Image Processing. IEEE, 2007:I-105-I-108.
[9] Figueiredo M A T, Nowak R D, Wright S J. Gradient Projection for Sparse Reconstruction: Application to Compressed Sensing and Other Inverse Problems[J]. IEEE Journal of Selected Topics in Signal Processing, 2008, 1(4):586-597.
[10] Ma J., Le Dimet F.-X.. Deblurring from highly incomplete measurements for remote sensing[J]. IEEE Transactions on Geoscience and Remote Sensing, 2009, 47(3): 792～802.

Single-image-based Rain Detection and Removal via CNN

Tianyi Chen[1, a] and Chengzhou Fu[1, b]

[1]School of Computer, South China Normal University, Guangzhou 510631, China
*Corresponding author. [a]tychen@m.scnu.edu.cn, [b]fucz@m.scnu.edu.cn

Abstract. The quality of the image is degraded by rain streaks, which have negative impact when we extract image features for many visual tasks, such as feature extraction for classification and recognition, tracking, surveillance and autonomous navigation. Hence, it is necessary to detect and remove rain streaks from single images, which is a challenging problem since we have no spatial-temporal information of rain streaks compared to the dynamic video stream. Inspired by the priori that the rain streaks have almost the same feature, such as the direction or the thickness, although they are in different types of real-world images. The paper aims at proposing an effective convolutional neural network (CNN) to detect and remove rain streaks from single image. Two models of synthesized rainy image, linear additive composite model (LACM model) and screen blend model (SCM model), are considered in this paper. The main idea is that it is easier for our CNN network to find the mapping between the rainy image and rain streaks than between the rainy image and clean image. The reason is that rain streaks have fixed features, but clean images have various features. The experiments show that the designed CNN network outperforms state-of-the-art approaches on both synthesized and real-world images, which indicates the effectiveness of our proposed framework.

1. Introduction

Clean image is necessary in computer vision research and application, such as feature extraction for classification or recognition, tracking, surveillance and autonomous navigation. Under rainy conditions, the qualities of image are often degraded by rain streaks. Effective methods to remove rain streaks from single image are becoming research hot in recent years. Single-image-based method and video-based method are two main tasks for rain detection and removal. Compared to video-based method, it is more difficult to do rain detection and removal by single-image-based method due to the following reasons.

- There are no temporal contents of rain streak in single image.
- Some details of the image are similar to the rain streak, which makes it hard to detect and remove the rain streak and keep the original image structure wholly.

In this paper, we propose an effective CNN network to detect the rain streak in the rainy image, thus by subtracting the rain streak, we can obtain the original clean image.

2. Background

For the rain streak in the dynamic video stream, it is easy to identify the rain streak by comparing the differences of the continuous video frames [1]. To detect and remove rain streak in single image is more challenging. J. H. Kim et al. [2] proposed a mean filtering and a kernel regression to detect and remove rain streaks. The patch-based approaches [3, 4] are used to remove rain streaks effectively. In recent years, the CNN [5] has been designed to do rain-removal and the satisfactory results have been gotten. Based on the deep CNN, the mapping from rainy image to clean image detail layer is learnt. We know that the mapping depends on each pair of the rainy image and clean image, which results in less robust of the CNN parameters. We notice that the rain streaks in different rainy images have the same features, such as the direction or the thickness. In this paper, we design a new CNN to get the

IOP Conf. Series: Journal of Physics: Conf. Series **1004** (2018) 012007 doi:10.1088/1742-6596/1004/1/012007

mapping between the rainy images and the rain streaks. Thus, the clean image is obtained by subtracting the rain streaks from the rainy images.

In this paper, we consider two models of synthesized rainy image, LACM model and SCM model.

- LACM model [1,3,5,6]

$$I = B + R, \tag{1}$$

- SCM model [4]

$$I = B + R - B \cdot R, \tag{2}$$

Where I denotes a rainy image, which is composited by two layers, de-rained image layer B and rain layer R. Here "·" denotes point-wise multiplication operator. Some examples are shown in Figure 1.

(a) Clean image (b) LACM (c) SCM

Figure 1 Some examples of clean image, LACM-model image and SCM-model image.

By guided image filter [7], a rainy image I could be decomposed to obtain the low-pass layer I_{base} and the high-pass layer I_{detail},

$$I = I_{detail} + I_{base}, \tag{3}$$

Where I_{base} and I_{detail} contain the smooth part and the detail part of I respectively. It is obvious that the rain layer R is high-pass, which is included in I_{detail}. Thus, the high-pass layer I_{detail} is mainly studied as shown in [5], which makes it easier to find the suitable mapping between I_{detail} and the rain streaks. Moreover, the computing cost is lower when only the I_{detail} is considered. A CNN network is trained in [5] instead of the raw rainy image to obtain the clean image.

Although the performance is better in [5] to remove the rain streaks from the rainy image. We know that the CNN network [5] is to directly find the mapping between I_{detail} and the clean image. There are so many kinds of rainy images, which make the CNN network unstable since it depends on each pair of the I_{detail} and the clean image. Thus, the effect is not robust when the rainy images, which are not like the training images, are tested. We notice that the rain streaks in different types of rainy images almost have the same features, such as the direction or the thickness of the rain streaks. In this paper, a new CNN network is proposed to find the mapping between the rainy images and the rain streaks, instead of the mapping between the rainy images and the clean images. The rain streaks are taken as the supervised object in our CNN network whose destination is to learn the features of the rain streaks. So, the result does not depend on each pair of the input images. Moreover, compared to the complex features of various clean images, the features of the rain streaks are more stable, which makes our CNN network more robust. Thus, our CNN network is suitable to different kinds of rainy images, including synthesized and real-world rainy images.

3. Our proposed method

3.1. Customized-filter rainy detection and removal network (CRDRN)

We propose a novel rain detection and removal model. The input images are the rainy image I and the rain streak R. A customized-filter rain detection and removal network(CRDRN) is designed to detect and remove the rain streak R in I, which minimizes the objective function as follows:

$$L = \sum_{n=1}^{N} ||f(I_{detail}^{n}, W, b) - R^{n}||_F^2, \tag{4}$$

Where N is the number of training images and n indexes the image, $f(\cdot)$ is CRDRN, W and b are network parameters that need to be learned, I^0 is the detail layer obtained by guided image filter, R is the supervised rain streaks. Our CRDRN network architecture is shown in Figure 2.

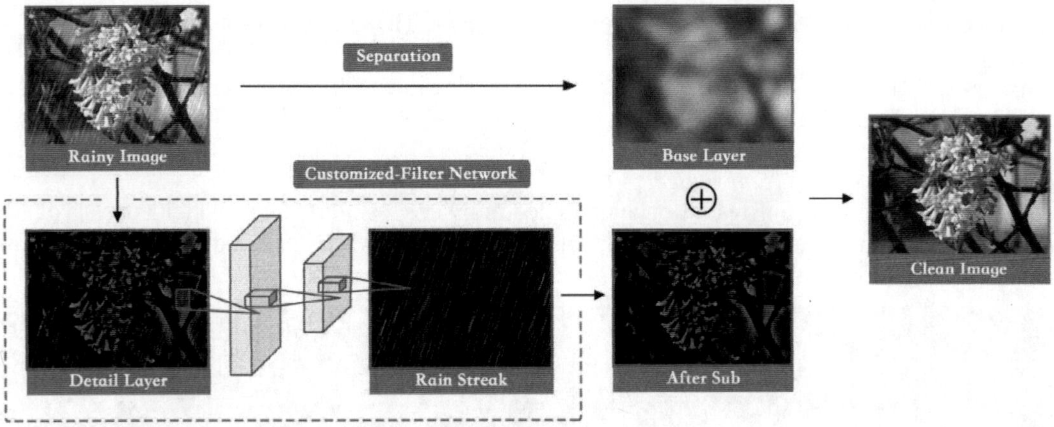

Figure 2 The proposed CRDRN network architecture for single-image-based rain detection and removal.

3.1.1. Rain detection and removal. Our CRDRN network consists of three CNN layers to detect and remove rain streaks, which can be expressed as the following operations:

$$I_{detail} = I - I_{base}, \tag{5}$$

$$I^l = \sigma(W^l * I_{detail} + b^l), \ l = 1, \tag{6}$$

$$I^l = \sigma(W^l * I^{l-1} + b^l), \ l = 2, \tag{7}$$

$$I^l = (W^l * I^{l-1} + b^l), \ l = 3, \tag{8}$$

$$O = I_{detail} - I^l + I_{base}, \ l = 3, \tag{9}$$

Where l =1,2,3 indexes the layer number, "*" is the convolution operation, W^l is the trained weight and b^l is the bias. The $\sigma(\cdot)$ is ReLU nonlinear activation function [8]. The ReLU is effective to truncate the parameters less than *0*, which makes the CRDRN network parameters take normal values. Two hidden layers are used in our CRDRN network and O is the output.

For the first layer in CRDRN, we use filters of size $s_1 \times s_1 \times b \times c$ to obtain c feature maps by b input image channels; s denotes filter size. For the second layer, the size of filters is $s_2 \times s_2 \times c \times c$, which extract the rain streaks from the rainy images. For the last layer, the size of filters is $s_3 \times s_3 \times c \times b$, which turn c feature map into b-dimension output image.

3.1.2. Training. We use Adam optimizer [9] and take the average of 50000 iterated times to minimize the objective function (4). We selected 350 clean images to generate training set as in [5]. Based on two generation models (LACM model and SCM model) and seven different streak orientations, each clean image was used to synthesize 14 pairs of rainy images and rain streaks. Thus, a dataset containing $350 \times 14 = 4900$ pairs of rainy images and rain streaks is set up. We randomly selected 40 $64 \times 64 \times 3$ rainy image/rain streak patch pairs from each pair. The final training dataset contains 196000 rainy image/rain streak patch pairs.

IOP Conf. Series: Journal of Physics: Conf. Series **1004** (2018) 012007 doi:10.1088/1742-6596/1004/1/012007

4. Experiment

4.1. Parameters setting

To evaluate the performance of our CRDRN, both synthesized and real-world rainy images are tested. Three recent high-quality methods [3,4,5] are compared with ours. Software implementations of our method were provided by Python based on Tensorflow. All experiments are carried on a personal computer with Intel Core CPU i7-7700, 32GB RAM and NVIDIA GeForce GTX 1080. Our CRDRN network consists of two hidden layers and one output layer. The kernel sizes are set as $s_1 = 5$, $s_2 = 1$ and $s_3 = 3$ respectively. The numbers of convolution kernels are $b = 3$ and $c = 512$. All these parameters s, b and c are defined in Sec. 3.1.1. The learning rate of Adam optimizer is set as 10^{-4}, and the mini-batch size is set as 50.

4.2. Results on synthesized rainy image

Firstly, we evaluate the testing results on synthesized rainy images. Rain12 [3], which consists of 12 synthesized rainy images and is a benchmark dataset for rain removal, is selected as the testing dataset. The synthesized rainy images in Rain12 are generated using photorealistic rendering techniques [10].

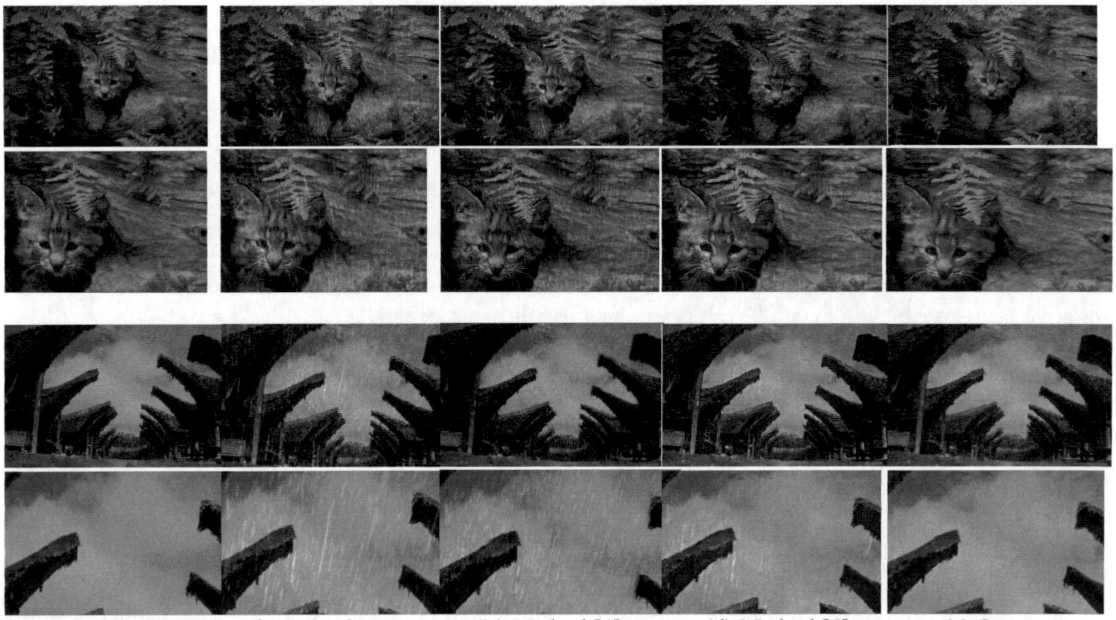

(a) Ground truth (b) Rainy image (c) Method [4] (d) Method [5] (e) Ours

Figure 3. Example results on synthesized rainy images "cat" and "temple" respectively, which were for testing and not for training. The odd row shows the full image and the even row shows one zoomed-in region.

Figure 3 shows visual comparisons for two synthesized rainy images "cat" and "temple" respectively. As can be seen, the outputs of method [4] still contain many short rain streaks although they are lighter. Method [5] is better, but some rain streaks are remained in the images. However, as shown in the last column, our CRDRN method can detect and remove rain streaks.

Since we have the ground truth, the SSIM [11] could be used for quantitative evaluation. A higher SSIM value means a better result in terms of image structural properties. As shown in Tab.1, our method has the highest SSIM values, which is in agreement with the visual results in Figure 3.

Table 1. Each quantitative comparison results on Rain12 [3] benchmark dataset using SSIM.

	Pic1	Pic2	Pic3	Pic4	Pic5	Pic6	Pic 7	Pic8	Pic 9	Pic 10	Pic 11	Pic 12	Average
DSC	0.86	0.92	0.86	**0.98**	0.93	0.97	0.97	0.94	0.91	0.86	0.90	0.87	0.91

IOP Conf. Series: Journal of Physics: Conf. Series **1004** (2018) 012007 doi:10.1088/1742-6596/1004/1/012007

GMMs	0.88	0.93	0.93	0.93	0.90	0.95	0.96	0.90	0.91	0.90	0.86	0.92	0.91
DerainNet	0.90	0.93	0.92	**0.98**	0.94	0.97	**0.98**	0.95	0.94	0.91	0.91	0.93	0.94
Ours	**0.91**	**0.95**	**0.95**	**0.98**	**0.95**	**0.98**	**0.98**	**0.98**	**0.96**	**0.93**	**0.94**	**0.95**	**0.96**

Moreover, we selected 100 images, which are never used in training set, from BSD300 [12] dataset. LACM model is used to generate rainy images as testing dataset with random directions. SSIM is used for quantitative evaluation. As shown in Tab.2, our method has the highest SSIM values compared to DSC method [4] and DerainNet method [5].

Table 2. Quantitative measurement results on BSD300 [12] using SSIM.

	DSC	DerainNet	Ours
SSIM	0.92	0.93	**0.94**

(a)Real-world image (b)Method [4] (c)Method [5] (d)Ours

Figure 4. Example results on real-world rainy image "soccer" and "street".

4.3. Results on real-world rainy image

For we have no ground truth for real-world rainy images, our CRDRN method is tested on real-world images using the network trained on the 196000 rainy image/rain streak patch pairs as described in Sec. 3. As shown in Figure 4, the outputs of method [4] still contain many rain streaks. There are less rain streaks in the images of method [5], but the images are blurred. However, as shown in the last column, our CRDRN method not only detects and removes rain streaks, but also make the image clearer, which indicates that our CRDRN method behaves well for rain detection and removal for real-world rainy image.

5. Conclusion

In this paper, we propose a CNN network architecture called CRDRN for rain detection and removal from single images. Based on CRDRN network our method learns the features of rain streaks of rainy image detail layers. Although the CRDRN network is trained on synthesized clean/rainy image pairs, it shows better performances on both synthesized dataset and real-world dataset, which indicates that our CRDRN network is effective and robust for rain detection and removal. The experiments show that our CRDRN method outperforms other state of-the-art methods regarding image quality.

In future, we will work on setting up an image dataset for better evaluating framework for single-image rain detection and removal. In addition, we will also investigate the applicability of the proposed method to other challenges in computer vision, for example, single-image snow detection and removal.

Acknowledgement
Tianyi Chen thanks the supports by the Special Funds for the Cultivation of Guangdong College Students' Scientific and Technological Innovation ("Climbing Program 2018" Special Funds, No.pdjha0129), and the Project of Golden Seed Cultivation (No.17JJKC06). Chengzhou Fu thanks the supports by Guangdong Science and Technology Project (No.2013B090800024), the Project of Young Education Research and Cultivation Fund of South China Normal University (No.17KJ16), and the Science and Technology Program of Guangzhou, China (No.201604016007).

References

[1] Barnum P C, Narasimhan S and Kanade T 2010 Analysis of rain and snow in frequency space *Int. Journal of Computer Vision* 86 256–274
[2] Kim J H, Lee C, Sim J Y and Kim C S 2013 Single-image deraining using an adaptive nonlocal means filter *Int. Conf. on Image Processing* 914-917
[3] Li Y, Tan R T, Guo X, Lu J and Brown M S 2016 Rain streak removal using layer priors *IEEE Conf. on Computer Vision and Pattern Recognition* 2736-2744
[4] Luo Y, Xu Y and Ji H 2015 Removing rain from a single image via discriminative sparse coding *Int. Conf. on Computer Vision* 3397-3405
[5] Fu X, Huang J, Ding X, Liao Y, Paisley J 2016 Clearing the skies: a deep network architecture for single-image rain streaks removal *IEEE Transactions on Image Processing* 26 2944-2956
[6] Kang L W, Lin C W and Fu Y H 2012 Automatic single-image-based rain streaks removal via image decomposition *IEEE Transactions on Image Processing* 21 1742–1755
[7] He K, Sun J and Tang X 2013 Guided image filtering *IEEE Transactions on Pattern Analysis and Machine Intelligence* 35 1397-1409
[8] Krizhevsky, Alex, Sutskever I and Hinton G E 2012 ImageNet classification with deep convolutional neural networks Annual Conf. on Neural Information Processing Systems 25 1097-1105
[9] Kingma D P and Ba J 2014 Adam: A method for stochastic optimization Preprint arXiv:1412.6980
[10] Garg K and Nayar S K 2006 Photorealistic rendering of rain streaks *ACM Transactions on Graphics* 25 996–1002
[11] Wang Z, Bovik A C, Sheikh H R and Simoncelli E P 2004 Image quality assessment: From error visibility to structural similarity *IEEE Transactions on Image Processing* 13 600–612
[12] Martin D, Fowlkes C and Tal Dand Malik J 2001 A database of human segmented natural images and its application to evaluating segmentation algorithms and measuring ecological statistics *Int. Conf. on Computer Vision* 2 416–423

IOP Conf. Series: Journal of Physics: Conf. Series **1004** (2018) 012008 doi:10.1088/1742-6596/1004/1/012008

Understanding of the Cyber Security and the Development of CAPTCHA

Yu Yang[1,*]

[1]School of Computer, Wuhan University, Wuhan, Hubei Province, P.R.China 430072

*562536365@qq.com

Abstract. CAPTCHA is the abbreviation of "Completely Automated Public Turing Test to Tell Computers and Humans Apart", which is a program algorithm for distinguishing between computers and humans. It is able to generate and evaluate tests that are easy for human to pass yet are not possible for computers to. Common CAPTCHA generally contains symbols, text, pictures, and even videos, which is mainly used for human-computer verification. With the popularization of the Internet and its related applications, many malicious attacks against websites, systems and servers gradually appear. Therefore, the research on CAPTCHA is especially important. This article will briefly summarize and introduce the existing CAPTCHA technology, and summarizes the common problems of network attacks and information security. After listing the common type of CAPTCHA, it will finally propose feasible suggestions for the development of CAPTCHA.

1. Introduction

With the popularization and development of information technology, the issue of information security has drawn more and more attention. Especially the importance of cyber security has become a hot issue in many countries. One of the many solutions to common cyber security problems is to provide protection against network attacks on the client side, of which the CAPTCHA technology is the most widely used.

Completely Automated Public Turing Test to Tell Computers and Humans Apart (CAPTCHA) is a program that generates and evaluates solvable tests yet which exceed the capabilities of current computer programs [1]. This technology is now almost a standard security mechanism for defense. A good human-machine identification system must not only be humanized, but powerful enough to automatically pass (or challenge) the CAPTCHA test so as to prevent the computer program written by an attacker.

The most widely used CAPTCHA is the so-called text-based approach which relies on the complex distortions of textual images with the aim of rendering them unable to be recognized by the current pattern recognition methods. The popularity of such schemes is due to the many advantages they have, such as being intuitive to global users (user tasks are character recognition only), few localization problems (people from different countries recognize Roman characters), and have good potential to provide strong security (for example, if there is a perfect strategy for anti-attack, the computing amount of violent attacks will be huge) [2].

Earlier studies show that even though these characters are highly distorted, computers are very good at recognizing individual characters [3]. If CAPTCHA produces a known position of a character in the challenge image, this is a purely recognition problem for computers. Standard machine learning techniques, such as neural networks, can achieve high recognition success rates [4]. However, existing techniques (including machine learning) methods do not perform well in locating roles when the character position in the CAPTCHA challenge is unknown. The problem of identifying a character's position in the correct order or segment is still a challenging issue in areas such as handwriting recognition and computer vision. In general, segmentation is computationally expensive and is usually a comprehensive issue [2].

This article discusses common cyber security issues and CAPTCHA, including a description of common types of malicious cyber attacks, common types of CAPTCHAs, and challenges that CAPTCHA has encountered and is about to encounter.

2. Background

CAPTCHA systems are sometimes referred to as "reverse Turing tests," for whose goal is to let the computer determine whether a remote client is a human. Although they are extremely important and widely used, there is no systematic way to design or evaluate them [5, 6, 7]. In fact, as we've demonstrated through in-depth research, many popular Web sites still rely on solutions that are vulnerable to match automated attacks. For instance, our program of automated DECAPTCHA tool breaking the Wikipedia. Among the 15 most widely used current scenarios, 13 are equally vulnerable to automated attacks by our tools. Therefore, a complete set of design and test principles will result in a significant need for more powerful CAPTCHAs. Previous work [5] showed that the security of the CAPTCHA depends on preventing the splitting, and we found that in our research, relying on segmentation alone does not provide a reliable defensive attack auto-attack.

Based on the techniques described in reference [8, 9], machine learning techniques [10, 11, 12] and efficient visual algorithms [10, 13, 14] can be used. We divide the automatic verification code solving process into five steps: Preprocessing, segmentation, general post-segmentation, recognition, and post-processing. While segmenting and separating a sequence of characters into single characters, identifying and recognizing these features is intuitive and generally understood, there are good reasons to consider extra pre-processing and post-processing steps as a standard procedure. For example, the preprocessing may remove the background pattern or eliminate other additional images that may interfere with the segmentation, while the segmented steps may "clean" the segmented output to regulate the size of each image, otherwise performing a different segmentation step. Recognition and post-processing can improve accuracy, for example, using spell checking for any verification code based on actual words (such as Slashdot).

3. Common attacks

In this section, the article will mainly introduce the main types of existing cyber attacks, and its attack principle. In general, there are four broad categories: read attacks, operational attacks, cheating attacks, and flood attacks.

3.1. Read Attacks

Read attacks primarily involve all relevant attacks that get information from the victim. Such attacks get the IP address of the organizational structure, perform port scanning and vulnerabilities scans in those address ranges, and finally gain access to compromised hosts.

3.1.1. Reconnaissance Attacks. Designed primarily to give attackers access to more information about victims, reconnaissance attacks can be proactive and reactive, and in almost all cases successful reconnaissance attacks greatly increase the likelihood of subsequent attacks succeeding since the attacker gets more information about the victim.

3.1.2. Sniffing attack. This can be referred to as some form of sniffing attack when an attacker catches a packet from the cable, or a packet passes through an attacker's system. The purpose of sniffing attacks is to read the information to obtain intelligence so that attackers understand the target system, so that the sniffed protocol information must be sent in plaintext rather than cipher text, then the sniffing is successful. The main information is as follows: Authentication information, network management information, confidential affairs and so on.

3.2. Operational Attacks

3.2.1. Network manipulation attacks. The most common is IP fragmentation, where attackers deliberately fragment traffic flow in an attempt to bypass network-based (IDS or firewall) or application-based security controls. In addition to IP fragmentation, attackers can also perform source routing attacks, where an attacker can use source routing to choose an attack path in the network. Source routing has almost no legal application and is off by default on most routers. There are many attackers who can exploit the IP, TCP, and UDP protocols to attack.

3.2.2 Application manipulation attacks. Application manipulation attacks are attacks that take place at the application level. Its main use of application design or implementation of the program flaws, the most famous application manipulation attack is a buffer overflow attacks. Such as web application attacks, and insecure Common Gateway Interface (CGI).

3.3. Cheating Attacks

A cheating attack is that an attacker could cause a user or a device in the system to believe that the information came from a source that did not actually send the message, which can happen in any virtually location that is not certified for weakness or network traffic.

3.3.1. MAC cheat. It is a very straightforward attack. Attack systems use this attack to change their MAC address to a trusted system address. In an Ethernet environment, the content addressable memory (CAM) table on the switch tracks MAC addresses, VLANs, and MAC addresses. When an attacker changes the MAC address to the address of another system connected to the switch, the CAM table will be updated and the switch master considers a machine to move from one location to another.

3.3.2. IP cheat. An attacker simply enters the system's original packet driver, and an attacker can send a packet containing the IP header.

3.3.3. Transmission cheat. Transmission cheat means that communication spoofing is successfully implemented at the transport layer, such as UDP spoofing, TCP spoofing, and identity spoofing.

3.4. Flood Attacks

The flood attack is when an attacker sends too much data to some network resources. These network devices can be routers, switches, applications, hosts, or network links.

3.4.1. MAC Flooding. MAC flooding refers to the fake source MAC address and destination address of the packet from the attacker's system sent to the Ethernet link for MAC address occupied on the exchange in the CAM position, the CAM table capacity is limited, When CAM is full, other hosts can only flood the local LAN, allowing attackers to sniff these frames.

3.4.2. Network flooding. Network flooding is generally designed to consume the available bandwidth of the network link. This attack sends legitimate traffic to cables flooded with fraudulent traffic, greatly reducing the amount of legitimate traffic on the available broadband, often targeting internet links on the network.

4. Common types of CAPTCHA

4.1. Graphic CAPTCHA

IOP Conf. Series: Journal of Physics: Conf. Series **1004** (2018) 012008 doi:10.1088/1742-6596/1004/1/012008

The Graphic CAPTCHA is the earliest and most widely used CAPTCHA. Common graphic CAPTCHA codes often contain symbols, numbers, letters, etc., and then the verification code can be enhanced by aided means to improve the recognition difficulty. Common aids as shown in Table 1 and Table 2 below.

Table 1. Common CAPTCHA reinforcement method

Fonts distortion	Background color interference
Fonts Adhesion	Background letters interference
Fonts hollow	Formula verification code
Mix of fonts	Addition and subtraction verification code
Main interference line	Logic verification code

Table 2. Example of CAPTCHA hardening method

Example				
Reinforcement method	Noise, fonts distortion	Different styles, alphabetic shading, alphabetical adhesion, background color interference	Main interference line, background color interference, interference background letters, font distortion, alphabetical adhesion	Background interference line, background color interference, background letter interference, font distortion, alphabetical adhesion, fonts hollow

4.2. SMS verification code

SMS verification code is a valid verification code system by sending a verification code to the phone. SMS verification code is commonly seen in registration, the stage of forgetting the password, confirming the order and other stages, especially when it comes to the user's personal sensitive behavior. In order to confirm the operation is conducted by the user, the SMS authentication code is usually used for secondary verification. At the same time, SMS verification code is also one of the most common types of CAPTCHA.

4.3. Voice verification code

Compared with the SMS verification code, the voice verification code has the following advantages: 1) the voice verification is more effective on the anti-brush single effect; 2) the voice verification code has a higher arrival rate; and 3) the user experience is more friendly

4.4. Sliding verification code

IOP Conf. Series: Journal of Physics: Conf. Series **1004** (2018) 012008 doi:10.1088/1742-6596/1004/1/012008

Now there are a lot of Internet companies begin to use the sliding verification code for verification. Seemingly a simple sliding operation, the wind control engine and the corresponding rules are behind as a guarantee, the specific verification process is shown in Figure 1.

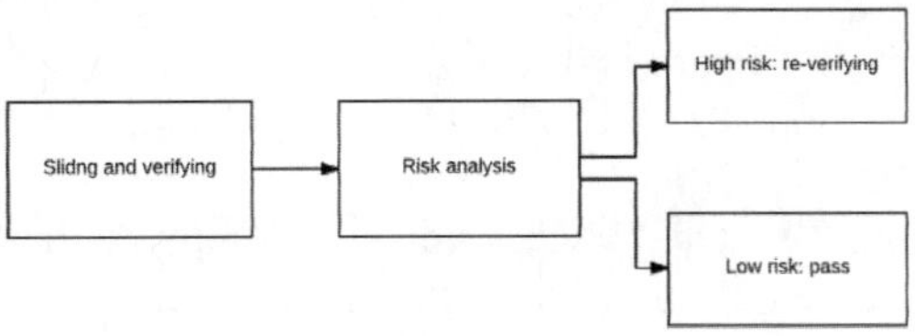

Figure 1. Sliding verification process

First, the user slides the CAPTCHA to the specified location. Upon completion, it will send back various encrypted information to the server. In order to judge whether there is any abnormality, the rules include user IP, operation path, UA, COOKIE, device fingerprint, etc. If the rule is not achieved, it will release verification, if achieved, it will pop up a secondary calibration, only after passing the verification can be released.

5. Main challenges

5.1. Database limitations

No matter what form of verification code is, it has a database to store pictures, text, voice and other information, if the database is limited or updated too slowly, malicious attackers will easily traverse the entire database, making them easily to master the crack code verification method.

5.2. Poor experience

Many users encounter this type of problem when entering a passcode, such as too obscured images, illegible numbers or letters, such as "0" and "O," or images that are too small. These problems directly increase the difficulty of the user identification code, when the user fails to verify the output of multiple answers and the system does not prompt, the user is likely to cease operation, so that the site will lose potential users.

5.3. New technology brings challenges

As mentioned before, the development of new technologies will bring new challenges to CAPTCHA technology. Machine learning, especially deep learning, greatly improves the computing power of the computer, reduces the computing time, and greatly reduces the difficulty of cracking the verification code.

6. Conclusion

With the development and popularization of information technology, the problem of cyber security has become increasingly prominent. This article focuses on the forms of network attacks that are common today, and discusses the types and features of CAPTCHAs according to one of the solutions.

The CAPTCHA design is an interdisciplinary topic where the expertise from multiple domains plays an important role. The design of CAPTCHAs with good robustness and usability has been the research topic for a long time. The current collective understanding of this topic is still in its infancy. In order to develop the CAPTCHA design from the arts to science still need a lot of research. Our experience shows that CAPTCHA will lead to the development of more robust systems through similar evolutionary processes, such as cryptography and digital watermarking, and through successful iterations.

References

[1] L von Ahn, M Blum and J Langford. "Telling Humans and Computer Apart Automatically", CACM, V47, No2, 2004.
[2] K Chellapilla, K Larson, P Simard and M Czerwinski, "Building Segmentation Based Human-friendly Human Interaction Proofs", 2nd Int'l Workshop on Human Interaction Proofs, Springer-Verlag, LNCS 3517, 2005.
[3] K Chellapilla, K Larson, P Simard, M Czerwinski, "Computers beat humans at single character recognition in reading-based Human Interaction Proofs", 2nd Conference on Email and Anti-Spam (CEAS), 2005.
[4] P Simard, D Steinkraus, J Platt. "Best Practice for Convolutional Neural Networks Applied to Visual Document Analysis", International Conference on Document Analysis and Recognition (ICDAR), IEEE Computer Society, Los Alamitos, pp.958-962, 2003.
[5] K. Chellapilla, K. Larson, P.Y. Simard, and M. Czerwinski. Computers beat humans at single character recognition in reading based human interaction proofs (hips). In CEAS, 2005.
[6] K Chellapilla and P Simard. Using machine learning to break visual human interaction proofs. In MIT Press, editor, Neural Information Processing Systems (NIPS), 2004.
[7] J. Yan and A.S.E. Ahmad. Breaking visual captchas with naive pattern recognition algorithms. In ACSAC 2007, 2007.
[8] K. Chellapilla and P.Y. Simard. Using machine learning to break visual hips. In Conf. on Neural Information Processing Systems, NIPS 2004, 2004.
[9] J. Yan and A.S. El Ahmad. A Low-cost Attack on a Microsoft CAPTCHA. In Proceedings of the 15th ACM conference on Computer and communications security, pages 543–554. ACM, 2008.
[10] H. Bay, T. Tuytelaars, and L. Van Gool. Surf: Speeded up robust features. Computer Vision–ECCV 2006, pages 404–417, 2006.
[11] Y. LeCun, L. Bottou, Y. Bengio, and P. Haffner. Gradient-based learning applied to document recognition. Proceedings of the IEEE, 86(11):2278–2324, 1998.
[12] R. Quinlan. Machine Learning. Morgan Kaufmann Pub, 1992.
[13] Duda R O, Hart P E. Use of the Hough transformation to detect lines and curves in pictures. Communications of the Acm, 15(1):11-15, 1972.
[14] S. Geman and D. Geman. Stochastic relaxation, Gibbs distributions and the Bayesian restoration of images*. Journal of Applied Statistics, 20(5):25–62, 1993.

IOP Conf. Series: Journal of Physics: Conf. Series **1004** (2018) 012009 doi:10.1088/1742-6596/1004/1/012009

Research on Stereo Matching Algorithm Based on Improved Steady-State Matching Probability

Jianye Zhang[1] and Yan Piao[1]

[1]Changchun University of Science and Technology, Changchun of JiLin, China

*E-mail: piaoyan@cust.edu.cn

Abstract. Aiming at the hole phenomenon generated by steady-state matching probability (SSMP) stereo matching algorithm in the test graph where the disparity is large,and the problem of correct disparity loss in the left disparity map caused by the error disparity in the right disparity map after using that algorithm. A stereo matching algorithm based on SSMP and semi-global matching (SGM) is proposed. Using the consistency detection method of the disparity maps obtained by SSMP and SGM are presented to fill the disparity map and the filling criterion based on hill-climbing algorithm is introduced. The experimental results demonstrate that the average bad pixels rates of the improved SSMP algorithm is reduced from 5.38% to 5.23% on the Middlebury benchmark. And the proposed algorithm can deal with the above problems well, improve the matching accuracy effectively and robustness.

1. Introduction

As one of the hot spots in the field of machine vision, stereo matching can be divided into two categories: local algorithms and global algorithms. Global algorithms include graph cut (GC) [1], dynamic programming (DP) [2], believe propagation (BP) [3], etc. Global algorithms have high precision but slow speed and local algorithms have fast speed but low precision.

In 2011, Rhemann used color graph as the guide filter to aggregate matching cost quickly [4]. Yang pioneered the cost aggregation using the minimum spanning tree method in 2012 [5]. In 2014 Ham proposed the SSMP to obtain disparity map with high accuracy [6]. However, the SSMP often leads to the loss of correct disparity information because of the occlusion. The hill-climbing algorithm is used to segment images based on the consistency of color, using the disparity of the same label page to fill the disparity inconsistency pixels and solving the hole problem of the disparity map. And the disparity consistency between the disparity map obtained by using the SSMP and another one derived by using the cost aggregation method in the SGM is used to solve the problem of the correct disparity loss in the left disparity map due to the inaccuracy of the right disparity map.

2. SSMP

The process of the SSMP is shown in figure 1. First, the initial matching cost function is constructed by using GRD. Then, using the RWR and winner takes all(WTA) to aggregate the matching cost and get an initial disparity map. And finally the refined disparity map is obtained by the empty pixels filling and median filtering.

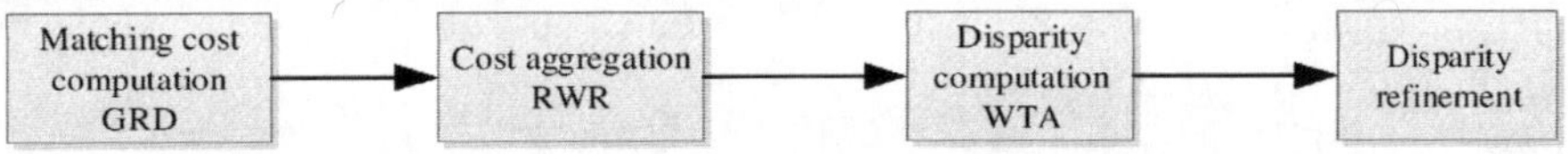

Figure 1 The flow chart of SSMP algorithm

2.1. Matching cost computation

The matching cost is constructed by using the combination of intensity and gradient information. For any pixel point $p(x,y)$ in the left image, the corresponding disparity is d, and in the right image $pd(x-d,y)$ is the corresponding matching point of p. The initial matching cost function is as follows:

$$C^0(p,d)=\delta\cdot\min\{\|\psi(p,d),\tau_1\|\}+(1-\delta)\cdot\min\{\|\nabla_x\psi(p,d)\|,\tau_2\} \tag{1}$$

Where $\psi(p,d)$ is the color difference between p and pd; and the gradient difference between p and pd along the X direction is $\nabla\psi_x(p,d)$; δ is weight coefficient; τ_1 and τ_2 are cut off value.

2.2. Cost aggregation

Firstly, the initial matching probability of the pixel point can be determined by its initial matching cost:

$$P^0(p,d)=\frac{1}{Z(p)}\exp\{-vC^0(p,d)\} \tag{2}$$

$$Z(p)=\sum_d \exp\{-vC^0(p,d)\} \tag{3}$$

Where $P^0(p,d)$ can be calculated with an initial matching cost $C^0(p,d)$; v is the positive factor that controls the growth of the matching probability.

And then, the RWR is used to obtain the stable matching probability:

$$P^{t+1}(p,d)=(1-\alpha)\frac{\sum_{q\in N_p} w(p,q)P^t(q,d)}{\sum_{q\in N_p} w(p,q)}+\alpha P^t(p,d) \tag{4}$$

Where t is the number of iterations. N_p denotes the 4-neighborhood of p. $w(p,d)$ is an edge weight between p and q.

$$w(p,q)=\exp\left(-\frac{\|I_l(p)-I_l(q)\|_2^2}{\gamma_c}\right) \tag{5}$$

Where γ_c represents the bandwidth parameter, typically set to the intensity variance.

After $t+1$ iteration, the matching probability reaches a stable state $P^{t+1}(m,d)=P^t(m,d)$. At this pixel point, $P^t(m,d)$ is the stable matching probability when the disparity is d.

2.3. Disparity computation

Using the WTA to derive the disparity :

$$d=\arg\max(P(p,1),P(p,2),\ldots,P(p,d_{\max})) \tag{6}$$

Where $d_{\max}$ represents the Parallax search range; $\arg\max(\ )$ is the value of d when $P(p,d)$ takes the maximum.

3. Proposed algorithm

3.1. Algorithm structure

Firstly, the SSMP is used to obtain the initial disparity map, and the filling criterion based on color similarity is added. Then, using SGM[7] to obtain another disparity map, and the consistency disparity information of the left disparity map obtained by the SSMP and the SGM is used to fill left disparity map which contains empty pixel points. Finally, the refined disparity map is obtained by the empty pixel points filling and median filtering of the SSMP. In figure 2, there is the diagram of the proposed algorithm and the dotted line area is the improvement of the SSMP.

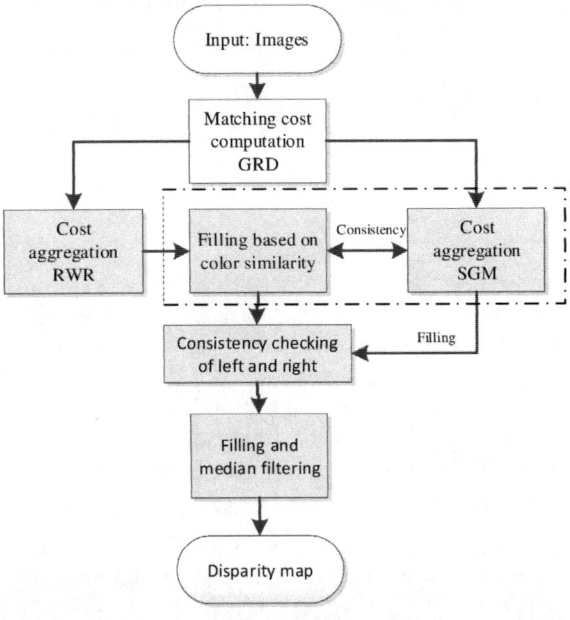

Figure 2 The improved SSMP algorithm diagram

3.2. The cost aggregation method in the SGM algorithm

The energy function is as follows in term of the cost aggregation algorithm of the SGM.

$$E(D)=\sum_{p}\left(C(p,d_p)+\sum_{q\in N_p}P_1T\left[\left|d_p-d_q\right|=1\right]+\sum_{q\in N_p}P_2T\left[\left|d_p-d_q\right|>1\right]\right) \tag{7}$$

Where $C(p,d_p)$ is initial matching cost; P_1 and P_2 are constant penalties, and P_2 should be far greater than P_1.

$$L_r(p,d)=C(p,d)+\min\begin{pmatrix} L_r(p-r,d), \\ L_r(pr,d-1)+P_1, \\ L_r(p-r,d+1)+P_1, \\ \min_i L_r(p-r,i)+P_2 \end{pmatrix}-\min_k L_r(p-r,k) \tag{8}$$

Where $L_r(p,d)$ is along a distance r of the pixel p at disparity d.

3.3. Filling criteria based on hill-climbing algorithm

Segment color image by using hill-climbing algorithm[8]. Establish 3D color histogram and count the number of pixels in each color at first. Then, select a number of pixels greater than 0 from the color histogram to start climbing the mountain, and count the number of pixels of the adjacent color. If the number of pixels in the adjacent color is different, then the neighbor as the current point, continue to climb the mountain of adjacent color pixels. When the mountain climbing can no longer be carried out, it is considered to be the peak. Then, select the number of other pixels more than 0 and have not been climbed, repeat the climb process until every color is climbed. The number of peaks selected represents the initial number of clusters. All colors which can climb to the same peak are classified as a class to segment color image.

After the image segmentation is completed, the different classes of the class of the pixels after the segmentation are labeled with different labels. The label page of any point p in the original image is expressed as $L_p, q \in N_p$, q is a point in eight neighborhood pixels of. The label page of q is represented as L_q. When the disparity $d_p < 1$, filling d_p with the following steps:

a) Initialization: $n = 0$, $S = 0$

b) When $d_q > 1$ and $L_p = L_q$: $n = n+1$, $\sum S + d_q$

c) Filling d_p: $d_p = \begin{cases} d_p & n < 3 \\ \left[S/n \right] & 3 \le n \le 8 \end{cases}$

4. Experimental results and analysis

The experiments of tsukuba, venus, teddy and cones on the Middlebury benchmark are carried out so as to demonstrate the effectiveness of the proposed algorithm. MATLAB programming language is used to realize the proposed algorithm. Computer hardware configuration for experiment is Intel(R)Core(TM)i7-6700HQ CPU@2.60GHz. Using the percent of bad pixels to evaluate the matching accuracy of the proposed algorithm. And the parameter values are set as follows:

$\{\delta, \tau_1, \tau_2, \gamma_c, \mathrm{v}, \alpha, P_1, P_2\} = \{0.11, 15/255, 2/255, 50, 3000, 0.003, 0.01, 1000\}$.

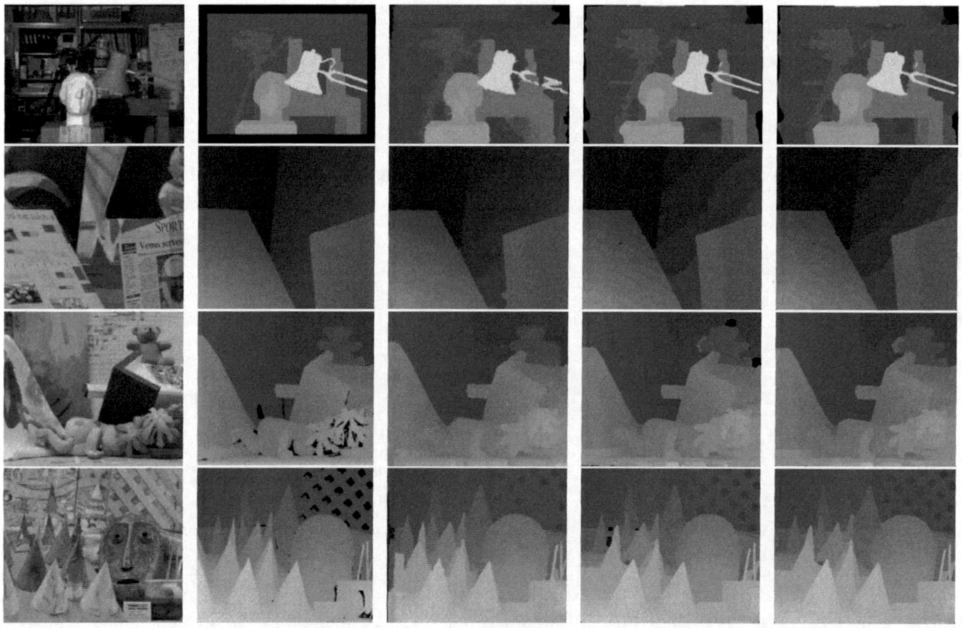

Figure 3 Experimental results of different algorithms on Middlebury2.0; (a)reference image; (b)standard disparity map;(c) SGM; (d)SSMP; (e) proposed

Figure 3 shows that the overall effect of the disparity map obtained by the SSMP is significantly better than the disparity map obtained by the SGM, but the SSMP has obvious cavity phenomenon in teddy and cones. The proposed algorithm has some improvement on tsukuba and venus, and the effect of teddy and cones are obviously improved. So as to compare the SSMP and the proposed algorithm better, the local magnification of the disparity maps are shown in figure 4.

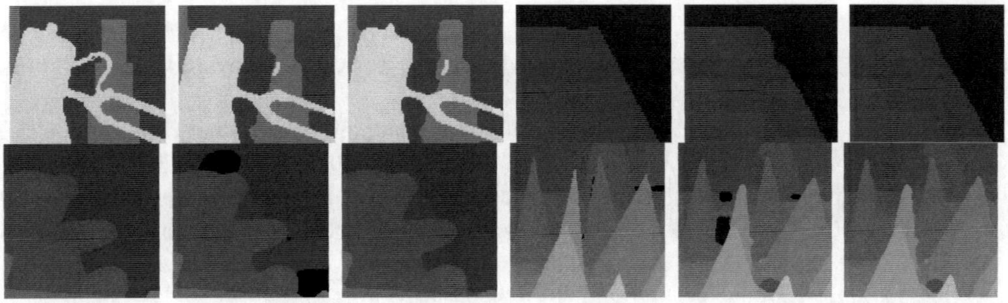

Figure 4 Local disparity map; (a)standard disparity map; (b)SSMP; (c) proposed

It is apparent that the difference between the SSMP and the proposed algorithm from figure 4, the proposed algorithm not only increases the correct information of the disparity map, but also can solve the hole problem well. With the purpose of comparing the proposed algorithm with others better, the bad pixels rates of different stereo matching algorithms are shown in table 1.

Table 1 Bad pixel rates of different stereo matching algorithms evaluated in all region (%) (Error threshold = 1)

Algorithm	Tsukuba	Venus	Teddy	Cones	Average
Proposed	1.87	0.36	11.1	7.60	5.23
SSMP	1.97	0.38	11.5	7.92	5.38
NL	1.88	0.42	11.6	8.45	5.48
ST[9]	2.76	0.60	15.2	7.86	6.82
SGM	3.96	1.57	12.2	9.75	7.50
BP	3.40	1.90	13.2	11.6	7.69
GC	4.12	3.44	25.0	18.2	11.4
DP	5.04	11.0	21.6	19.1	14.2

As shown in figure 3 and table 1, the SSMP has obvious advantages over traditional global algorithms on some images (such as the tsukuba and venus), but the advantages on some other images (such as teddy and cones) are less obvious. On the basis of SSMP, the proposed algorithm not only keeps the advantages of the SSMP, but also significantly reduces the false matching points in the teddy and cones. Therefore, from the average bad pixels rates of the above algorithms, the effect of the proposed algorithm is better than others, and the matching accuracy is improved.

5. Conclusion

An improved SSMP stereo matching algorithm is proposed. Hill-climbing segmentation algorithm based on color similarity and the consistency detection method of the disparity map obtained by SGM and SSMP are used to add more accurate disparity to the disparity map. The experimental results show that the proposed algorithm reduces the average percent of bad pixels of the images on the second version of Middlebury benchmark from 5.38% to 5.23%. It can solve the problem of missing Correct disparity in the left disparity map because of the inaccuracy of the right disparity map well, and effectively improve the matching accuracy. And further optimization will be made for the details area and edge in future research.

References

[1] Kolmogorov V, Zabih R. Computing visual correspondence with occlusions using graph cuts[C]. ICCV, 2001:508-515.
[2] Veksler O. Stereo correspondence by dynamic programming on a tree[C]. CVPR, 2005:384-390.
[3] Felzenszwalb P F, Huttenlocher D P. Efficient belief propagation for early vision[J]. International Journal of Computer Vision, 2006,70(1):41-54.
[4] Hosni A, Rhemann C, Bleyer M, et al. Fast cost-volume filtering for visual correspondence and beyond[J]. TPAMI, 2013, 35(2):504-511.

IOP Conf. Series: Journal of Physics: Conf. Series **1004** (2018) 012009 doi:10.1088/1742-6596/1004/1/012009

[5] Yang Q. A non-local cost aggregation method for stereo matching[C]. CVPR, 2012:1402-1409.
[6] Ham B, Min D, Oh C, et al. Probability-based rendering for view synthesis[J]. IEEE Transactions on Image Processing, 2014,23(2):870-884.
[7] Hirschmüller H. Stereo processing by semi-global matching and mutual information.[J]. IEEE Transactions on Pattern Analysis & Machine Intelligence, 2008, 30(2):328-341.
[8] Ohashi T, Aghbari Z, Makinouchi A. Hill-climbing algorithm for efficient color-based image segmentation[C], IASTED Conf. On Signal Processing, Pattern Recognition and Applications (SPPRA 2003). June 2003:17-23.
[9] X.Mei, X.Sun, W.Dong ,et al. Segment-tree based cost aggregation for stereo matching[C]. Computer Vision and Pattern Recognition. IEEE, 2013,313–320.

Multi-Stage Target Tracking with Drift Correction and Position Prediction

Xin Chen[1], Keyan Ren[1,*] and Yibin Hou[1]

[1]Beijing University of Technology, Beijing Engineering Research Center for IoT Software and Systems, Beijing, China.

*keyanren@bjut.edu.cn

Abstract. Most existing tracking methods are hard to combine accuracy and performance, and do not consider the shift between clarity and blur that often occurs. In this paper, we propound a multi-stage tracking framework with two particular modules: position prediction and corrective measure. We conduct tracking based on correlation filter with a corrective measure module to increase both performance and accuracy. Specifically, a convolutional network is used for solving the blur problem in realistic scene, training methodology that training dataset with blur images generated by the three blur algorithms. Then, we propose a position prediction module to reduce the computation cost and make tracker more capable of fast motion. Experimental result shows that our tracking method is more robust compared to others and more accurate on the benchmark sequences.

1. Introduction

VOT (Visual object tracking) is an essential topic in computer vision with many conceptually or methodologically various object-tracking algorithms being proposed recently [1,2,3,4]. The state-of-art target tracking methods are divided into two categories: Convolutional Neural Networks (CNN) [4,3,5,6] and Correlation Filter [1,2,7,8].

The advantage of correlation filtering method is faster than CNN, but the accuracy of correlation filter is lower. At the same time, the method of general correlation filtering needs to mark the object artificially at the first frame, and the bounding box of target is usually prone to drift due to the changes of object appearance changes. The application method of CNN is general detect each frame of the video in the target tracking field, which lead to a significant decrease in speed. Currently, universal trackers do not consider the blurry situation explicitly, which is general on the realistic scene when the targets move fast or photographic equipment shake unexpectedly.

To address the above practical problems, we proposed a multi-stage tracking framework which conduct tracking by correlation filter with a corrective measure by CNN detection at key frame to increase the performance, accuracy and meet the practical requirements. The whole framework is shown in Figure 1. In this work, to solve the blur situation in the realistic scene and the drift problem, we apply deep CNN for detection and correlation filter for tracking. And we train the VGG16 [9] module provided by SSD [10] with dataset to extracting object proposal with classification and confidence score. The dataset combines VOC (Visual Object Classes) with blurry images that blurred by Gaussian blur, motion blur, and defocus blur to increasing accuracy. We obtain the fi, si and bi from detection and track by discriminative scale space tracking [2]. We also design a corrective

IOP Conf. Series: Journal of Physics: Conf. Series **1004** (2018) 012010 doi:10.1088/1742-6596/1004/1/012010

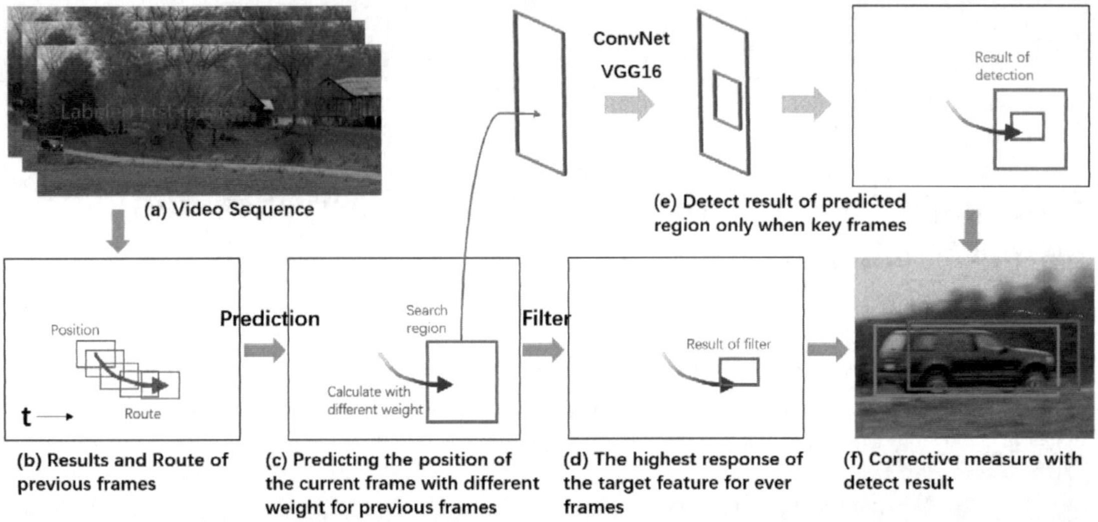

Figure 1. Object tracking framework of our proposed method. (a)Video sequence with labeled first frame. (b) Results and Route (center point of target box) of previous frames. (c) Predicting the position of the current frame by position predicting module and determine the search region. The detailed description is in section 3.2. (d) The result of filter which is the highest response of the target on the search region. (e) Search the pink region by ConvNet and obtain the detect result only when key frames. The detailed description is in section 3.3. (f) Correct the filter result when drift. The detailed description is in section 3.4.

measure module to solve the drift problem, and a position prediction module to reduce the computation cost.

The contributions of our research are: (1) An efficient method proposed for target accurate tracking when visual tracking scene contains clear, blur or illumination variation stages. (2) Different from existing methods, the proposed tracking method has a corrective measure module to relocation the target when tracker drifts. (3) The proposed tracker is capable of scale accommodation in different challenging factors. We assess the proposed tracking framework on OTB-100 (Visual Tracker Benchmark) [5], Extensive experimental results on the sequences of OTB-100 show that our framework performs good tracking efficiency and is effective on full scene stage, which can restrain the tracking drift. Thus our framework can increase the performance of the general tracker especially in the situations of scale transformation, motion blur and fast motion.

2. Related work

2.1. Convolutional neural network

Since CNN is applied to the computer vision field, highest level methods of object detection are principally based on deep CNN detection. Girshick et al. [11] proposed a scalable detection algorithm called R-CNN (Regions with Convolutional Neural Networks features) by training deep CNN with extracted region proposals which are localized and segmented by Selective Search [12]. The whole object detection system is decomposed into several stages including extracting region proposals, computing CNN features, classifying with class-specific linear SVMs. At that time, this framework showed prominent performance and many methods had based on it [13,14].

As a result of massive repeated calculations in CNN features computation of extracted region, RCNN faces a serious speed bottleneck. To accelerate the training and detecting process, Fast R-CNN [15] is proposed. A SPP-NET [16] network layer was used in CNN. Each region patch has a fixed dimension feature representation and is classified by the normal softmax instead of wrapping to a fixed size for CNN features computation. And it skillfully put the bounding box regression into the neural network combine with region classified to a multi-task model. In [17], the selective search algorithm for region proposal was replaced by a Region Proposal Network (RPN) and put into the neural network. The framework can be train as a full convolution network end-to-end that achieved training and update synchronous.

The real-time detection method proposed by W. Liu [10] eliminates the process of bounding boxes and pixel or feature resampling in the middle. It ensures both the speed and the detection accuracy even when the resolution of input image is low by predicting object and box offsets at different levels of feature maps. Training of the convolutional neural networks has become simpler. And a better trade-off is obtained between the detection speed and the detection precision. The mAP of the result can be comparable to the method using region proposals (such as Faster R-CNN).

2.2. Correlation filter-based tracking

The research of object tracking mainly has two categories: generative and discriminant model. The method of the former is to find the best matching window, and the latter is to distinguish the target from the surroundings. In discriminative model, correlation filter performs better that generate high response (correlation peak) to each interested object in the scene and conversely to the background. The main process is mark the object artificially at the first frame, predict the position by the target initial position.

KCF [8], the typical and high-performance Correlation filter-based tracking algorithm. It used multiple channels instead of the original work, cyclic matrix to acquire samples and ridge regression to train target detector, has reduced the computational complexity and improved the computational speed. Yang, Y. et al [6] proposed a multiple object tracking system based on KCF for tracking and background subtraction for targets detecting. Martin et al [1] designed an efficient tracker based on correlation filter named DSST which designs two consistent correlation filters relatively independent to training or testing with different feature types and computing methods. This algorithm can be transplanted to any algorithm for target location and scale estimation. fDSST [2] and ECO [18] were proposed to reduce the computational cost without sacrificing its robustness and accuracy.

3. Multi-stage target tracking with drift corrective and position prediction

We proposed a multi-stage tracking framework which conduct tracking by correlation filter with a corrective measure by CNN detection at key frame to increase the performance, accuracy and meet the practical requirements. The framework has been designed with two particular modules: Position predicting module and Corrective measure module that will be elaborated in this section. Figure 1 shows the steps of our framework.

3.1. Tracking by correlation filter

The tracker we choose is the fast discriminative scale space tracker from [2], the correlation scores at the search region Z_t^l are computed by $\mathcal{F}^{-1}\{Y_t\}$. [2] shows more robust performance than [18] in our experiment condition by standard translation filter and learned scale filter in the situation of object deformation and scale changes. The correlation filter-based tracker is more suitable and have comparable performance under the consideration of real word tracking conditions, taking into account the performance of deep CNNs-based tracker rely on hardware facilities.

$$Y_t = \frac{\sum_{l=1}^{d} \overline{A_{t-1}^l} Z_t^l}{B_{t-1} + \lambda} \tag{1}$$

We use the A_{t-1}^l and B_{t-1} to update the filter in the previous frame same as [2], and the search region Z_t^l is generated by equation (2) in section 3.2.

The correlation filter used in object tracking has a drift problem. We have designed a module named corrective measure to handle this problem, in which we use CNN detection to corrective the

IOP Conf. Series: Journal of Physics: Conf. Series **1004** (2018) 012010 doi:10.1088/1742-6596/1004/1/012010

position of target when tracker drifts. And the module position predicting is designed for increasing speed to counterpoise the speed drop caused by detection.

3.2. Position prediction module

The main procedure of the tracking method we used is to extract the HOG feature of the previous frame, then to set up a search region to matching characteristics. The search region is obtained from expending the target size two times with the midpoint of the target as the center. We find that even if search range is already reduced to two times of the target size from the whole image after numerous experiments, there is still a large disinterested area. Therefore, we propose a position prediction method to reduce the redundant cost of computation. The basis of this method is that no matter what kind of movement the object does, in a small time range, it can be regarded as a curve or even a straight line. If we fit the motion trend of the target in current frame according to the previous data, we can predict the moving direction and distance of the target in next frame. The search range Z_t in equation (1) can be saved nearly half in terms of the computational cost and can be better applied into the fast motion situation.

We predict the position $(\hat{X}_t, \hat{Y}_t)$ of the current frame t by fitting the central point of the target in the front N frames. The information to be predicted includes the motion direction (angle) and the motion distance. Because the motion vector of arbitrary angle can be decomposed into sum of two vectors that directions in horizontal and longitudinal, we decompose the position information vector $\vec{a} = (x, y)$ on two dimensions into two one-dimensional information vectors $\vec{a} = x\vec{\imath} + y\vec{\jmath}$, which reduces the computational complexity. We use a N-th-order difference equation based on Taylor series expansion. To predicting the position $\hat{X}_t$ and $\hat{Y}_t$ of the target at frame t with the center position of the target X_{t-n+i} and Y_{t-n+I} in front N frames, which $i \in (0, 1, \cdots, N-1)$ by:

$$\left(\hat{X}_t, \hat{Y}_t\right) = \sum_{j=0}^{n-1} \sum_{i=0}^{j} \frac{(-1)^i C_j^i (X_{t-n+i}, Y_{t-n+i})}{j\,!} + \varepsilon \tag{2}$$

Where ε is the error offset to reduce the error of the prediction. Then we obtain the motion vectors $\vec{a} = \mathrm{D}_x^t\vec{\imath} + \mathrm{D}_y^t\vec{\jmath}$ in which $\mathrm{D}_x^t = (\hat{X}_t - X_{t-1})$, $\mathrm{D}_y^t = (\hat{Y}_t - Y_{t-1})$we use the prediction $\hat{X}_t$ and $\hat{Y}_t$ that obtain from equation (2) to compute the predicted motion direction $\hat{\theta}$ and distance $\hat{d}$, and calculate the size of the search range:

$$\Lambda = \begin{cases} \left(\mathrm{D}_x^t, \mathrm{D}_y^t\right) + (w, h) \;; & others \\ (w, h)/3 + (w, h) \;; & \left(\mathrm{D}_x^t, \mathrm{D}_y^t\right) = 0 \end{cases} \tag{3}$$

Where Λ is the extended length of search range based on target center at frame t-1, w and h are the width and height of the target size at frame t-1, respectively. D_x^t and D_y^t are the predicted distance of vertical and horizontal direction.

Figure 2 is an example of our position prediction that shows the search region is more accurate

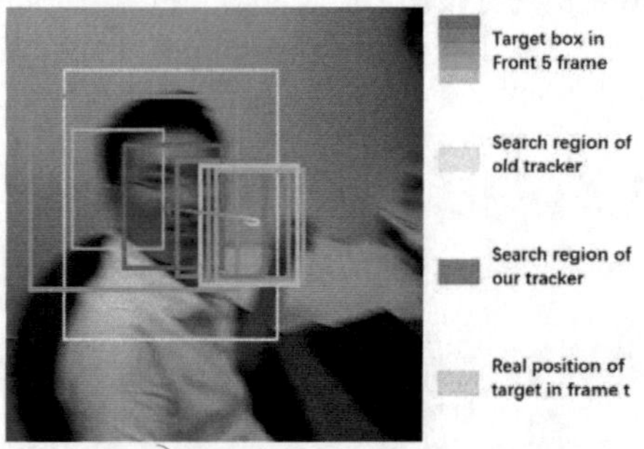

Figure 2. An example of prediction. The image is extracted from sequence named BlurFace in OTB-100 at frame 302

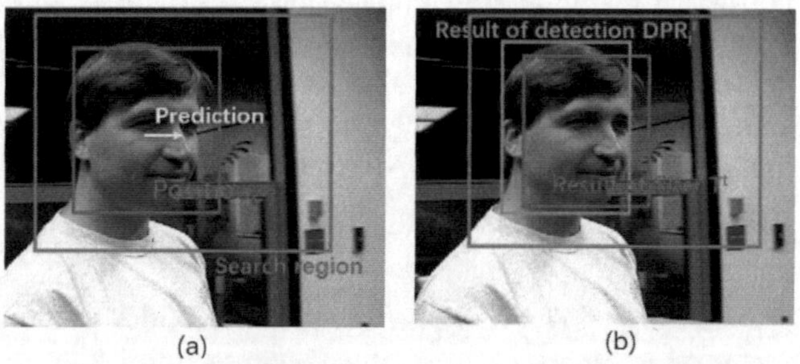

Figure 3. An example of corrective measure. The images are extracted from sequence named Dudek in OTB-100 dataset at frame 925 and 926

appropriate. This approach can reduce the computation cost and make tracker more capable even in fast motion situation. When object moves rapidly, there is a real possibility that the search region of old tracker could not combine the real position of target at the next frame, leading to the failure of tracking. Our method effectively solves this problem.

We also give an example in Figure 3(a) about equation (3), On the basis of $(\hat{X}_t, \hat{Y}_t)$, the extended width and height is based on predicting motion direction. The D_y in Figure 3(a) is zero, so the Λ of vertical direction is $(w,h)/3 + (w,h)$.

3.3. Detecting key frames by CNN Net

We use the detection algorithm based on neural network to provide object proposal for the corrective measure module. The algorithm needs to generate a set of parameter: frame number fi for matching the frame at tracking module, and confidence score si for showing result and filter negative objects, and bounding box bi for locating object.

3.3.1. Dataset preparation. For the selection of dataset, we choose the VOC (Visual Object Classes) and OTB-100 (Visual Tracker Benchmark) [5] dataset. For increasing the performance of algorithm when the image suddenly fuzzy because of the suddenly moving of object or shaking of equipment, the dataset has been adapted by adding moderate blur image in data sample. And we obtained the blur images by change the clear image with three different blur algorithm.

3.3.1.1. Gaussian blur. Assigning the weights with Gauss function and weighted average to calculating the transformation of each pixel in each images. Central pixel receives the highest weights. And the greater the distance to central pixel is, the smaller weight is received. We use Gaussian blur to imitate the lack of detail information of the target in the image due to insufficient resolution or the short distance between target and equipment.

3.3.1.2. Motion Blur. Motion blur is used to simulate the blur image as a result of suddenly moving of target or shaking of device, that weighted average to calculating the transformation of each pixel in each images and assign the weights with equation (4) when $1 - |ycos\theta - xsin\theta| \geq 0$:

$$\mathrm{G(x,y)} = 1 - |ycos\theta - xsin\theta| \;\; ; \; x \in [0, cos\theta len/2], y \in [0, sin\theta len/2], \theta \in [0,\pi] \quad (4)$$

Where x is the pixel's horizontal distance to the central pixel, y is the pixel's vertical distance to the central, θ is the angle in which the picture rotates counter-clockwise, and len is the motion length. The result matrix is symmetric about the central pixel that we can only calculate quarter element of the matrix we decrease the speed. As the central pixel received the highest weight, the further the distance perpendicular to the motion vector and the distance to the central pixel, the lower weight is received.

3.3.1.3. Defocus Blur. Contrary to Gaussian Blur we used defocus blur to simulate the image when the distance from object to device is far relatively and cause the blur. This method designs the weights with a circular template, in which weights is related to the position of the pixel on the circular. The pixels outside the circle receive the weights 0, the pixels inside the circle receive the weights 1, and the pixels on the circle receive the weight with parameter m1 and m2 by:

$$\mathrm{G(x,y)} = \frac{\mathrm{r}^2}{2}\left(\arcsin\left(\frac{m2}{r}\right) - \arcsin\left(\frac{m1}{r}\right)\right) + \frac{\mathrm{r}^2}{4}\left(\sin\left(2\arcsin\left(\frac{m2}{2}\right)\right) - \sin\left(2\,arcsin\left(\frac{m1}{2}\right)\right)\right)$$
$$- (\max(\mathrm{x,y}) - 0.5) * (\mathrm{m2} - \mathrm{m1}) + \mathrm{m1} - (\min(\mathrm{x,y}) - 0.5) \quad (5)$$

Where x and y are the pixel's horizontal and vertical distance to the central pixel, r is the radius of circle template, m1 and m2 are the vertical distance between the intersection of the circle and the pixel to the central pixel.

3.3.2. Structure of detection network. We use the VOC dataset pre-trained VGG16 [9] model provided by SSD and fine-tuning with the OTB-100 dataset to extracting object with classification and confidence score. There are five classes: person, biker, dog, bird, car, which are enough to tracking at general scene and can be extend for special scene. And removing negative object proposal which the confidence scores of VGG detection and bounding box information are below a certain threshold. We use 0.75 as confidence score threshold. And the objects with confidence score below the threshold are considered as disinterested objects.

3.4. Corrective measure module

The scale change and drift problem is a basic and common problem in correlation filter-based tracking field. When the size of target decreases, the filter will learn a lot form background information. If the target is expanding, the filter will go with the local texture. These two situations are likely to lead to unexpected results, drift and failure. Even fDSST [2] uses learned scale filter, the accuracy of tracking algorithm with the drift problem had been improved relative to filters like KCF [8] without scale update. But eventually cause drift or failure because of the accumulation of small error when the tracking task for a long time.

We design a corrective measure module in our framework shown in Figure 1(f) to solve this problem. Contrasting the results between tracker and detection algorithm at discontinuous key frame to determine whether the drift occurred, and correcting the track result by the detection algorithm while the drift occurring. The problem for that is 2-fold: how to correctly extract the bounding box of the target in detection results at the current frame; how to determine the drift occurred and update the bounding box for a new tracker. The position prediction proposed a search region like Figure 1(c) has narrowed the detection range from whole image to a predicting region. It increases the speed and filtering out the disinterested targets. Then we compare the detection proposal region DPR_j^t with the tracking target T^t and design the corrective measure by overlap region OR_j^t between tracking result and bounding box of detection algorithm. We divide the problem into following situations (with a premise that there is at least one DPR_j^t obtained from detection):

- (situation 1) If the tracking target T^t is not overlapping with any proposal region DPR_j^t it means tracker drifts seriously.
- (situation 2) If the tracking target T^t is only overlapping with one proposal region DPR_j^t.
- (situation 3) If the tracking target T^t is overlapping with two or more proposal regions DPR_j^t,

At the situation 1, because the tracking method of correlation filter is based on target information in prior frame T^{t-1} and evaluate the target T^t in current frame with filter, we chose the DPR_j^t as the result of current frame t and build a new filter for tracking. At the situation 2, Since drift may happen, we set up a threshold Thr to determine whether to correct the filter result. When the overlapping of the DPR_j^t and the T^t is bigger than Thr, we consider the tracking performance is good enough. Because generally the detection of bounding box will contain some background information, we use the tracking results, and use the detection results contrarily. And at the situation 3, we choose the DPR_j^t which has the maximum overlapping area with the target T^t and compare with Thr like situation 2.

$$T = \begin{cases} \mathrm{DPR}_j^t \; ; & n = 0(\text{situation 1}) \\ \mathrm{DPR}_j^t \; ; & n = 1(\text{situation 2}), \mathrm{OR}_j^t \le \mathrm{Thr} \\ T^t \; ; & n = 1(\text{situation 2}), \mathrm{OR}_j^t > \mathrm{Thr} \end{cases} \tag{6}$$

Where n is the number of the overlap between tracking target T and proposal region DPR_j^t, T is the result of the tracker in current frame t.

IOP Conf. Series: Journal of Physics: Conf. Series **1004** (2018) 012010 doi:10.1088/1742-6596/1004/1/012010

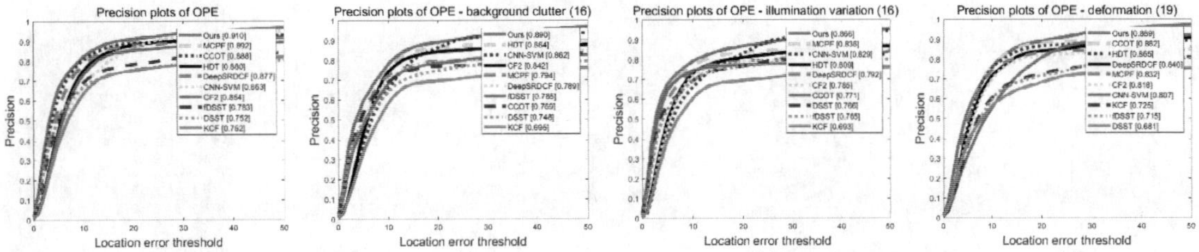

(a) (b) (c) (d)

Figure 4. (a) Precision plots on 10 trackers of fifty sequences using one-pass evaluation on the OTB-100 Dataset. (b)(c)(d) Precision plots on 10 trackers of background clutter, illumination variation and deformation.

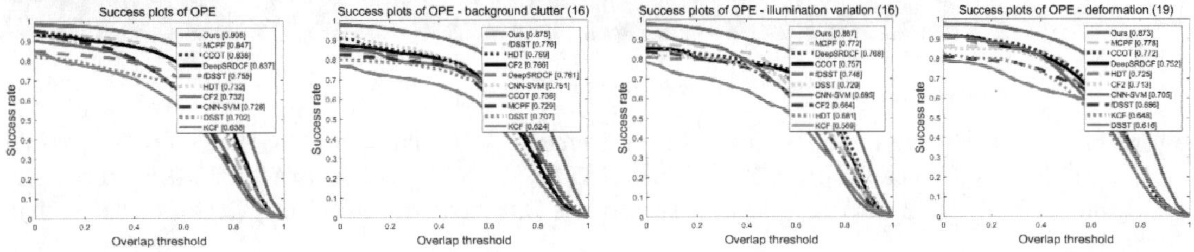

(a) (b) (c) (d)

Figure 5. (a) Success plots on 10 trackers of fifty sequences on the OTB-100 Dataset. (b)(c)(d) Success plots on 10 trackers of background clutter, illumination variation and deformation.

Figure 3 shows an example of the corrective measure processing that target position at frame t-1 and prediction result of frame t in Figure 3(a). And the green box in Figure 3(b) is the search region's result of detection and the red box is the filter result.

4. Experiments

In this chapter, we provide the experiment results to verify the performance of our tracking framework. Comparing with nine current and state-of-the-art tracker including MCPF [7], HDT [19], CF2 [20], DeepSRDCF [21], CNN-SVM [22], KCF [8], DSST [1], fDSST [2] and CCOT [23]. The tested fifty sequences we choose are extracted from OTB-100 mainly including person, animal and car as target.

We analyse the performance based on OTB-100 sequences in terms of 11 challenging factors, and show the representative results. The results showed in Figure 4 and Figure 5 using the distance precision and overlap success rate, illustrate the mean distance and overlap precision over all fifty sequences. The figures contain the AUC (area-under-the-curve) score and precision score of average distance for each tracker. Our tracker performs better compared with all nine compared trackers in both precision and success plots. Column (b) in Figure 4 and 5 illustrate precision and success plots of background clutter, column (c) illustrates the results of illumination variation and the column (d) illustrates the results of deformation. The results of success plots show that our method is capable of scale accommodation in different challenging factors.

Column (a) in Figure 4 and Figure 5 illustrate the success and precision plots on 10 trackers of fifty sequences on the OTB-100 Dataset. The average distance precision score for each tracker is showed in the figure 4. The top method MCPF [7] runs on real-time with an AUC score of 84.7%. Ours tracking method runs with score of 90.8%, significantly outperforming MCPF [7] by 6.1%. The overlapping rate is best without sacrificing too much location precision, which shows the more robust when size and appearance changing. Summarily, the precision plot proves that our framework is outperform in robustness compared with others, and the success plot demonstrates that our tracker tracks scale more accurately.

Figure 6. The visualization tracking result of our framework compare with the state-of-the-art visual trackers MCPF [7], CCOT [23], fDSST [2] and DSST [1] on (a)-(e) five benchmark sequences are. Gym MotorRolling Bolt2 Bird1 CarScale, respectively. Our method significantly performs better than the others.

We also provide a comparison of our framework with several current trackers MCPF [7], CCOT [23], fDSST [2] and DSST [1] on representative video sequences named Gym, MotorRolling, Bolt2, Bird1 and CarScale. Shows the comparative result in various situation in Figure 6.

Figure 6(a), (b) and (e) illustrate sequences with significant scale variations and deformation. MCPF are capable in-plane rotation, but incompetent to significant scale changes. MCPF is bigger and others is smaller than real scale. Our tracker performance better on both the scale and position in spite of these challenges.

Figure 6(c) and (d) illustrate the results on sequences with partial occlusions by similar objects named Bolt2 and Bird1. Especially, the target is completely blocked for a while in the sequence Bird1 showed in Figure 6(d). The compared trackers are incompetent to the significant clutter and occlusions in the both two sequences. Figure 6(b) and (c) illustrate the results on sequences with fast motion and Figure 6(b) and (e) illustrate the results on sequences with blur motion. Our tracker demonstrates the best performance in all five representative sequences. As speed of tracking framework, our tracker shows average 30.7 fps with visualization in our experimental environment.

5. Conclusion

In this paper, we propose a multi-stage tracking framework that tracking by correlation filter. Our method consists of a corrective measure based on CNN detection at key frame to increase the accuracy and meet the practical requirements. To handle the shift between clarity and blur problem, we design the blur dataset based on VOC and OTB-100 generated by three blur methods to simulate the blur caused by different reasons. Experiments are executed by the challenging benchmark sequences with 11 attributes such as scale variation, deformation, motion blur, fast motion, and background clutter. Compared with above state-of-the-art tracking scheme, the results clearly certify that our method outperforms them on both success and precision plot of OPE. Since our detection and tracking methods are respectively independent that can be utilized in any multi-stage tracking framework for different realistic scenes.

Further studies beyond this work include improving the accuracy of position prediction and test the performance of our method in occlusion. And improve our tracker for multi target tracking.

Acknowledgments

This work was supported by The Aeronautical Science Foundations of China (No.20161375002) through a grants for our project.

References

[1] Danelljan M, Häger G, Khan F S and Felsberg M 2014 Accurate scale estimation for robust visual tracking *Proceedings of the British Machine Vision Conference* (Nottingham)

[2] Danelljan M, Häger G, Khan F S and Felsberg M 2017 Discriminative scale space tracking *IEEE Transactions on Pattern Analysis & Machine Intelligence* (vol 39 n 8) pp 1561-75

[3] Held D, Thrun S and Savarese S 2016 *Proc. European Conf. on Computer Vision* (vol 9905 LNCS) pp 749-65

[4] Redmon J, Divvala S, Girshick R and Farhadi A 2016 You only look once: unified, real-time object detection *IEEE Computer Vision and Pattern Recognition* pp 779-88

[5] Wu Y, Lim J and Yang M H 2013 Online object tracking: a benchmark *IEEE Conf. on Computer Vision and Pattern Recognition* (vol 9) pp 2411-8

[6] Yang Y and Bilodeau G A 2016 Multiple object tracking with kernelized correlation filters in urban mixed traffic

[7] Zhang T Z, Xu C S and Yang M H 2017 Multi-task correlation particle filter for robust visual tracking *Computer Vision and Pattern Recognition.*

[8] Henriques J F, Caseiro R, Martins P and Batista J 2015 High-speed tracking with kernelized correlation filters *IEEE Transactions on Pattern Analysis and Machine Intelligence* (vol 37 n 3) pp 583-96

[9] Simonyan K and Zisserman A 2014 Very deep convolutional networks for large-scale image recognition *Computer Science*

[10] Liu W et al. 2016 SSD: Single shot multibox detector *Proc. European Conf. on Computer Vision* (vol 9905 LNCS) pp 21-37

[11] Girshick R, Donahue J, Darrell T and Malik J 2014 Rich feature hierarchies for accurate object detection and semantic segmentation *Proc. Computer Society Conf. Computer Vision Pattern Recognit* pp 580-7

[12] Uijlings J R R, Sande K E A V D, Gevers T and Smeulders A W M 2013 Selective search for object recognition *International Journal of Computer Vision* (vol 104 n 2) pp 154-71

[13] Szegedy C et al. 2015 Going deeper with convolutions *Proc. IEEE Computer Society Conf. Computer Vision Pattern Recognition* (vol 07-12-June-2015) pp 1-9

[14] Ouyang W et al. 2016 DeepID-Net: Deformable deep convolutional neural networks for object detection *Proc. Computer Society Conf. Computer Vision Pattern Recognition* pp 2403-12

[15] Girshick R 2015 Fast R-CNN *Int. Conf. on Computer Vision* pp 1440-8

[16] He K, Zhang X, Ren S and Sun J 2014 Spatial pyramid pooling in deep convolutional networks for visual recognition *European Conf. on Computer Vision* (vol 8691 LNCS) pp 346-61

[17] Ren S, He K, Girshick R and Sun J 2015 Faster R-CNN: towards real-time object detection with region proposal networks *IEEE Transactions Pattern Analysis Machine Intelligence* (vol 39 n 6) pp 1137-49

[18] Danelljan M, Bhat G, Khan F S and Felsberg M 2017 Efficient convolution operators for tracking *Proc. IEEE Conf. Computer Vision and Pattern Recognition*

[19] Qi Y et al. 2016 Hedged deep tracking Proc. IEEE Computer Society Conf. Computer Vision Pattern Recognition pp 4303-11

[20] Ma C, Huang J B, Yang X and Yang M H 2015 Hierarchical convolutional features for visual tracking *Proc. IEEE Int. Conf. Computer Vision* pp 3074-82

[21] Danelljan M, Häger G, Khan F S and Felsberg M 2015 Convolutional features for correlation filter based visual tracking *Proc. Int. Conf. Computer Vision* (vol 2015-February) pp 621-9

[22] Hong S, You T, Kwak S and Han B 2015 Online tracking by learning discriminative saliency map with convolutional neural network *International Conference on Machine Learning* (vol 1) pp 597-606

[23] Danelljan M, Robinson A, Khan F and Felsberg M 2016 Beyond correlation filters: learning continuous convolution operators for visual tracking *European Conf. on Computer Vision*

IOP Conf. Series: Journal of Physics: Conf. Series **1004** (2018) 012011 doi:10.1088/1742-6596/1004/1/012011

JANUS: A Compilation System for Balancing Parallelism and Performance in OpenVX

Hossein Omidian[1,*] and Guy G. F. Lemieux[1]

[1]Department of Electrical and Computer Engineering, University of British Columbia, Vancouver, BC, Canada

E-mail: *hosseino@ece.ubc.ca

Abstract. Embedded systems typically do not have enough on-chip memory for entire an image buffer. Programming systems like OpenCV operate on entire image frames at each step, making them use excessive memory bandwidth and power. In contrast, the paradigm used by OpenVX is much more efficient; it uses image tiling, and the compilation system is allowed to analyze and optimize the operation sequence, specified as a compute graph, before doing any pixel processing. In this work, we are building a compilation system for OpenVX that can analyze and optimize the compute graph to take advantage of parallel resources in many-core systems or FPGAs. Using a database of prewritten OpenVX kernels, it automatically adjusts the image tile size as well as using kernel duplication and coalescing to meet a defined area (resource) target, or to meet a specified throughput target. This allows a single compute graph to target implementations with a wide range of performance needs or capabilities, e.g. from handheld to datacenter, that use minimal resources and power to reach the performance target.

1. Introduction

In recent years both industrial and academic communities have focused on implementing Computer Vision (CV) applications on different embedded platforms. For the most computationally intensive CV algorithms, using parallelism techniques for implementing them is considered a must. Although in theory CPU and GPU platforms can perform such intensive computations, power remains the main challenge. Previous studies suggest custom hardware implementations [1, 2] as well as programmable many-core systems [3, 4, 5, 6, 7, 8] can be good alternatives since they are able to perform such intense computations within the power limits. Moreover, most CV applications can be define as streaming applications (i.e. each stage receives stream of image pixels, rows or frames, processes them, and sends the results as stream of data to the next stage). This means we can describe CV applications as compute graphs or Synchronous Data Flow Graphs (SDFGs) [9]. Previous studies have shown custom hardware implementations on FPGA as well as reconfigurable many-core systems have the potential to dramatically increase the performance/Watt for implementing computationally intensive SDFGs. The target embedded systems (for CV applications) can have on-chip resources (e.g. computational, power, area, etc) as large as an autonomous car or as small as a battery operated device. In both cases the final goal for developers is maximizing performance while decreasing the power. Previous studies such as Chen et. al. [10] report the off-chip DRAM traffic dominates the energy consumption and show eliminating the DRAM access reduces energy by up to 150.31x. This means in order to save power, we need to keep data locally on chip as long as possible and avoid dumping/restoring data to/from main memory. Traditional CV programming systems such as OpenCV [11] operate on entire image frame at each step, which means for every node in the compute graph (e.g. ConvertColor node), the application needs to read the entire frame from off-chip memory, do the required process on the whole frame and then provide the results for the next node by dumping the results back to the off-chip memory. In contrast, OpenVX [12] has been introduced as a cross-platform standard which a computation can be tiled and the workload can be run on those tiles independently while keeping several tiles in on-chip memory concurrently.

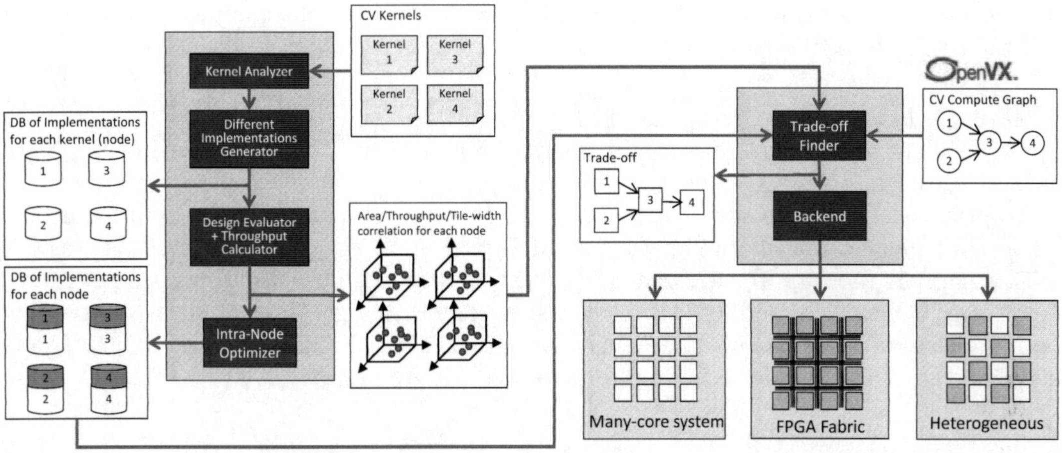

Figure 1. Tool flow

In this work we introduce a compilation system for OpenVX style compute-graphs. Our tool is able to automatically explore the space/time tradeoff problem for these compute graphs and find optimum solutions satisfying different throughput targets and area budgets while targeting a wide range of FPGAs and programmable many-core systems.

Every node (kernel) in the compute graph is analyzed to find all degrees of parallelism based on different loop transformation strategies [13, 14], pipeline opportunities and data dependencies. Based on those, our tool generates different implementations with different area, IO and throughput characteristics for each node in the compute graph. Moreover, using different optimization techniques such as node combining and node replication, our tool is able to widen the problem space by covering more possible solutions. After finding variety of different solutions for each node in the compute graph, the tool needs to find an optimum overall solution based on defined area budget or throughput target. Previous research has shown finding the optimum solution for this space/time tradeoff is an NP-complete optimization problem. It can be defined as Integer Linear Programming (ILP) problem and solved using ILP solver [15].

Although the ILP approach can be useful in some cases, using the ILP optimization model within the tool prohibits the use of certain optimizations. In contrast, using heuristic approaches allows us to perform object coalescing which cannot be done as an ILP formulation. Using heuristic approach leads to area saving and less runtime compared to the ILP approach [16].

There have been several recent studies on implementing image processing and OpenVX applications on reconfigurable platforms and exploring the area/throughput trade-off for them [17, 18, 19, 20]. These existing approaches either use a specific programming model which requires the user to learn a new programming language, or implement a soft multi-core platform on FPGA and then schedule and run the application on it. Automatically exploring area/throughput tradeoffs, automatically finding optimum implementations for different area budgets or throughput targets as well as achieving better results in faster runtime compared to the existing ILP approaches, make our approach unique. Moreover our approach can be used for both FPGA and many-core platforms as well as heterogeneous platforms such as GRVI Phalanx with accelerators[8].

2. SDFG based HLS for CV applications

Figure 1 illustrates the detailed flow of the proposed tool. Our tool targets two different backends: 1) a Java-based language inspired by Ambric's KPN programming model and many-core architecture [7], and 2) a C-based OpenVX library system built using Xilinx's Vivado HLS tool. Since Synchronous Data Flow (SDF) is a restriction of Kahn process networks (KPN) [21], it is possible to transform any SDFG to a KPN model which means our tool works on SDFG for both of these backends.

2.1. OpenVX Programming Model

OpenVX is a cross-platform C-based API standard for Computer Vision applications. OpenVX can be a great programming model for embedded systems since it enables performance and power-optimized CV processing. It specifies a higher level of abstraction which allows us targeting different computing architectures. Most CV applications can be described as a set of vision kernels (nodes) which communicate through input/output data dependencies (e.g. AXI stream). OpenVX describes this set of vision kernels in a graph-oriented execution model (Compute Graph) based on Directed Acyclic Graphs (DAGs). Figure 2 shows an OpenVX code example (Canny) and Figure 3 shows the corresponding graph for it.

```
//Canny example
vx_node nodes[] = {
  vxColorConvertNode(graph,rgb, gray),
  vxGaussian3x3Node(graph,gray,gauss),
  vxSobel3x3Node(graph,gauss,gradx,grady
  vxMagnitudeNode(graph,gradx,grady,mag)
  vxPhaseNode(graph,gradx,grady,phase),
  vxNonMaxima(graph,mag,phase,nm),
  vxThreshold(grpah,nm,output)
};
```

Figure 2. OpenVX source code.

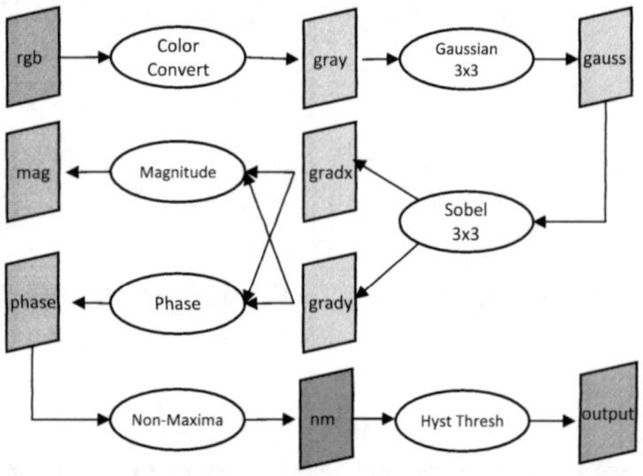

Figure 3. Canny edge detection graph representation.

2.2. Finding Different Implementations

Consider an application described as a compute graph with N nodes $f_1, f_2, ..., f_N$. For each node f_m our tool tries to find different implementations $P_m^1, P_m^2, ..., P_m^{S_m}$ where each implementation P_m^s can perform functionality of f_m with area cost $A(P_m^s)$, number of pixels it can consume/produce $NP(P_m^s)$ each firing, initial interval $II(P_m^s)$ and "inverse throughput" $\vartheta_{in/out}(P_m^s)$ for each input/output channel [16]. For most CV kernels, their input/output channels have matched throughput, so we can use a single variable $\vartheta_{IO}(P_m^s)$ which allows us to define "kernel throughput" $\Theta(P_m^s)$ as number of pixels consumed/produced in each clock cycle [2]. The initiation interval is the number of time units before another firing of the inputs is permitted; it can be just 1 for fully pipelined execution or > 1 for partially pipelined execution.

The Different Implementation Generator (DIG) module in our tool automatically finds the above mentioned implementations. The DIG needs to be able to automatically find a wide range of different implementations with different area cost, throughput and tile-width to cover the solution space as much as possible. DIG uses different loop transformations, data dependency analysis, pipeline opportunities as well as changing the image tile-size (in both kernel IO and kernel core). In addition, an Intra-node Optimizer step in the tool generates a wider range of implementations. Intra-node Optimizer replicates and combines existing implementations of nodes in order to widen the solution space. Node replication can be used to either increase the throughput or widen the tile-width [2]. All of the abovementioned techniques are used find a wide range of implementations for each kernel which, widens the solution space for the area/throughput scaling problem. Below we discuss our trade-off formulation and solutions.

3. Trade-off Finding Problem Definition and Solutions

The trade-off finding problem can be defined in two different ways.

- Given a throughput target Θ_{tgt}, and different implementations for each node f_j, which implementation P_j^i should be selected and how many replicas nr_j^i are needed in order to minimize area cost A_A subject to the constraint the application throughput Θ_A is bigger than Θ_{tgt}.
- Given an available area on chip A_C and different implementations for each node f_j, which implementation P_j^i should be selected and how many replicas nr_j^i are needed in order to maximize application throughput Θ_A subject to the constraint the application area cost A_A is not bigger than A_C.

For our baseline, we modeled the optimization problem as ILP formulations similar to previous studies such as Cong et.al. [15] and used the GLPK open-source ILP solver [22]. Although ILP solvers are able to find an optimum solution for the defined problem, we need to define the problem beforehand which prohibits the use of certain optimization such as graph manipulation (node combining/splitting) during runtime. Moreover, as ILP finds the optimum solution by going through all different possible solutions, finding an optimum solution for a problem with a large search space is quickly becomes highly time inefficient. Our proposed heuristic approach addresses both of these shortcomings.

3.1. *Using Heuristic Approach for Trade-off Finding Problem*

Trade-off finding problem can be solved using heuristic approaches such as Omidian et.al. [16]. Our heuristic approach uses there following steps to find a good tradeoff:

- Throughput Analysis
- Application Throughput Propagation and Balancing
- Bottleneck Optimizer

In contrast to the ILP approach, a heuristic approach enables us to use additional optimization opportunities such as Inter-node optimization. Using heuristic approach also can improve the runtime for trade-off finding problem.

4. Experimental Results

Our experiments are carried out in two parts. We evaluate our heuristic approach for both different throughput target and area budgets (different FPGA architectures as well as different throughput targets for many-core system) and compare it to the ILP approach. Then we evaluate the throughput results by implementing CV benchmarks on Zedboard development platform with a Xilinx Zynq SOC. To show the capabilities of our approach, we have utilized the following benchmarks implemented as OpenVX compute graph:

- *Sobel* is a Sobel-filter based edge detection with 5 nodes.

IOP Conf. Series: Journal of Physics: Conf. Series **1004** (2018) 012011 doi:10.1088/1742-6596/1004/1/012011

- *Canny* implements Canny edge detector with 7 nodes.
- *Harris* implements Harris corner detector with 6 nodes.
- *JPEG* implements JPEG encoding with 4 nodes.

Table 1. Implementation Library for JPEG encoder.

module	Color Conversion				DCT					Quantization					Encoding
Version	v1	v2	v3	v4	v1	v2	v3	v4	v5	v1	v2	v3	v4	v5	v1
Inverse Throughput	1	2	4	8	1	2	4	6	32	1	2	4	8	128	512
Area	512	256	128	64	800	400	224	160	50	512	256	128	64	4	22

Table 2. Heuristic vs ILP for many-core system.

Method	Inverse Throughput	Color Conversion		DCT		Quantization		Encoding		Fork/join Overhead	Total Area
		impl	rep	impl	rep	impl	rep	impl	rep		
ILP	1	v1	1	v1	1	v1	1	4	512	10880	23968
Heuristic		v1	1	v5	32	v5	128	224	512	640	13888
ILP	2	v2	1	v2	1	v2	1	v1	128	5376	11920
Heuristic		v2	1	v5	16	v5	64	v1	128	256	7456
ILP	4	v3	1	v3	1	v3	1	v1	64	2688	5984
Heuristic		v3	1	v5	8	v5	32	v1	64	128	3600
ILP	8	v4	1	v4	1	v4	1	v1	32	1280	2976
Heuristic		v4	1	v5	4	v5	16	v1	32	0	1736

We evaluated our tool by implementing JPEG on the many-core system targeting different throughput targets. Our tool is able to find 11 different implementations for Color Conversion and Quantization modules, 17 different implementations for DCT, and only one implementation for Encoding. Table 1 shows a selection of these implementations. Both ILP and Heuristic approaches have been used by our tool in order to find a trade off between area and throughput for different inverse throughput targets for JPEG. Table 2 shows the results generated by these two approaches for given throughput targets. We list the selected implementation and number of replicas for each module. As we can see our heuristic approach finds better area/throughput trade-off compare of the ILP approach. For example, for an inverse throughput target of 2, our heuristic approach used 37% less area compare to ILP.

We also evaluated our tool with different Xilinx 7-series FPGAs. Both heuristic and ILP approaches could fill over 95% of the chip area on average. The heuristic approach however is able to achieve better throughput while improving the runtime up to 3.6x compared to the ILP approach. Figure 4 shows the runtime result improvement achieved through using our heuristic approach compared to the ILP for different Xilinx FPGAs. In addition, in order to examine the area saving achieved through using heuristic approach (using inter-node optimization), we first used the ILP approach to fill the chip and achieved the best throughput possible for different Xilinx FPGAs, then we used the achieved throughput on each FPGA as a throughput target for the heuristic approach. Figure 5 shows the area cost comparison between the heuristic and the ILP. The heuristic approach saves 19% area on average while decreasing the throughput only by less than 2%. Figure 6 shows the percentage of throughput reduction for heuristic compared to ILP and Figure 7 shows the results for setting different throughput targets and implement the tradeoff solutions on Zedboard development platform. As it's shown our tool is able to hit all the throughput targets while increasing the area cost linearly. The tool is able to achieve up to 5.5GigaPixel/sec throughput for implementing Sobel on a small size FPGA.

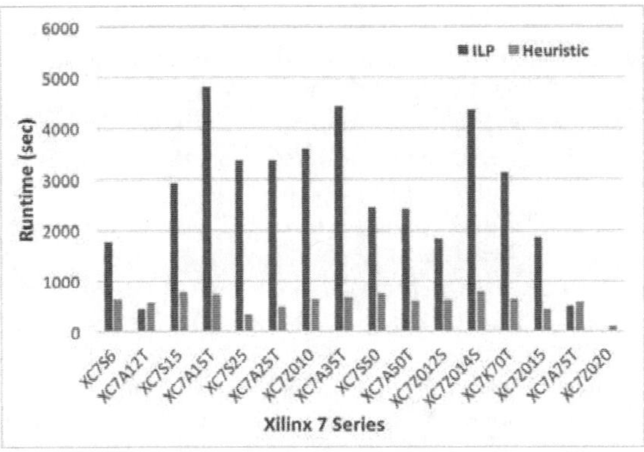

Figure 4. Heuristic vs ILP runtime speedup for Harris corner detection.

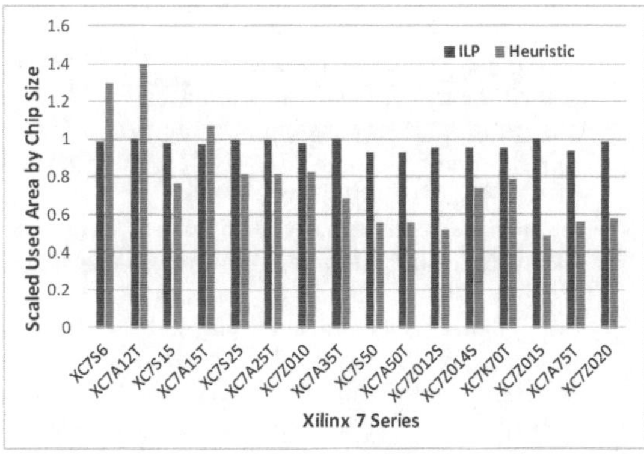

Figure 5. Area cost results for Harris using Heuristic and ILP approaches.

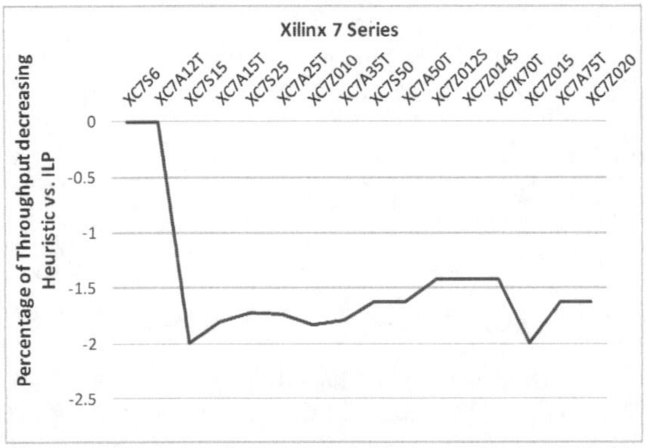

Figure 6. Heuristic vs ILP throughput results for Harris corner detection.

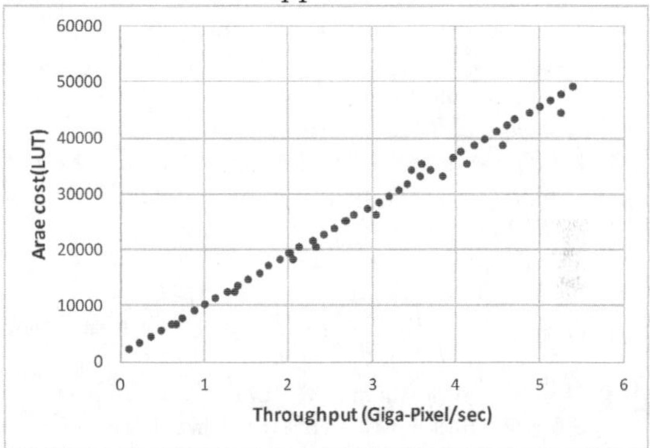

Figure 7. Area/Throughput results for implementing Sobel on Xilinx Zedboard.

5. Conclusion

In this work we introduce a compilation system for OpenVX style compute-graphs. Our tool is able to automatically explore the space/time tradeoff problem for these compute graphs and find optimum solutions satisfying different throughput targets and area budgets while targeting a wide range of FPGAs and programmable many-core systems. Our approach is differentiated from existing approaches as it automatically investigates finding different implementations, and it also combines module selection and replication methods as well as changing tile-size with node combining and splitting in order to automatically find a better area/throughput tradeoff. This approach was verified with different OpenVX benchmarks targeting different FPGA sizes. Our tool is able to achieve over 95% of the target area budget while improving the throughput. Using Inter-node Optimizer step, our heuristic tradeoff finder is able to hit the same throughput targets while reducing the area cost by 19% on average compared to existing ILP approaches. Hitting different throughput targets as well as getting up to 5.5 GigaPixel/sec for a small FPGA target shows our tool is efficient in managing on chip resources efficiently in order to hit different throughput targets.

References

[1] R. Hameed, W. Qadeer, M. Wachs, O. Azizi, A. Solomatnikov, B. C. Lee, S. Richardson, C. Kozyrakis, and M. Horowitz, "Understanding sources of inefficiency in general-purpose chips," in *Proceedings of the 37th Annual International Symposium on Computer Architecture*, 2010, pp. 37–47.

[2] H. Omidian and G. G. F. Lemieux, "Exploring automated space/time tradeoffs for openvx compute graphs," in *2017 International Conference on Field-Programmable Technology (FPT)*, Dec 2017.
[3] A. Severance, J. Edwards, H. Omidian, and G. Lemieux, "Soft vector processors with streaming pipelines," in *Proceedings of the 2014 ACM/SIGDA International Symposium on Field-programmable Gate Arrays*, ser. FPGA '14, 2014, pp. 117–126.
[4] Adapteva-Inc, "Epiphany-iv 64-core 28nm microprocessor," 2014. [Online]. Available: http://www.adapteva.com/products/silicon-devices/e64g401/
[5] L. Benini, E. Flamand, D. Fuin, and D. Melpignano, "P2012: Building an ecosystem for a scalable, modular and high-efficiency embedded computing accelerator," in *2012 Design, Automation Test in Europe Conference Exhibition (DATE)*, 2012, pp. 983–987.
[6] KALRAY-Corporation, "Massively parallel processor array." [Online]. Available: http://www.kalray.eu/
[7] M. Butts, A. M. Jones, and P. Wasson, "A structural object programming model, architecture, chip and tools for reconfigurable computing," in *15th Annual IEEE Symposium on Field-Programmable Custom Computing Machines (FCCM 2007)*, 2007, pp. 55–64.
[8] J. Gray, "Grvi phalanx: A massively parallel risc-v fpga accelerator accelerator," *International Workshop on Overlay Architectures for FPGA (OLAF)*, vol. abs/1606.03717, 2016.
[9] E. A. Lee and D. G. Messerschmitt, "Synchronous data flow," *Proceedings of the IEEE*, pp. 1235–1245, 1987.
[10] Y. Chen, T. Luo, S. Liu, S. Zhang, L. He, J. Wang, L. Li, T. Chen, Z. Xu, N. Sun, and O. Temam, "Dadiannao: A machine-learning supercomputer," in *2014 47th Annual IEEE/ACM International Symposium on Microarchitecture*, 2014, pp. 609–622.
[11] *The OpenCV Reference Manual*, 2nd ed., Itseez, April 2014.
[12] Khronos-Group, "Openvx," 2017. [Online]. Available: https://www.khronos.org/openvx/
[13] A. W. Lim and M. S. Lam, "Maximizing parallelism and minimizing synchronization with affine transforms," in *Proceedings of the 24th ACM SIGPLAN-SIGACT Symposium on Principles of Programming Languages*, ser. POPL '97, 1997, pp. 201–214.
[14] V. Sarkar and R. Thekkath, "A general framework for iteration-reordering loop transformations," *SIGPLAN Not.*, pp. 175–187, 1992.
[15] J. Cong, M. Huang, B. Liu, P. Zhang, and Y. Zou, "Combining module selection and replication for throughput-driven streaming programs," in *2012 Design, Automation Test in Europe Conference Exhibition (DATE)*, 2012, pp. 1018–1023.
[16] H. Omidian and G. G. F. Lemieux, "Automated space/time scaling of streaming task graph," *International Workshop on Overlay Architectures for FPGA (OLAF)*, vol. abs/1606.03717, 2016.
[17] G. Tagliavini, G. Haugou, A. Marongiu, and L. Benini, "Adrenaline: An openvx environment to optimize embedded vision applications on many-core accelerators," in *2015 IEEE 9th International Symposium on Embedded Multicore/Many-core Systems-on-Chip*, 2015, pp. 289–296.
[18] J. Hegarty, J. Brunhaver, Z. DeVito, J. Ragan-Kelley, N. Cohen, S. Bell, A. Vasilyev, M. Horowitz, and P. Hanrahan, "Darkroom: Compiling high-level image processing code into hardware pipelines," *ACM Trans. Graph.*, pp. 144:1–144:11, 2014.
[19] J. Hegarty, R. Daly, Z. DeVito, J. Ragan-Kelley, M. Horowitz, and P. Hanrahan, "Rigel: Flexible multi-rate image processing hardware," *ACM Trans. Graph.*, pp. 85:1–85:11, 2016.
[20] J. Pu, S. Bell, X. Yang, J. Setter, S. Richardson, J. Ragan-Kelley, and M. Horowitz, "Programming heterogeneous systems from an image processing dsl," *ACM Trans. Archit. Code Optim.*, pp. 26:1–26:25, 2017. [Online]. Available: http://doi.acm.org/10.1145/3107953
[21] G. Kahn, "The Semantics of a Simple Language for Parallel Programming," in *Information Processing '74: Proceedings of the IFIP Congress*, 1974, pp. 471–475.
[22] GNU-project, "Gnu linear programming kit," 2017. [Online]. Available: https://www.gnu.org/software/glpk/

IOP Conf. Series: Journal of Physics: Conf. Series **1004** (2018) 012012 doi:10.1088/1742-6596/1004/1/012012

Multi-viewpoint Image Array Virtual Viewpoint Rapid Generation Algorithm Based on Image Layering

Lu Jiang[1] and Yan Piao[1]

[1]Changchun University of Science and Technology, Changchun of JiLin, China

*E-mail: piaoyan@cust.edu.cn

Abstract. The use of multi-view image array combined with virtual viewpoint generation technology to record 3D scene information in large scenes has become one of the key technologies for the development of integrated imaging. This paper presents a virtual viewpoint rendering method based on image layering algorithm. Firstly, the depth information of reference viewpoint image is quickly obtained. During this process, SAD is chosen as the similarity measure function. Then layer the reference image and calculate the parallax based on the depth information. Through the relative distance between the virtual viewpoint and the reference viewpoint, the image layers are weighted and panned. Finally the virtual viewpoint image is rendered layer by layer according to the distance between the image layers and the viewer. This method avoids the disadvantages of the algorithm DIBR, such as high-precision requirements of depth map and complex mapping operations. Experiments show that, this algorithm can achieve the synthesis of virtual viewpoints in any position within 2×2 viewpoints range, and the rendering speed is also very impressive. The average result proved that this method can get satisfactory image quality. The average SSIM value of the results relative to real viewpoint images can reaches 0.9525, the PSNR value can reaches 38.353 and the image histogram similarity can reaches 93.77%.

1. Introduction

In the process of three-dimensional integrated imaging, traditional lens array recording systems have problems of low resolution and crosstalk that have long been difficult to solve [1]. At the same time, multi-view high-resolution image recording system (such as camera array) has a very broad application prospect in the acquisition of three-dimensional information of large scenes and the acquisition of high quality elemental images [2] because of independent non-interfering imaging units. Therefore, the combination of sparse scene viewpoint information and virtual viewpoint generation technology to achieve multi-view stereoscopic display has become the focus of the integrated imaging research. Currently, the mainstream algorithm for generating the virtual viewpoint image rendering technology is DIBR (Depth-Image-Based Rendering) [3]. However, its 3D mapping operation is very complicated and requires a large amount of computation, and the generating image quality depends on the depth image accuracy.

Aiming at the existing problems of virtual viewpoint technology, this paper presents a method which generate virtual viewpoint of camera array fast based on image layering. This method does not require the input of depth data and camera array parameters. Experiments show that this method can achieve the rendering of viewpoints anywhere within 2×2 viewpoint range.

IOP Conf. Series: Journal of Physics: Conf. Series **1004** (2018) 012012 doi:10.1088/1742-6596/1004/1/012012

2. Three-dimensional information acquisition based on multi-view image array

With the continuous development of three-dimensional display technology, the lens array recording system has been unable to meet the demand of obtaining large-area three-dimensional information due to its low resolution and crosstalk problems[4]. Therefore, people began to research the multi-view image array recording system to achieve the recording of the three-dimensional information recording.

As shown in the figure 1, camera array after the virtual viewpoint synthesized and display lens array has the same dimension $n \times m$. The focal length of the neutron camera in the camera array is shown as f_1 and the interval is a_1. The focal length of the lens element in the lens array is shown as f_2 and the interval is a_2. The center coordinates of the sub-camera 0 and the sub-camera n in the same column are respectively set as $(0,0,0,)$, $(na_1,0,0)$. The center coordinates of the lens element m and the lens element n in the same columnin of the display lens array are respectively set as $(0,0,0)$, $(na_2,0,0)$. Take any two points on three-dimensional objects P and Q, they can be imaged as pixel points P_1 and Q_1 on the elemental image array through the camera array. And then through the display lens array integration, we can get two points in the space of the integrated three-dimensional phase points P_3 and Q_3.

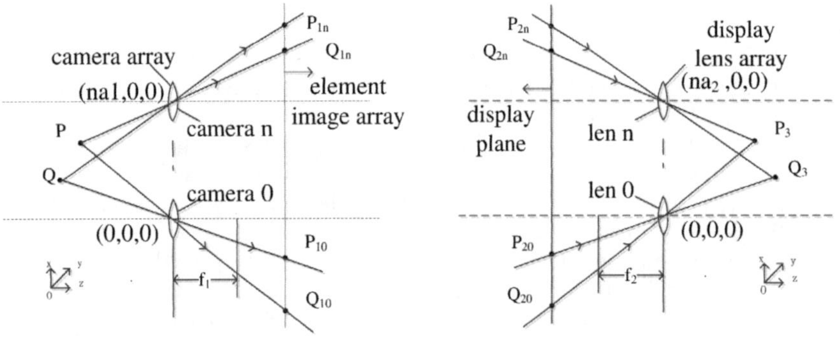

(a)Recording process (b) Display process

Figure 1.Any two points on a three-dimensional object recording and display process

In practical application, the parameters of recording camera and display side lens are often difficult to match[5]. One-time recording of object three-dimensional information can not be applied to different display lens array with different parameters. If we use a sparse camera array to capture reference viewpoint information, the problem will be solved effectively.

3. Virtual viewpoint synthesis algorithm

3.1. Synthetic structure of two-dimensional viewpoint

The contribution of the two reference views to the virtual viewpoint synthesis effect is very limited, the background information occluded in the scene is difficult to complete. Therefore, this method of virtual viewpoint synthesis is based on multi-viewpoint, which can improve the virtual viewpoint information and the rendering image quality. Reference point of view images used in this article and their relative position shown as figure 2(a), the size of array is 4×4, the adjacent viewpoints have the same horizontal and vertical spacing, image resolution of 320×240.

Create a virtual view synthesis coordinate system with *0,1,4,5* reference view as a plane(solid line selected area shows in figure 2(b), the viewpoints and their coordinate positions as shown in figure 2 (c).

3.2. Obtain the depth information

Acquire scene depth based on *0* and *1* viewpoint images, this method requires less depth of image accuracy. Therefore, the SAD similarity measure is adopted as the stereo matching, which has the advantage of high speed [6]. The proceed shows as follows:

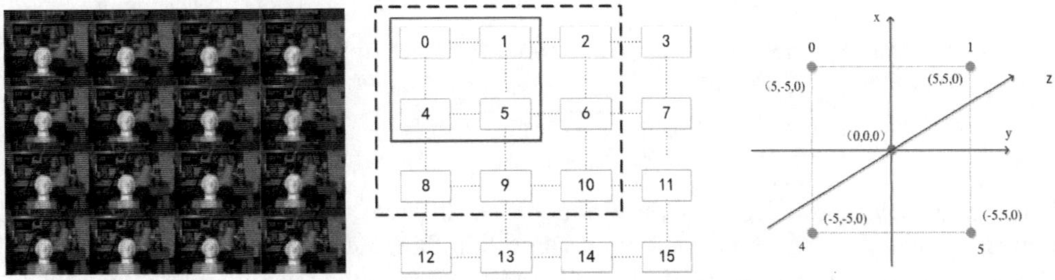

(a) image array (b) relative position (c) coordinate system

Figure 2. Structure of virtual viewpoint synthesis

- Enter the original image pair, select a pixel in the upper left corner of the base map (right) to create a fixed 15×15 size window.
- Calculate the similarity between the corresponding pixels within the search range according to the similarity measure function SAD. The point with high similarity is the matching point, then calculate their disparity value.
- Determine whether the pixel matches the lower right corner of the image. If it is transferred go ahead, otherwise turn back to last step.
- Output disparity, complete the match.

3.3. Two-dimensional image layering and disparity calculation

The obtained depth map is set as $I_D(x,y)$, the depth layer index value *Sign (x,y)* corresponding to each pixel in the depth map is calculated according to the set number of layers *N* (*N* is a positive integer greater than 1). The formula is as follows

$$Sign(x,y) = \langle\ I_D(x,y)/(256/N+0.5)\ \rangle \tag{1}$$

Where $\langle a \rangle$represents the largest integer operation less than or equal to a. Obtain the layered depth map $I_D^S\ (x, y)$ of the two-dimensional image according to the depth layer flag *Sign (x, y)*. The index of the stratified depth map $I_D^S\ (x, y)$ is *sign*, which is smaller the value, the closer the depth layer is to the viewer. When *Sign (x, y)* equals to *sign*, $I_D^S\ (x, y)$ equals to $I_D(x, y)$, otherwise equals to 255.

Select a depth layer in the layered depth image as the focal layer of the camera array. Determine the foreground layer and background layer according to the focus layer. If the index value $sign > focus$, set it as the background layer; If the index value $sign < focus$, set it as the background layer. The reference viewpoint images are similarly stratified. Calculate the layered image $I_C^S\ (x, y)$ based on the layered depth image, at the same time record each layer images assignment flag $F_c^s\ (x, y)$. When *Sign (x, y)* equals to *sign*, $I_C^S\ (x, y)$ equals to $I_D^S\ (x, y)$, $F_c^s\ (x, y)$ equals to 1. Otherwise, $I_C^S\ (x, y)$ equals to 255 and $F_c^s\ (x, y)$ equals to 0.

Let the maximum disparity value between the reference viewpoint foreground layer and its adjacent viewpoint fore-ground layer be D_M (D_M is an integer greater than or equal to one), calculate the maximum disparity value R_M between the foreground and background layers

$$R_M = (D_M / focus) \bullet (N-1) \tag{2}$$

Equation(6) shows how to compute the corresponding dis-parity value of each depth layer combine the distance of each layer to the focal layer

$$Dis^{sign} = -|(focus - sign) * (R_M /(N-1)) + 0.5| \tag{3}$$

Select the number of layers $N = 5$, the maximum disparity value $D_M = 30$, get layered result as shown in figure 3.

3.4. Virtual viewpoint rendering

Assuming that the size of virtual viewpoints in the virtual parallel multi-view array is $M \times M$ (*M* is an integer equal to or greater than 2)and the zero reference viewpoint is the central viewpoint of the multi-view array. For the virtual viewpoint i with coordinates *(a, b, 0)*, its index value of sub_i can be count by the formula(7)

$$sub_i = (M-b)M - | \; M/2 - a \; | \quad i = 0,1,2...,M^2 \tag{4}$$

The relative distance from the virtual viewpoint *(a, b, 0)* position to the reference viewpoint position *(x, y, 0)* is

$$D_{sub_i} = \sqrt{(x_j - a)^2 + (y_j - b)^2} \tag{5}$$

Where *j* denotes the position of the reference viewpoint, *j = 0, 1, ... 16*

According to each layer of images corresponding to different disparity values, render the virtual viewpoint image based on weighted translation algorithm. Suppose we want to generate the image of the *i* position in *M* virtual viewpoints, we should perform a weighted panning operation on the image layer and the assignment flag according to the the relative distance between the virtual viewpoint position and the reference viewpoint position and the disparity corresponding to each image layer. The equations are as follows

$$I_H^S(x,y) = \begin{cases} 0 & (x + Dis^{sign} \bullet D_{sub_i}) < 0 \;\; or \;\; (x + Dis^{sign} \bullet D_{sub\_i}) \ge H \\ I_C^S(x + Dis^{sign} \bullet |x_j - a|, y + Dis^{sign} \bullet |y_j - b|) & 0 \le (x + Dis^{sign} \bullet D_{sub_i}) \le H \end{cases} \tag{6}$$

$$F_H^S(x,y) = \begin{cases} 0 & (x + Dis^{sign} \bullet D_{sub\_i}) < 0 \;\; or \;\; (x + Dis^{sign} \bullet D_{sub_i}) \ge H \\ F_C^S(x + Dis^{sign} \bullet |x_j - a|, y + Dis^{sign} \bullet |y_j - b|) & 0 \le (x + Dis^{sign} \bullet D_{sub_i}) \le H \end{cases} \tag{7}$$

Where, *H* represents the image horizontal resolution.Initialize the virtual viewpoint's assignment flag with position *i* $F^i_{vir}(x,y)$to 0.

The translated images $I_H^S(x,y)$ are superimposed layer by layer follow the order of putting the foreground in the background, that is, the order of *sign* from big to small. For Each time superimposed image layer, update the image assignment flag $F^i_{vir}(x,y)$. After the overlay operation, a virtual view point image with position *i* is obtained, the formula is as follows

$$F_{vir}^i(x,y) = \begin{cases} 0 & F_{vir}^i(x,y) = 0 \;\; and \;\; F_H^S(x,y) = 0 \\ sign + 1 & F_H^S(x,y) = 1 \\ sign + 1 + N & F_{vir}^i(x,y) = 0 \;\; and \;\; F_H^S(x,y) = 2 \end{cases} \tag{8}$$

$$I_{vir}^i(x,y) = \begin{cases} 0 & F_{vir}^i(x,y) = 0 \\ I^{F_{vir}^i(x,y)-1}(x,y) & F_{vir}^i(x,y) > 0 \;\; and \;\; F_{vir}^i(x,y) - N \le 0 \\ I^{F_{vir}^i(x,y)-1-N}(x,y) & F_{vir}^i(x,y) - N > 0 \end{cases} \tag{9}$$

This rendering order, effectively ensure that the foreground object and the background object between the correct occlusion relationship. Perform the above operations on the *M* virtual viewpoints in turn, we can get *M*×*M* virtual view of the image. The composite view at *(0,0,0)* position of coordinate system shows in figure 4 below.

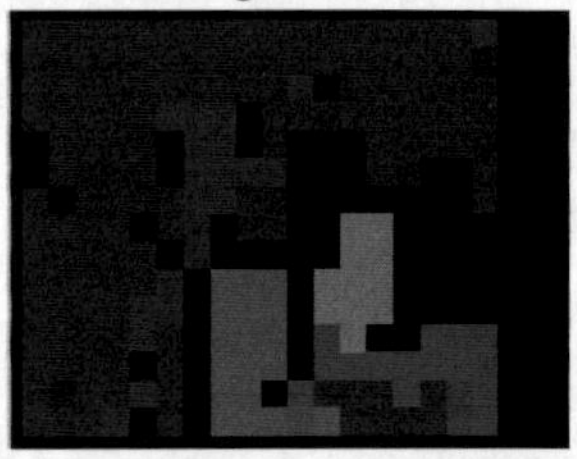

Figure3. Layered schematic of depth images

Figure4. Virtual viewpoint image of (0,0,0)

4. Experimental results

For the coordinates in figure 2(c), the synthesized view point image at some position is shown in the figure 5 using the above method.

IOP Conf. Series: Journal of Physics: Conf. Series **1004** (2018) 012012 doi:10.1088/1742-6596/1004/1/012012

(a)coordinate points（2,0,0） (b)coordinate points（-2,2,0） (c)coordinate points (-3,4,0)

Figure 5. Partial position virtual viewpoint image

In order to verify the effect of the synthesized viewpoint and the true viewpoint, in the same way, viewpoint *5's* image is rendered as a virtual viewpoint using reference view *0,2,8,10*(As shown in the dotted line area in figure 2(b), the experimental results shown in figure 6.

(a) real point of view and render point of view (b)detail comparison chart

Figure 6.Viewpoint *5* image contrast

In the subjective evaluation, the virtual viewpoint rendered image is already very close to the true viewpoint image, basically retained the complete image information. Only a slight blur appears in the foreground plaster and the lamp. In the objective evaluation, firstly, use color histograms as measures for image similarity comparison. The method of this paper is used to synthesize viewpoint *5, 6, 9*and*10*. The virtual viewpoint images are shown in figure 7, and the data results shown in figure 8.

(a)virtual viewpoint *5* (b)virtual viewpoint *6* (c) virtual viewpoint *9* (d) virtual viewpoint *10*

Figure 7.Virtual viewpoint image

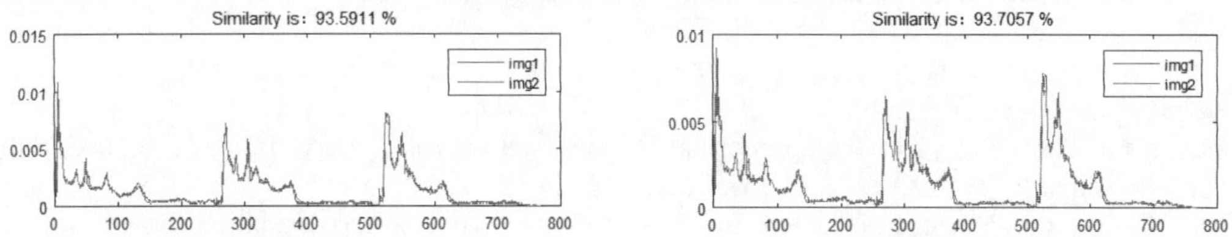

(a)virtual viewpoint *5* (b)virtual viewpoint *6*

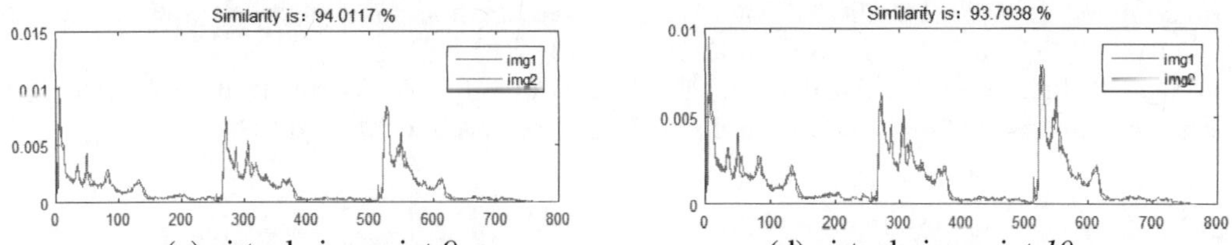

(c) virtual viewpoint *9* (d) virtual viewpoint *10*

Figure 8.Color histograms similarity result

As can be seen from the figure 9, virtual point of view and the true point of view have a high degree of similarity and the mean of similarity is 93.77%.

IOP Conf. Series: Journal of Physics: Conf. Series **1004** (2018) 012012 doi:10.1088/1742-6596/1004/1/012012

Then, PSNR(peak signal to noise ratio)[7], SSIM(structural similarity index)[8] and *t*(time)are used to objectively compare the quality of generated virtual viewpoint image, the unit of *t* is *s*. Among them, the higher the PSNR value, the closer the SSIM value is to 1, the better the image quality is. The synthesis time for each viewpoint was also measured. Synthetic viewpoint images and the real viewpoint images are compared separately and the table below shows the data.

It can be seen in the table that all the numerical values are basically acceptable and the objective synthetic quality is satisfactory.

Table 1. Image quality data

	PSNR	SSIM	t
5	38.3369	0.9533	6.7895
6	38.3296	0.9544	6.9209
9	38.4082	0.9511	7.0619
10	38.3374	0.9510	6.8302
average	38.3530	0.9525	6.8998

5. Conclusion

Based on the principle of integrated imaging, a virtual view rendering method based on image hierarchical algorithm is proposed in this paper. Firstly, the view point image is layered according to the depth information of the scene, and then the weighted viewpoint translation of the layered image is used to obtain the virtual viewpoint image. The synthesis effect of different viewpoint is validated, experiments show that we can quickly get any position of virtual view images effectively, which laid a good foundation for the synthesis of elemental image array of lens array display system.

References

[1] JIAO Xiao-xue, ZHAO Xing, YANG Yong, FANG Zhi-liang, YUAN Xiao-cong 2012 Pick-up system for three-dimensional integral imaging with camera array *Optics and J.Precision Engineering*. **8**. 1653-60

[2] YUAN Xiao-cong, XU Yu-ping, YANG Yong, ZHAO Xing, BU Jing 2011 Design parameters of elemental images formed by camera array for crosstalk reduction in integral imaging *J.Optics and Precision Engineering*. **9** .2050-56

[3] ZENG Yao-Xian 2012 A Dissertation Submitted to Shanghai Jiao Tong University for Master Degree of Engineering *D.Shanghai Jiao Tong University*

[4] Jiao Xiao-Xue, Zhang Lei, Sun Yu, Zhou Li-Qiu, Zhao Xing 2015 Three-dimensional reconstruction for a large scene of integral imaging based on color-position characteristics. *J.Optoelectronics Letters*. **4** .268-272

[5] Wang Yu, Piao Yan 2014 Computational Reconstruction for Integral Imaging with Sampled Elemental Images *J.ACTA OPTICA SINICA*. **5**.70-75

[6] Yue Chen-ping, Meng Liya 2013 Implementation og the stereo matching based on SAD algorithm. *J.Image Processing and Multimedia Technology*. **6** .41-43

[7] PINSON M H, WOLFS 2004 A new standardized method for objectively measuring video quality *J .IEEE Transactions on Broadcasting* .**50(3)**. 312-322

[8] WANG Z, BOVIK A C, SHEIKHHR 2004 Image quality assessment: from error visibility to structural similarity *J.IEEE Transactions on Image Processing* .**13(4)**. 600-612

IOP Conf. Series: Journal of Physics: Conf. Series **1004** (2018) 012013 doi:10.1088/1742-6596/1004/1/012013

Fast Deep Tracking via Semi-Online Domain Adaptation

Xiaoping Li[1, a], Wenbing Luo[1, *, b], Yi Zhu[1, c], Hanxi Li[1, d] and Mingwen Wang[1, e]

[1]Jiangxi Normal University, 99 Ziyang Road, Nanchang, Jiangxi, China

[a]xp.li@jxnu.edu.cn; [*, b]lwb@jxnu.edu.cn; [c]deke.zhu@jxnu.edu.cn; [d]hanxi.li@jxnu.edu.cn; [e]mwwang@jxnu.edu.cn

Abstract. Deep tracking has been illustrating overwhelming superiorities over the shallow methods. Unfortunately, it also suffers from low FPS rates. To alleviate the problem, a number of real-time deep trackers have been proposed via removing the online updating procedure on the CNN model. However, the absent of the online update leads to a significant drop on tracking accuracy. In this work, we propose to perform the domain adaptation for visual tracking in two stages for transferring the information from the visual tracking domain and the instance domain respectively. In this way, the proposed visual tracker achieves comparable tracking accuracy to the state-of-the-art trackers and runs at real-time speed on an average consuming GPU.

1. Introduction

Visual tracking is one of the most fundamental computer vision tasks. The visual tracker is required to predict the motion state of the tracking target at each frame, given the target state at the first frame.

During the last decade, as the surge of deep learning, more and more tracking algorithms benefit from deep neural networks, e.g. Convolutional Neural Networks[1][2] and Recurrent Neural Networks [3][4]. Despite the well-admitted success, a dilemma still existing in the community is that, deep learning increases the tracking accuracy, while at the cost of high computational complexity. As a result, most well-performing deep trackers usually suffer from low efficiency[5][6]. Recently, some real-time deep trackers were proposed[7][8]. They achieved very fast tracking speed, but can not beat the shallow methods in some important evaluations, as we illustrate latter.

Wang *et.al* [9] illustrated an effective way to make the deep trackers real-time. They try to transfer the deep features from the classification domain to the tracking domain and meanwhile reduce the feature channels. The yielded tracking algorithm, termed MSDAT, achieves real-time tracking speed and still maintains the high tracking accuracy. Figure 1 compares the network structures and the learning strategy of MD-Net and the MSDAT method. Both of them successfully transfer the knowledge from the classification domain to the tracking domain and thus both of them achieve high tracking accuracies[6][9].

However, the 30 FPS speed of MSDAT is obtained on a high-end GPU. For normal computers in the market, the MSDAT can not achieve real-time. A straightforward way to accelerate the MSDAT further is to reduce the network size while this usually leads to serious performance drop. In this work, we propose to enhance the original MSDAT method via adding a semi-online learning procedure on MSDAT. The semi-online learning stage only covers the first 10 frames of the video sequence and the network is fixed after that. In this way, we further transfer the knowledge from the visual tracking domain to the current video sequence. The experiment part of this paper verifies the superiority of the

IOP Conf. Series: Journal of Physics: Conf. Series **1004** (2018) 012013 doi:10.1088/1742-6596/1004/1/012013

proposed two-stage transfer learning which achieves 60 FPS on a high-end GPU and 23 FPS on a average consuming GPU.

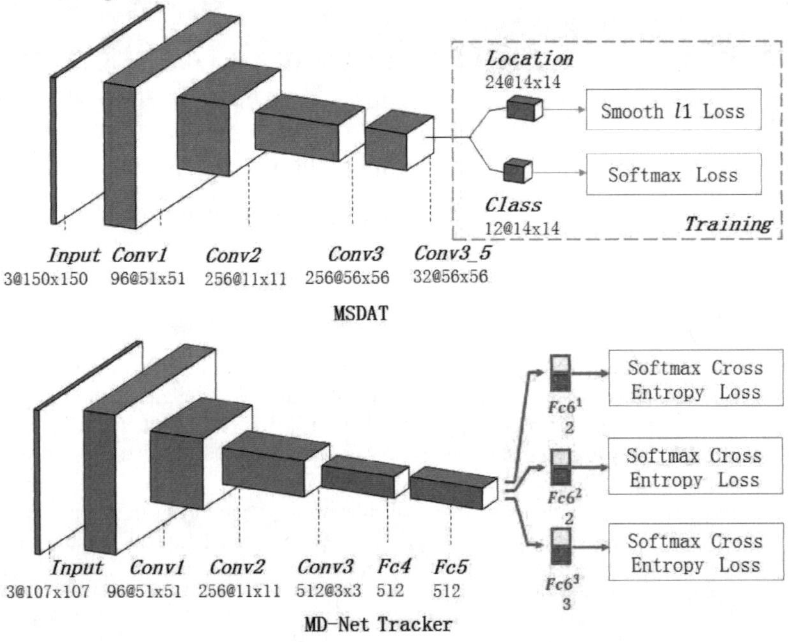

Figure 1. The flow-charts of the training process of MSDAT and MD-net. Note that the network parts inside the dashed blocks are only used for training and will be abandoned in tracking. Better view in color.

2. The proposed method

The MSDAT algorithm performs well at a real-time speed. However, the 30 FPS speed is obtained on a high-end GPU (NVIDIA GTX1070 as described in [9]). For normal computers in the market, the MSDAT can not achieve real-time. A straightforward way to accelerate the MSDAT further is to reduce the network size while this usually leads to serious performance drop. In this work, we propose to split the domain adaptation into two stages: an offline stage that transfers the knowledge (parameterized in the CNN model) from the image classification domain to the visual tracking domain; an semi-online stage that further transfers the visual tracking knowledge to the current video sequence. In this way, we obtain a faster deep tracker compared with MSDAT while maintain its high accuracy.

An brief illustration of the proposed two-stage learning is shown in Figure 2.

2.1. Network structure

Considering that the semi-online learning is time consuming, especially for those median size network. In this work, we modify the PVA net, which was originally proposed for object detection for visual tracking. The network structure is shown in Figure 3.

As we can see in the figure, the network structure is mainly inherited from the PVA net [10]except the three "tails" mounted on the network, where the dimension of the deep feature is reduced. The branches are set following the convention of HCF[5] and MSDAT[9] When tracking, the second stage of the domain adaptation, *i.e.*, the knowledge transfer from the visual tracking domain to the object instance domain, is also conducted in those branches. That is to say, we update the CNN parameters of the branches online, namely, with the image samples captured in the current test sequence (see Figure 2).

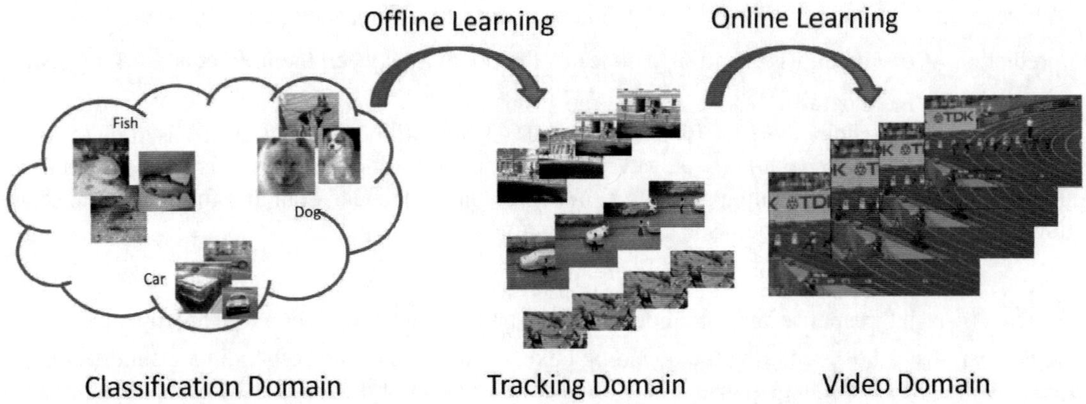

Figure 2. The two-stage learning strategy of this work. We can see that the transfer learning from the classification domain to the visual tracking domain is performed offline while when testing, the second stage, *i.e.*, the transfer learning from the visual tracking domain to the particular instance domain is conducted.

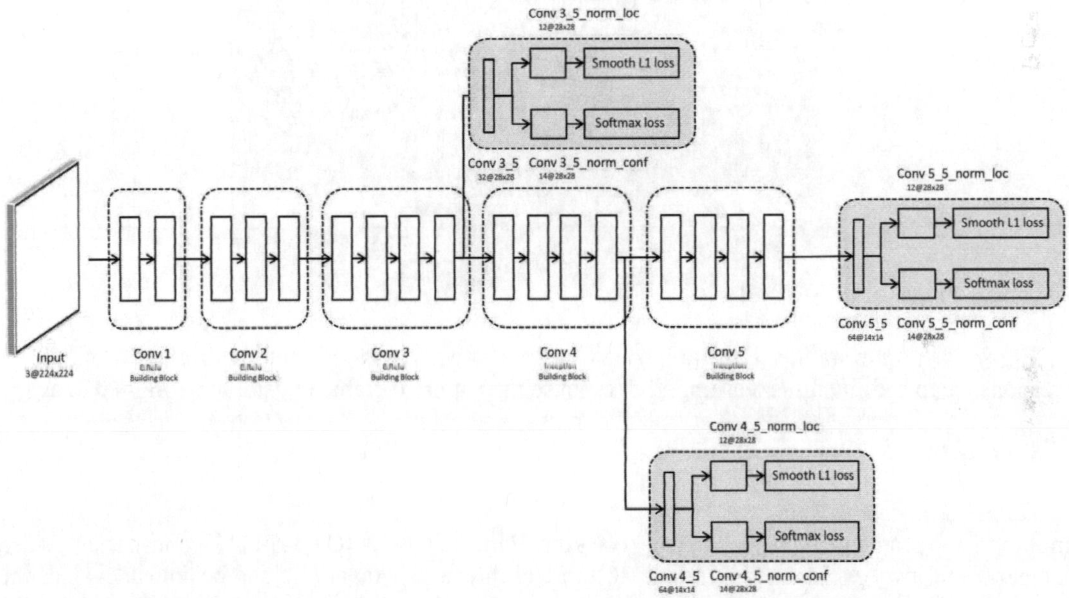

Figure 3. The network structure of this work. We can see that the network is "narrower" than that used in MSDAT and there are also three branches mounted on to the original PVA net. The KCF trackers are performed based on the deep features extracted from the three branches. The CNN parameters of the three branches are also updated in the test video for the first 10 frames.

2.2. Gaussian response loss function

As described in[9], the MSDAT algorithm modified the original loss function of the SSD algorithm[11] as

$$L_{loc}^{i,j}(l,g) = \sum_{\mu \in \{x,y\}} m_{i,j} \cdot smooth_{L_1}\left(l_i^u - \hat{g}_j^u\right) \tag{1}$$

Where $\hat{g}_j^u, u \in \{x, y, w, h\}$ is one of the geometry parameter of normalized ground-truth box, l_i^u is the prediction of one default box and $m_{i,j}$ indicates the relation between them. Readers are suggested to refer to [9] for more details.

In this work, we adopt the loss function of MSDAT for offline training as it has proved itself successfully for offline adaptation. However, we empirically find that loss function defined in Equation 1 is not suitable for online adaptation. We thus employ the Gaussian loss for the second stage of domain adaptation in this paper. The new loss is defined as

$$L_{gaussian} = \left\| M - \hat{M} \right\|_{l2} \quad (2)$$

Where M is the response map predicted by one branch (note that here we directly predict the response map instead of predicting the bounding boxes) and $\hat{M}$ is the corresponding ground-truth. To better illustrate the Gaussian response loss, we show the learning objective *i.e.*, the ground-truth maps, the original response maps and the learned response maps in Figure 4.

In practice, when testing, we update the parameters in the three branches during the first 10 frames, with learning rate 5e-5 and decay ratio 1e-3. The KCF tracker is restarted when the online learning is finished.

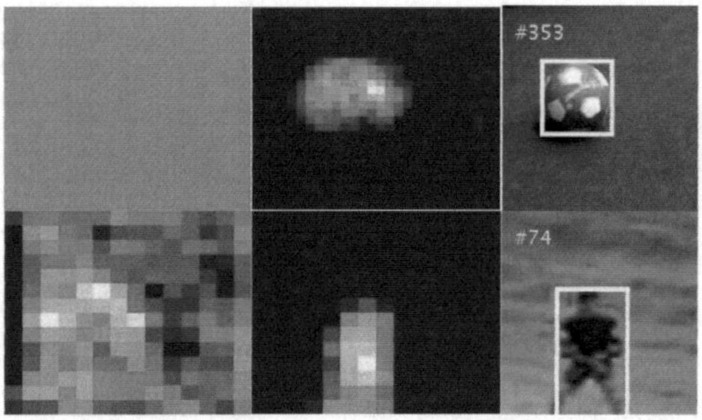

Figure 4. Two examples of the proposed Gaussian response loss. From left to right: the initial response map before online learning; the response map after 10-frame update; the original image.

3. Experiment

3.1. Dataset and competitors

Similar to its prototype[12], the Object Tracking Benchmark 50 (OTB-50)[13] consists 50 video sequences and involves 51 tracking tasks. It is one of the most popular tracking benchmarks since the year 2013, The evaluation is based on two metrics: center location error and bounding box overlap ratio. The one-pass evaluation (OPE) is employed to compare our algorithm with some well-performing shallow visual trackers including the KCF tracker[14], TGPR[15], Struck[16], MIL[17], TLD[18] and SCM[19]. Also, some recently proposed deep trackers including MD-net[6], HCF[5], MSDAT[9], GOTURN[7] and the Siamese tracker[8] are also compared.

3.2. Tracking results

The comparison result curves are shown in Figure5, note that we also report the running speeds (in FPS) of the comparing methods. From the curves we can see that the proposed method achieves comparable tracking accuracies to the state-of-the-art visual trackers while performs at 61 FPS on a high-end GPU (Nvidia GTX-1070). The number nearly double the speed of MSDAT. More importantly, on an average consuming GPU, Nvidia GTX-1050, our method performs at 27 FPS which is still real-time.

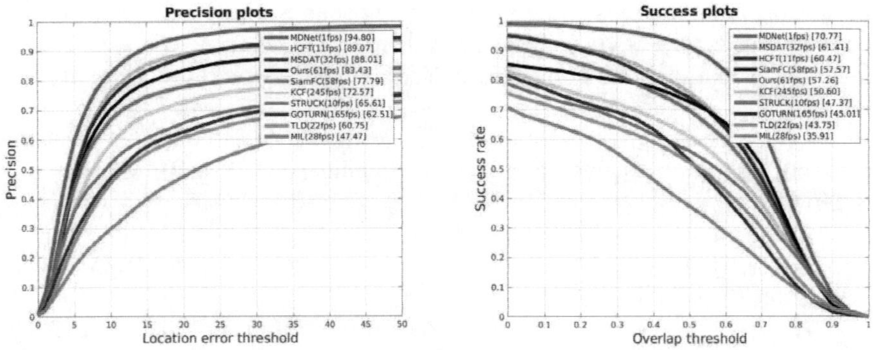

Figure 5. The location error plots and the overlapping accuracy plots of the involving trackers, tested on the OTB-50 dataset.

4. Conclusion

We propose a novel two-stage learning strategy for visual tracking. Firstly, a lightened CNN network is adopted for transferring the knowledge from the image classification domain to the visual tracking domain. When testing, the Gaussian response loss is used to update the "tracking branches" of the CNN model for the current tracking target. In this way, we obtain a deep tracker which runs real-time on average consuming GPUs while still maintain high tracking accuracy.

Acknowledgments

The work described in this paper was fully supported by a grant from Jiangxi Province Science Foundation for Youths (No. 20151BAB217019) and National Natural Science Foundation of China for Youths (No. 61503168)

References

[1] Naiyan Wang and Dit-Yan Yeung, "Learning a deep compact image representation for visual tracking," in NIPS, pp. 809–817. 2013. More references.

[2] Hanxi Li, Yi Li, and Fatih Porikli, "Deeptrack: Learning discriminative feature representations online for robust visual tracking," IEEE Transactions on Image Processing (TIP), vol. 25, no. 4, pp. 1834–1848, 2016.

[3] Anton Milan, Seyed Hamid Rezatofighi, Anthony Dick, Konrad Schindler, and Ian Reid, "Online multi-target tracking using recurrent neural networks," arXiv preprint arXiv:1604.03635, 2016.

[4] Guanghan Ning, Zhi Zhang, Chen Huang, Zhihai He, Xiaobo Ren, and Haohong Wang, "Spatially supervised recurrent convolutional neural networks for visual object tracking," arXiv preprint arXiv:1607.05781, 2016.

[5] Chao Ma, Jia-Bin Huang, Xiaokang Yang, and Ming-Hsuan Yang, "Hierarchical convolutional features for visual tracking," in ICCV, 2015, pp. 3074–3082.

[6] Hyeonseob Nam and Bohyung Han, "Learning multi-domain convolutional neural networks for visual tracking," arXiv preprint arXiv:1510.07945, 2015.

[7] David Held, Sebastian Thrun, and Silvio Savarese, "Learning to track at 100 fps with deep regression networks," in European Conference on Computer Vision, 2016, pp. 749–765.

[8] Luca Bertinetto, Jack Valmadre, Joao F Henriques, Andrea Vedaldi, and Philip HS Torr, "Fully-convolutional siamese networks for object tracking," in ECCV, 2016, pp. 850–865.

[9] Xinyu Wang, Hanxi Li, Yi Li, Fumin Shen, and Fatih Porikli, "Robust and real-time deep tracking via multi-scale domain adaptation," 2017.

[10] Kye Hyeon Kim, Sanghoon Hong, Byungseok Roh, Yeongjae Cheon, and Minje Park, "Pvanet: Deep but lightweight neural networks for real-time object detection," 2016.

IOP Conf. Series: Journal of Physics: Conf. Series **1004** (2018) 012013 doi:10.1088/1742-6596/1004/1/012013

[11] Wei Liu, Dragomir Anguelov, Dumitru Erhan, Christian Szegedy, and Scott Reed, "Ssd: Single shot multibox detector," arXiv preprint arXiv:1512.02325, 2015.
[12] Yi Wu, Jongwoo Lim, and Ming-Hsuan Yang, "Online object tracking: A benchmark," in CVPR, 2013, pp. 2411–2418.
[13] Yi Wu, Jongwoo Lim, and Ming-Hsuan Yang, "Object tracking benchmark," IEEE Transactions on Pattern Analysis and Machine Intelligence (TPAMI), vol. 37, no. 9, pp. 1834–1848, 2015.
[14] Joao F Henriques, Rui Caseiro, Pedro Martins, and Jorge Batista, "High-speed tracking with kernelized correlation filters," IEEE Transactions on Pattern Analysis and Machine Intelligence (TPAMI), vol. 37, no. 3, pp. 583–596, 2015.
[15] Jin Gao, Haibin Ling, Weiming Hu, and Junliang Xing, "Transfer learning based visual tracking with gaussian processes regression," in ECCV, pp. 188–203. 2014.
[16] Sam Hare, Amir Saffari, and Philip HS Torr, "Struck: Structured output tracking with kernels," in ICCV, 2011, pp. 263–270.
[17] Boris Babenko, Ming-Hsuan Yang, and Serge Belongie, "Visual tracking with online multiple instance learning," IEEE Transactions on Pattern Analysis and Machine Intelligence(TPAMI), pp. 1619–1632, 2011.
[18] Zdenek Kalal, Jiri Matas, and Krystian Mikolajczyk, "Pn learning: Bootstrapping binary classifiers by structural constraints," in CVPR, 2010, pp. 49–56.
[19] Wei Zhong, Huchuan Lu, and Ming-Hsuan Yang, "Robust object tracking via sparsity-based collaborative model," in CVPR, 2012, pp. 1838–1845.

IOP Conf. Series: Journal of Physics: Conf. Series **1004** (2018) 012014 doi:10.1088/1742-6596/1004/1/012014

Flu Diagnosis System Using Jaccard Index and Rough Set Approaches

Riswan Efendi[1,*], Noor Azah Samsudin[1], Mustafa Mat Deris[1] and Yip Guan Ting[1]

[1]Faculty of Computer Science and Information Technology, Universiti Tun Hussein Onn Malaysia, 86400 Batu Pahat, Malaysia
[2]Mathematics Department, UIN Sultan Syarif Kasim, 28293 Pekanbaru, Indonesia

*E-mail: riswan@uthm.edu.my

Abstract. Jaccard index and rough set approaches have been frequently implemented in decision support systems with various domain applications. Both approaches are appropriate to be considered for categorical data analysis. This paper presents the applications of sets operations for flu diagnosis systems based on two different approaches, such as, Jaccard index and rough set. These two different approaches are established using set operations concept, namely intersection and subset. The step-by-step procedure is demonstrated from each approach in diagnosing flu system. The similarity and dissimilarity indexes between conditional symptoms and decision are measured using Jaccard approach. Additionally, the rough set is used to build decision support rules. Moreover, the decision support rules are established using redundant data analysis and elimination of unclassified elements. A number data sets is considered to attempt the step-by-step procedure from each approach. The result has shown that rough set can be used to support Jaccard approaches in establishing decision support rules. Additionally, Jaccard index is better approach for investigating the worst condition of patients. While, the definitely and possibly patients with or without flu can be determined using rough set approach. The rules may improve the performance of medical diagnosis systems. Therefore, inexperienced doctors and patients are easier in preliminary flu diagnosis.

1. Introduction

In this paper, we are using two different approaches in diagnosing flu system, namely Jaccard index and rough set. All approaches is formed using set operations application, such as, intersection and subsets. In medical diagnostic domain, Venn diagram has been used to demonstrate the intersections between true positive, true negative, false positive and false negative results in a test for micro albuminuria [1]. While, Jaccard index has been also implemented to quantify similarity between hereditary diseases at molecular level [2]. However, the data reduction between conditional attributes and decision attribute cannot be solved using Venn and Jaccard approaches for medical diagnostic applications. Therefore, rough set theory was introduced and has been widely applied to solve complex problems by researchers in emergency room diagnostic medical. The rough set approaches can be used to assist such inexperienced doctors in diagnosing based on clinical decision support model of disease symptoms [3, 4].

There are existing rough set applications in medical diagnostic procedure to detect diseases, such as, dengue [3], diabetes mellitus [3, 4], chikungunya [4], and other. However, the step-by-step procedure

IOP Conf. Series: Journal of Physics: Conf. Series **1004** (2018) 012014 doi:10.1088/1742-6596/1004/1/012014

in determining suitable rules for the medical diagnostic applications remains an interesting issue since the ultimate goal is to achieve accurate prediction results. Motivated by application of set theory and its operations in various medical diagnostic applications [1-4], we are interested to investigate the dependency of conditional attributes (flu's symptoms) and decision attribute (flu) based on set operations from two different approaches. Furthermore, the symptom dependency values can be used for preliminary prediction purpose of the flu patients.

2. Fundamental Theories of Jaccard Index and Rough Set

In this paper, we explain two different approaches and its theories, namely, Jaccard index and rough set in determining the flu diagnosis system. While, both approaches and its theories will be explained in Sections 2.1 and 2.2.

2.1. Jaccard Index Theory

Motivated by Venn diagram and set operations, The Jaccard index, also known as intersection over union and the Jaccard similarity coefficient (originally coined coefficient de communauté by Paul Jaccard), is a statistic used for comparing the similarity and diversity of sample sets. The Jaccard coefficient measures similarity between finite sample sets, and is defined as the size of the intersection divided by the size of the union of the sample sets [5]:

$$J(A,B) = \frac{|A\cap B|}{|A\cup B|} = \frac{|A\cap B|}{|A|+|B|-|A\cap B|}, \quad (1)$$

The Jaccard distance, which measures dissimilarity between sample sets, is complementary to the Jaccard coefficient and is obtained by subtracting the Jaccard coefficient from 1, or, equivalently, by dividing the difference of the sizes of the union and the intersection of two sets by the size of the union:

$$dJ(A,B) = 1 - J(A,b) = \frac{|A\cup B|-|A\cap B|}{|A\cup B|}, \quad (2)$$

2.2. Rough Set Theory

The rough set theory has been introduced by Pawlak (1982) and well divided by researchers into information systems, indiscernibility relation, set approximations, rough clustering, and others. An information system $S = (U, \Omega, V_q, f_q)$ consists of [6, 7, 8]:

U : a nonempty, finite set called the universe;
Ω : a nonempty, finite set of attributes;
Ω : $C \cup D$, in which C is a finite set of condition attributes and D is a finite set of decision attributes; for each $q \in \Omega$, V_q is called the domain of q;
f_q : an information function $f_q: U \to V_q$.

The cases, states, processes, patients, companies, and observations can be interpreted as objects or elements of rough sets. The attributes of each element can be assumed as symptoms, factors, and characteristic information. A relationship between conditional attributes and decision attribute can be explained using information table. In this table, the row and column correspond to objects and attributes, respectively. The starting point of rough set theory is the indiscernibility relation, generated by information about objects of interest.

Let $S = (U, \Omega, V_q, f_q)$ be an information system, then any subset B of A determines a binary (equivalence) relation IND(B) on U, which will be called B-indiscernibility relation, and is defined as follows:

$$\text{IND}(B) = \{(x,y) \in U^2 : \forall a \in B, a(x) = a(y)\}, \quad (3)$$

Where $a(x)$ denotes the value of attribute a for element x in U. The collection of all equivalence

IOP Conf. Series: Journal of Physics: Conf. Series **1004** (2018) 012014 doi:10.1088/1742-6596/1004/1/012014

classes determined by IND(B), denoted by U/B. An equivalence class of U/B, containing x, is denoted by $[X]_B$.

Let $S = (U, \Omega, V_q, f_q)$ be an information system and let $B \subseteq A$ and $X \subseteq U$. We can approximate X using only the information contained in B by constructing the B-lower and B-upper approximations of X. Both approximations are denoted as:

$$\underline{B}(X) = \{\{x \in U | [x]_B \subseteq X\}\}, \tag{4}$$

And

$$\bar{B}(X) = \{x \in U | [x]_B \cap X \neq \emptyset\}, \tag{5}$$

Where $[x]_B$ is an equivalence class containing x. While, the difference between both approximations and its accuracy can be written:

$$\text{BND}(X) = \underline{B}(X) - \bar{B}(X), \tag{6}$$

3. Implementation

In this section, we discuss how to build the flu diagnosis system using Jaccard index and rough set approaches in sections 3.1 and 3.2.

3.1. Flu Diagnosis Using Jaccard Index Approach

This sub-section presents the implementation of set operations using Jaccard index in diagnosing flu system. A number data sets of patient flu [9] is presented in Table 1.

Table 1. Information of patient's flu and its symptoms

Patient's code	Conditional attributes			Decision attribute
	Headache	Muscle Pain	Temperature	Flu
p_1	No	Yes	High	Yes
p_2	Yes	No	High	Yes
p_3	Yes	Yes	Very High	Yes
p_4	No	Yes	Normal	No
p_5	Yes	No	High	No
p_6	No	Yes	Very High	Yes

Step 1: Based on table 1, define sets of conditional attributes and decision attribute using patient's code as presented in table 2.

Table 2. Sets attributes and its elements

Set Attributes	Set elements
Headache = H	= {{Yes}, {No}}, = {{p_2,p_3,p_5}, {p_1,p_4,p_6}.
Muscle Pain = MP	= {{Yes}, {No}}, = {{p_1,p_3,p_4,p_6}, {p_2,p_5}}.
Temperature = T	= {{Normal}, {High}, {Very High}, = {{p_4}, {p_1,p_2,p_5}, {p_3,p_6}}.
Decision (Flu) = F	= {{Yes, {No}}, = {{p_1,p_2,p_3,p_6}, {p_4,p_5}}.

Step 2: Find set intersections between conditional attributes and decision attribute.

Intersection between sets H (Yes) and MP (Yes), H = {p_2,p_3,p_5}, and MP = {p_1,p_3,p_4,p_6}. Then, H ∩ MP = {p_3}.

Intersection between sets H (Yes), MP (Yes) and T (very high), respectively. H = {p_2,p_3,p_5}, MP = {p_1,p_3,p_4,p_6}, T = {p_3,p_6}. Then, H ∩ MP ∩ T = {p_3}.

IOP Conf. Series: Journal of Physics: Conf. Series **1004** (2018) 012014 doi:10.1088/1742-6596/1004/1/012014

Intersection sets H (Yes), MP(Yes), T(Very high) and F(Yes), respectively. H = $\{p_2,p_3,p_5\}$, MP = $\{p_1,p_3,p_4,p_6\}$, T = $\{p_3,p_6\}$, F = $\{p_1,p_2,p_3,p_6\}$. Then, $H \cap MP \cap T \cap F = \{p_3\}$. It is presented in figure 1.

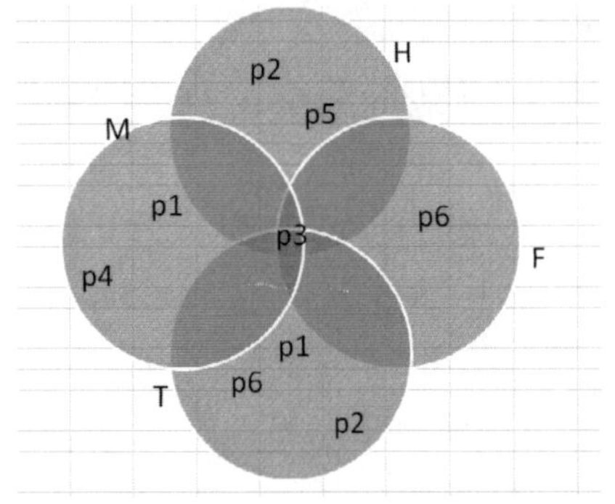

Figure 1. Intersections of sets H, MP, T and F

Step 3: Based on Step 2, calculate Jaccard similarity index (J) for each figure as presented in table 3.

Table 3. Similarity and dissimilarity based on Jaccard index

Intersection of sets	Jaccard similarity and dissimilarity indexes (J)
$H \cap MP = \{p_3\}$	= 1/6,
	= 0.1667 = 16.67%,
	Jaccard similariry index is: J_s(H, MP) = 16.67% similar.
	Jaccard dissimilarity index is:
	= 1 – J(H, MP),
	= 1 – 0.1667,
	= 0.8333.
	J_d(H, MP) = 83.33% dissimilar.
$H \cap MP \cap T = \{p_3\}$	= 1/7,
	= 0.1429 = 14.29%,
	Jaccard similariry index is: J_s(H, MP, T) = 14.29% similar.
	Jaccard dissimilarity index is:
	= 1 – J(H, MP, T),
	= 1 – 0.1429,
	= 0.8571.
	J_d(H, MP, T) = 85.71% dissimilar
$H \cap MP \cap T \cap F = \{p_3\}$	= 1/9,
	= 0.1111 = 11.11%,
	Jaccard similariry index is: J_s(H, MP, T, F) = 14.29% similar.
	Jaccard dissimilarity index is:
	= 1 – J(H, MP, T, F),
	= 1 – 0.1111,
	= 0.8889.
	J_d(H, MP, T, F) = 88.89% dissimilar

Step 4: Based on step 3 and table 3, determine the definitely patient have flu. In this case, we obtain that the patient p_3 is the worst condition based on similarity and distance of J_s(H, MP), J_s(H, MP, T), and J_s(H, MP, T, F). Therefore, the patient p_3 should be treated immediately.

3.2. Flu Diagnosis Using Rough Set Approach

Based on tables 1, 2 and [9], the rough set can be implemented to determine the dependency between conditional symptoms and decision attribute by following

Step 1: Determine lower, upper approximations and boundary regions as shown in tables 4 and 5.

Table 4. Lower and upper approximations

Lower approximation (LA)	Upper approximation (UA)
The patients that are definitely have flu = {p1, p3, p6}	The patients that possibly have flu = {p1, p2, p3, p6}.
The patients does not have flu = {p4}.	The patient that possible does not have flu = {p4, p5}.

Table 5. Boundary regions (BR)

BR for definitely have flu	BR for possibly have flu
BR = { p1, p3, p6}- { p1, p2, p3, p6} = {p2}.	BR = { p4}- { p4, p5} = {p5}.

By following all steps given in data reduction and extraction [9], we only present the final result of intersection data with symptoms and decision attributes on table 6.

Table 6. Final intersection data and information

Patient code	Conditional attribute			Decision attribute
	Headache	Muscle Pain	Temperature	Flu
p_3	Yes	Yes	Very High	Yes
p_6	No	Yes	Very High	Yes
p_1	No	Yes	High	Yes
p_4	No	Yes	Normal	No

Based on table 6, we generate the decision support rules for flu diagnosis (prediction) as presented in table 7, and we can also find the final intersection and information in table 8:

Table 7. Proposed decision support rules

Rule	Condition
Rule 1	If a patient with Symptom Headache: "Yes", and Symptom Muscle Pain: "Yes", and Symptom Temperature: "Very High". Then decision of flu: "Yes".
Rule 2	If a patient with Symptom Muscle Pain: "Yes", and Symptom Temperature: "High or Very High". Then decision of Flu: "Yes".
Rule 3	Otherwise, "No Flu".

Table 8. Final intersection data and information

Patient code	Conditional attribute			Decision attribute	Prediction attribute	
	Headache	Muscle Pain	Temperature		Headache	Muscle Pain
p_1	No	Yes	High	p_1	No	Yes
p_2	Yes	No	High	p_2	Yes	No
p_3	Yes	Yes	Very High	p_3	Yes	Yes
p_4	No	Yes	Normal	p_4	No	Yes
p_5	Yes	No	High	p_5	Yes	No
p_6	No	Yes	Very High	p_6	No	Yes

IOP Conf. Series: Journal of Physics: Conf. Series **1004** (2018) 012014 doi:10.1088/1742-6596/1004/1/012014

Table 7 shows the implementation of proposed rules in flu predicting and comparison result with rules proposed in [9]. Our proposed rules are able to predict almost all patients correctly, except the patient p2. While, the previous rule [9] is only able to predict the patient p1. In this case, the refining of rules are very important to be considered in order to improve the prediction accuracy.

4. Conclusion

In this paper, we implemented Jaccard index and rough set approximation for medical diagnosis systems, namely, flu diagnosis system. In the application, Jaccard index approach can be used to determine the worst condition based on the similarity and distance between conditional attributes and decision attribute. While, the rough set approach can be applied to determine the data reduction and decision support rules. Both approaches can be implemented to build the decision support systems for medical diagnosis. Thus, the inexperience doctors may use the systems for preliminary diagnosis of the patients.

Acknowledgement

This paper work is financially supported by ORICC Vot number D003 under Universiti Tun Hussein Onn Malaysia (UTHM).

References

[1] Stralen K J V, Stel V S, Reitsma J B, Dekker F W, Zoccali C, Jager K J 2009 *Kidney Inter.* **75** 1257
[2] Caniza H, Romero A E, Paccararo A 2015 *Sci. Report* **5** 17658
[3] Ali R, Hussain J, Siddiqi M H, Hussain M, Lee S 2015 *Sensors* **15** 15921
[4] Thivigar M L, Richard C, Paul N R 2012 *Inter. J. Inform. Sci.* **2** 33
[5] Jaccard P 1912 *New Phytologist* **11** 37
[6] Pawlak Z 1982 *Inter. J. Inform. Comp. Sci.* **11** 341
[7] Hampton J. 1997 *J. Comp. Intel. Fin.* **5** 25
[8] Tay F E H, Shen L 2002 *Euro. J. Operation. Res.* **141** 641
[9] Rissino S, Torres G L 2009 *Data Mining Knowledge Dis. in Real Life App. Inform.* 35

IOP Conf. Series: Journal of Physics: Conf. Series **1004** (2018) 012015 doi:10.1088/1742-6596/1004/1/012015

Plant Species Identification by Bi-channel Deep Convolutional Networks

Guiqing He[1], Zhaoqiang Xia[1,*], Qiqi Zhang[1], Haixi Zhang[1] and Jianping Fan[2]

[1]School of Electronics and Information, Northwestern Polytechnical University, Shaanxi, China.

[2]Department of Computer Science, University of North Carolina at Charlotte, Charlotte, USA

*The corresponding author's e-mail: zxia@nwpu.edu.cn

Abstract. Plant species identification achieves much attention recently as it has potential application in the environmental protection and human life. Although deep learning techniques can be directly applied for plant species identification, it still needs to be designed for this specific task to obtain the state-of-art performance. In this paper, a bi-channel deep learning framework is developed for identifying plant species. In the framework, two different sub-networks are fine-tuned over their pretrained models respectively. And then a stacking layer is used to fuse the output of two different sub-networks. We construct a plant dataset of Orchidaceae family for algorithm evaluation. Our experimental results have demonstrated that our bi-channel deep network can achieve very competitive performance on accuracy rates compared to the existing deep learning algorithm.

1. Introduction

Plants are tremendously important to human beings because they are source of food, clothing, medicines and other materials. In the past, plant species identification was the exclusive domain of professionals who identified the plants of interest by comparing them with previously collected specimens or by using books or identification manuals [1]. Obviously, learning about unknown plants is also an interesting adventure for amateur gardeners and outdoor enthusiasts. In recent years, image-based plant identification has been an area of growing interest. Plant classification and identification are generally based on the color, shape, texture and other characteristics of plants as the main basis, it always judges from the flowers, fruits, stems, leaves and other organs of the morphological characteristics. Leaf texture, color and morphological structure of plants are distinct, so they are the main basis for the distinction between plant species [2].

Due to the longer survival time of plant leaves and the convenience of image acquisition, the classification of plants by leaves has become the research direction of many scholars at present. In China, Wang et al. developed a method of identification of foliage from plants based on extraction of multiple features of leaf images. The system extracts the color, shape and texture features of the image of the foliage and implements identification based on the Support Vector Machine (SVM), achieving an identification rate of 91.41% [3]. Yang et al. analyzed the identification of seven Salix species using digital information analysis of leaf image characteristics and compared them with multiple features based on leaf-related features. The comprehensive identification accuracy reached 90.8% [4]. Based on the geometric and texture features of leaves, Zhang et al. proposed a classification method based on clonal selection algorithm and *K* nearest neighbors, which effectively improved the identification rate [5]. Besides, the CLEF (Cross Language Evaluation Forum) has organized multiple plant image identification competitions since 2011 and added identification conditions for different organs and complicated photographing situations in the 2014 competition. In 2015, the plant species database is up to 1000. In Image CLEF 2011, Villena-Román et al. used SIFT keypoints for object detection in the field of plant image identification [6]. Backes et al. used a complex network to analyze leaf characteristics and identify them, which has good effects on leaf image identification with different resolutions and noisy [7]. A smart phone-based plant identification system called Leafs-nap have been developed recently [8], which involves digitizing the plant image of a solitary flattened leaf and comparing it with a database of previously scanned leaves from some known plant species. By comparing the test plant image with the template images in the database, Leafsnap does not pay enough attention to both large intra-species visual variations and strong inter-species visual similarities which are typical in the botanical domain [9].

Although many advances have been made in plant identification research, there remain lots of problems. For example, in the process of leaf identification based on image analysis, the feature of extraction is usually determined by manual analysis. The differences among plant species are not solved, and the differences among plant data sets with the same features will be produced. In the case of complex background shooting, the identification accuracy of conventional methods decreased significantly. In addition, traditional leaf identification also has some disadvantages. The images contain too little information about the plants and do not have complex backgrounds. To this end, we have created a dataset of 27 different Orchidaceae plants that contain more complex information about the flowers, fruits, branches, stems and leaves.

Deep learning algorithm based on convolutional neural network can learn image features autonomously and reduce manual intervention. It can eliminate noise interference and improve image identification efficiency for leaf image with complex background. Therefore, we construct a bi-channel deep learning framework based on the convolutional neural network, train the model by using the Orchidaceae dataset that we created, and then conduct the test of identifying the plants. In order to verify the validity of the proposed framework, we will use some traditional methods to do the same experiment for comparative analysis. Experiments show that our proposed bi-channel deep learning framework has achieved good effect of identification.

2. Proposed method

2.1. Bi-channel Deep Learning Framework

The bi-channel deep learning framework consists of two sub-network layers: (1) VGG16Net based layer; (2) SqueezeNet based layer. The Figure 1 shows the architecture of our proposed network.

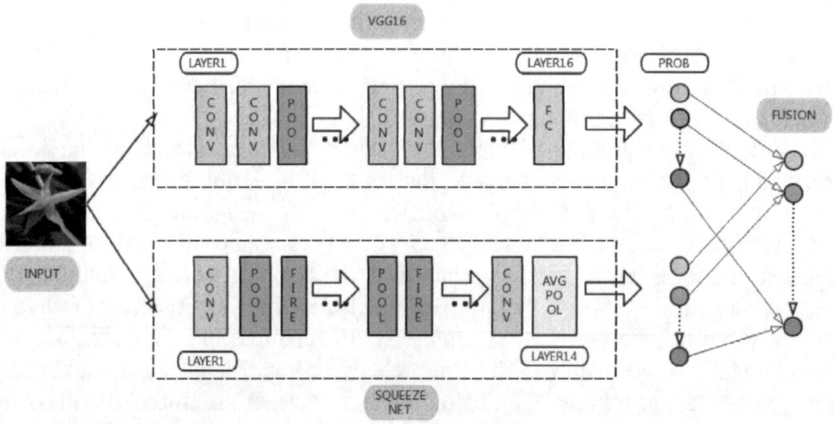

Figure 1. The architecture of bi-channel deep learning framework

IOP Conf. Series: Journal of Physics: Conf. Series **1004** (2018) 012015 doi:10.1088/1742-6596/1004/1/012015

2.1.1. VGG16Net. The VGG16Net was trained on the ImageNet1000 dataset and used for image classification. The format of input data is 244×224×3 pixels, after 13 rounds' processing of convolutional neural networks and pooling networks, we can get the feature data of 4096-dimension, and then through the 3 fully connected layers of neural network processing , finally we can obtain the standard softmax classification results. The size of overall convolution kernels is 3×3, concluding some special convolution kernels with 1×1, all the activation units in hidden layers are RELU. The pre-trained VGG16Net network is used while the output of the final layer is changed for the plant identification.

In order to adapt to the VGG16Net model for our plant identification problem, we fine tune the model using the constructed Orchidaceae dataset. In the case of plant identification, we have 27 classes of plants. So the output of the last fully connected layer is changed from 1000 to 27, which is change the number of classified species. After making these changes, we resume the pretrained parameters on ImageNet1000 for all layers as initialized parameters and then continue training on our dataset to fine tune all the parameters.

2.1.2. SqueezeNet. SqueezeNet was trained on the ImageNet1000 dataset and used for image classification. There are totally 9 fire modules of SqueezeNet, interspersed with some of the max pooling layers, the last global average pooling layer (AVG POOL) instead the fully connected layer (FC) of the traditional network. These changes make the parameters greatly reduced. In the beginning and the end of the network, there are two simple conv layers to ensure that the input and output size can be mastered. Fire module is a special module which splits the original conv layer into two layers: squeeze layer and expand layer, each with Relu activation layer. There are 1x1 convolution kernels in the squeeze layer, 1x1 and 3×3 convolution kernels in the expand layer. Behind the expand layer, the convolution of 1×1 and 3×3 output feature maps are spliced together in the channel dimension.

In order to adapt to the SqueezeNet model for our plant identification problem, we fine tune the model using the constructed Orchidaceae dataset. So we also have to change the output of the last average pooling layer from 1000 to 27, which is change the number of classified species. After making these changes, we resume the pretraining parameters on ImageNet1000 for all layers as initialized parameters and then continue training on our dataset to fine tune all the parameters.

2.2. Classification and Learning

As shown in Figure 1, the common classification layer with fully-connections is used to fuse the predictions from two networks. The *softmax* function is employed to predict the probability of recognizing plants for two deep models. It is defined by:

$$y_c = \frac{exp\,(\boldsymbol{W}_c^T \boldsymbol{V})}{\sum_{c=1}^{C} exp\,(\boldsymbol{W}_c^T \boldsymbol{V})} \tag{1}$$

Where y_c is the predicted probability of c th plants, and $\boldsymbol{V}$ denotes the output feature vector of last pooling layer.

The fusion operation is performed on the outputs of two deep models and the fusion results are used to predict the images. It is defined by:

$$c_i = \gamma_i a_i + (1 - \gamma_i) b_i, i = 1,2, \ldots ,27 \tag{2}$$

Where c_i is the final output of the ith plants. a_i and b_i are the output of the two sub-networks respectively. γ_i is the fusion weights and can determined by grid search.

Two deep models based on VGG16Net and SqueezeNet are pre-trained in the ImageNet1000 database and then fused for the use in our Orchidaceae dataset. When training the model, the *logloss* function is used to learn the parameters by the Stochastic Gradient Decent (SGD) algorithm. For the test, we calculate the top1 and top5 accuracy to measure the final identification results. *Top1* accuracy is the accuracy that the type of the highest output value is the correct one, and *Top5* accuracy is the performance that the types of five highest output values contain the correct one.

IOP Conf. Series: Journal of Physics: Conf. Series **1004** (2018) 012015 doi:10.1088/1742-6596/1004/1/012015

3. Experiments

3.1. Experimental Setup

In our experiment, we have created a dataset of 27 different types of Orchidaceaes. The training set contains 16,455 images, and the test set contains 1756 images. Images are of different backgrounds and different resolutions and contain a wealth of information, such as flowers, fruits, branches, stems, leaves and so on. All the images in this dataset are crawled from search engines such like Flickr, Google and Bing through keywords for plant species interpretation. These keywords are collected from a professional plant taxonomy website called "The Plant List" and shown in Table 1.

Table 1. The keywords for constructing the Ochidaceae dataset.

	Keywords
Type	Angraecum-calceolus, Bulbophyllum-ecornutum, Bulbophyllum-guttulatum, Calanthe-furcata, Cattleya-dormaniana, Cleisostoma-crochetii, Coelogyne-corymbosa, Cycnoches-cooperi, Dendrobium-acerosum, Dendrobium-furcatum, Dendrobium-tortile, Diuris-aurea, Epidendrum-stamfordianum, Eulophia-graminea, Lepanthes-bradei, Leptotes-bohnkiana, Masdevallia-racemosa, Mormodes-warscewiczii, Ophrys-vetula, Paphiopedilum-druryi, Phaius-wallichii, Phalaenopsis-pulcherrima, Platystele-compacta, Psychopsis-sanderae, Serapias-orientalis, Thelymitra-ixioides, Vanilla-planifolia

We first crawl all the species names under Orchidaceae from the "The Plant List", and then we use a python-based crawler to get images from different search engines through these keywords of Orchidaceae species. After that, a clustering-based rubbish image filter algorithm is used to filter out the noisy images. Finally all images are manually cropped and labelled following [10]. Some exemplar images are shown in Figure 2.

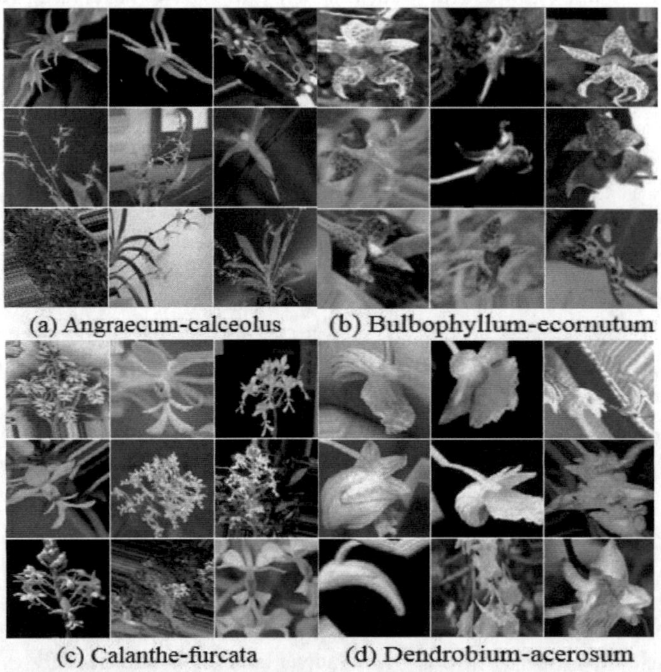

Figure 2. Exemplar images from the dataset

In order to evaluate the validity of the proposed framework, we use several traditional networks to conduct the same experiment and compare the experimental results. The first part of the comparison contains the traditional VGG16Net network [11] without any pretraining parameters (denoted as VGG16Net-1), the pretrained VGG16Net network without pretraining parameters updated (denoted as VGG16Net-2), and the fine-tune VGG16Net network with pretraining parameters updated (denoted as VGG16Net-3). The pretrained VGG16Net-2 network without pretraining parameters updated is set up as follows. First of all, each image in the training set and the test set is taken as the input data of the VGG16, and then we can get a 4096-dimensional feature vector as the output, directly replacing all the original data as the input of the behind layer. We add a 256-dimensional fully connected layer, a 27-dimensional fully connected layer and a *softmax* layer. The fine-tune VGG16Net-3 network updates the pretraining parameters of all the layers on our Orchidaceae dataset. The second part of the comparison contains AlexNet [12], VGG16 network [11], SqueezeNet (the three traditional networks are all fine-tuned) [13], and our bi-channel deep learning framework. The AlexNet network has 8 layers including 5 convolutional layers and 3 fully connected layers. The excitation function RELU and Local Response Normalization (LRN) are included in each convolutional layer.

We divide the training set into the training set and the validation set, using all the methods to train model on the training set, and validate the accuracy on the validation set after every training epoch. Finally, we conduct test on the test set, obtaining the accuracy as the assessment of these methods.

3.2. Experimental Results

As Table 2 shows, when using the traditional CNN network for classification, if the pre-training is not performed, the identification effect is poor at the same number of iterations. If feature extraction is performed on the basis of a pre-trained model and then new layers are added for training, good results will be achieved. The advantage of this method is allowing us to use the existing model with a small amount of data and training time to achieve good results, which is more convenient to be used for simple classification problems with a small dataset. Moreover, the parameters of this network have been reduced, which cut down the training difficulty. But it's worth noting that if we update the pretraining parameters of all the layers, the identification result will be great. The benefit of fine-tuning is that it does not have to completely retrain the model, which increases efficiency, as the accuracy of new training models generally increases slowly from very low values, but fine-tuning allows us to get a better result with less iteration.

Table 2. The identification accuracy of VGG16 approaches for comparison on the Ochidaceae dataset.

	VGG16Net-1	VGG16Net-2	VGG16Net-3
Top1 accuracy	75.00%	88.55%	95.39%

As Table 3 shows, In most configures, the deep method outperforms the shallow models, while in some cases the shallow models are competitive with deep models, that might be due to the insufficient samples. Furthermore, our proposed method outperforms the other two deep models. Although VGG16Net achieved better performance than the SqueezeNet in ImageNet1000 classification, it can be seen from the experimental results that the performance of SqueezeNet and VGG16Net are quite for our small dataset, which shows that even a very small network can achieve the good CNN identification accuracy. By combining the two networks into a unified framework, the performance of plant identification can be improved.

4. Conclusion

We proposed a bi-channel deep learning framework for plant identification. In this model, the CNNs based on VGG16Net and SqueezeNet are employed to automatically extract the features. These two models are then learned jointly with shared classification layer. Models are pre-trained in the ImageNet1000. Experimental results assessed the effectiveness of our proposed approach which outperformed existing conventional networks.

IOP Conf. Series: Journal of Physics: Conf. Series **1004** (2018) 012015 doi:10.1088/1742-6596/1004/1/012015

Table 3. The identification accuracy of some fine-tuned CNNs and proposed Bi-channel Framework for comparison.

	AlexNet	**VGG16Net**	**SqueezeNet**	**Bi-channel Framework**
Top1 accuracy	94.82%	95.39%	95.73%	96.81%
Top5 accuracy	99.51%	99.66%	99.26%	99.54%

Acknowledgments
This work is partly supported by the National Aerospace Science and Technology Foundation and the National Nature Science Foundation of China (No. 61402368 and No. 61702419).

References
[1] Cowan R, Chase M, Kress W, Savolainen V 2006 Taxon 55(3).
[2] Zhang S, Huai Y J 2016 J. Beij. Fore. Univ. 38(9).
[3] Wang L J, Huai Y J, Peng Y C 2015 J. Beij. Fore. Univ. 37(1).
[4] Yang T T, Pan X X, Mu L Q 2014 J. North. Fore. Univ. 42(12).
[5] Zhang N, Liu U W P 2013 J. Comp. Appl. 33(7).
[6] Villena Roman J,Lana Serrano S,Cristobal J C G 2011 Amsterdam: Clef Labs & Workshop.
[7] Backes A R, Casanova D, Bruno O M 2009 Patt. Reco. 42(1).
[8] Kumar N, Belhumeur P N, Biswas A, Jacobs D W, Kress W J, Lopez I C, Soares J V 2012 ECCV.
[9] Tera Botanica www.tela-botanica.org.
[10] Zhang H X, Fan J P 2017 IEEE MIPR.
[11] Parkhi O M, Vedaldi A, Zisserman A 2015 BMVC.
[12] Krizhevsky A, Sutskever I, Hinton G E 2012 NIPS 60(2).
[13] Iandola F N, Han S, Moskewicz M W, Ashraf K, Dally W J, Keutzer K 2016 preprint (arXiv:1602.07360).

IOP Conf. Series: Journal of Physics: Conf. Series **1004** (2018) 012016 doi:10.1088/1742-6596/1004/1/012016

Face Hallucination with Linear Regression Model in Semi-Orthogonal Multilinear PCA Method

Krissada Asavaskulkiet[1,*]

[1]Department of Electrical Engineering, Mahidol University, Nakhonpathom, Thailand

*Krissada.asa@mahidol.ac.th

Abstract. In this paper, we propose a new face hallucination technique, face images reconstruction in HSV color space with a semi-orthogonal multilinear principal component analysis method. This novel hallucination technique can perform directly from tensors via tensor-to-vector projection by imposing the orthogonality constraint in only one mode. In our experiments, we use facial images from FERET database to test our hallucination approach which is demonstrated by extensive experiments with high-quality hallucinated color faces. The experimental results assure clearly demonstrated that we can generate photorealistic color face images by using the SO-MPCA subspace with a linear regression model.

1. Introduction

In digital signal processing, video surveillance is often of small size because of the great distance between the camera and the objects. Image resolution is a potential factor affecting face recognition performance. For face identification, especially by human, it is desirable to render a high-resolution face image from the low-resolution one. This technique is called face hallucination or face super-resolution [1]. They infer the high frequency components from a parent structure by recognizing the local features from the training set, but there exists some noise in certain area. The nearest neighbor or cubic spline is the simplest way to increase image resolution because it is a direct interpolation of input images. On the other hand, the performance of direct interpolation is usually poor since no new information is added in the process. For this reason, face super-resolution methods which are based on learning from the training set are proposed [2]–[4]. Some methods with the assumption that high-resolution images are Markov random fields (MRFs) are more suitable for synthesizing local texture, and are usually applied to generic images without special consideration of the property of face images [5,6].

A novel hallucination method based on eigentransformation is proposed [7]. It is similarly related to the work in [8] and this approach is developed for sketch recognition. To go beyond the current super-resolution techniques which only consider face images under fixed imaging conditions in terms of pose, expression and illumination, these factors are crucial to face analysis and synthesis. Then, Vasilescu et al. introduce multilinear analysis to face modelling [9] and demonstrate its promising application in computer vision. In the method, equipped with tensor algebra, the multiple factors are unified in the same framework with the coordination between factors expressed in an elegant tensor product form. Next, Wu et al. propose a novel regression model to use tensor principal component analysis (PCA) subspace as the face representation [10], which is a special case of the concurrent subspace analysis. In addition, Multi-linear principal component analysis (MPCA) is a general extension of traditional linear method such as matrix singular value decomposition (SVD) or PCA

[11]. The idea of MPCA is extended to use Tensor-to-Vector Projection (TVP) that learns low-dimensional vectors from high-dimensional tensors in a successive way [12].

We focus on color face hallucination in a SO-MPCA method. Motivated by the problem in the studies involving the hallucination method in the HSV color space [13] and we use color face images from the FERET database [14]. We proposes a novel hallucination technique in color face images of HSV color space with a linear regression model in SO-MPCA formulation for tensor object. In this paper, a novel technique can capture more variance and learn more features than full orthogonality. Then, the bias can be increased and the variance of learning model be reduced. The rest of this paper is organized as follows. In Section 2, we explain the basic notation. We introduce the basic idea of the MPCA in Section 3. In Section 4, we propose a color face hallucination with SO-MPCA for HSV model. Some experimental results are shown in Section 5. Finally, conclusions are summarized in Section 6.

2. Basic Notation

In this paper, scalars are denoted by lower case letters $(a, b, \ldots)$, vectors by italic upper-case case letters $(A, B, \ldots)$, matrices by bold upper-case letters (**A,B,...**), and higher-order tensors by calligraphic upper-case letters $(\mathcal{A}, \mathcal{B})$. $(\bullet)^T$ denotes the transpose of a matrix. $(\bullet)^{\dagger}$ denotes the pseudo-inverse.

3. Basic Idea in Multi-linear Principal Component analysis (MPCA)

The basic idea in MPCA solution to the problem of dimensionality reduction for tensor objects is introduced [11]. In Figure. 1, we show the basic idea in MPCA, an n-mode unfolding of a tensor. An *N*th-order tensor is denoted as $\mathcal{A} \in \mathrm{R}^{I_1 \times I_2 \times \ldots \times I_N}$ for $I_n = 1, \ldots, N$. The *n*-mode vectors of $\mathcal{A}$ are defined as the I_n-dimensional vectors obtained from $\mathcal{A}$ by varying the index i_n while keeping all the other indices fixed. The unfolding of $\mathcal{A}$ along the *n*-mode is denoted as $\mathrm{A}_{(n)} \in \mathrm{R}^{In \times (I_1 \times I_2 \times \ldots \times I_N)}$ and the column vectors of $\mathrm{A}_{(n)}$ are the *n*-mode vectors of $\mathcal{A}$. Let the set of tensors be $\{\mathcal{A}_m, m = 1, \ldots, M\}$ and the total scatter of these tensors is defined as

$$\psi_{\mathrm{A}} = \sum_{m=1}^{M} \| A_m - \bar{A} \|^2 \tag{1}$$

Where $\bar{\mathcal{A}}$ is the mean tensor calculated as

$$\bar{\mathcal{A}} = (1/\mathrm{M}) \sum_{m=1}^{M} \mathrm{A}_{\mathrm{m}}. \tag{2}$$

Then, the total scatter matrix of these samples can be defined as

$$C_A = \sum_{m=1}^{M} \left(A_{m(n)} - \bar{A}_{(n)} \right) \left(A_{m(n)} - \bar{A}_{(n)} \right)^T, \tag{3}$$

Where $\mathrm{A}m(n)$ is the *n*-mode unfolded matrix of $\mathrm{A}m$.

The main objective of MPCA is to define a multi-linear transformation $\tilde{\mathrm{U}}^{(n)}$ which denoted $I_n \times P_n$ matrix containing the orthornormal *n*-mode basis vectors and the matrix $\tilde{\mathrm{U}}^{(n)}$ is *n*th projection matrix, *n* = 1,...,*N*. It can map the original tensor space $\mathrm{R}^{I_1} \otimes \mathrm{R}^{I_2} \ldots \otimes \mathrm{R}^{I_N}$ into a tensor subspace $\mathrm{R}^{P_1} \otimes \mathrm{R}^{P_2} \ldots \otimes \mathrm{R}^{P_N}$ with ($Pn < In$, for n = 1,..., *N*). We can define the projection of *n*-mode vector of $\mathcal{X}_m$ as

$$y_m = \mathcal{X}_m \times_1 \tilde{\mathrm{U}}^{(n)^{\mathrm{T}}} \times_2 \tilde{\mathrm{U}}^{(n)^{\mathrm{T}}} \ldots \times_N \tilde{\mathrm{U}}^{(N)^{\mathrm{T}}}. \tag{4}$$

The tensor y_m can capture most of the variations observed in the original tensor objects, assuming that these variations are measured by the total tensor scatter. The objective of MPCA is the determination of the *N* projection matrices $\mathrm{U}^{(n)}$ that maximize the total tensor scatter ψ_{A} as

$$\{\tilde{U}^{(n)}, n = 1, 2, ..., N\} = \underset{\tilde{U}^{(1)}, \tilde{U}^{(2)}, ..., \tilde{U}^{(N)}}{\arg\max} \psi_y \tag{5}$$

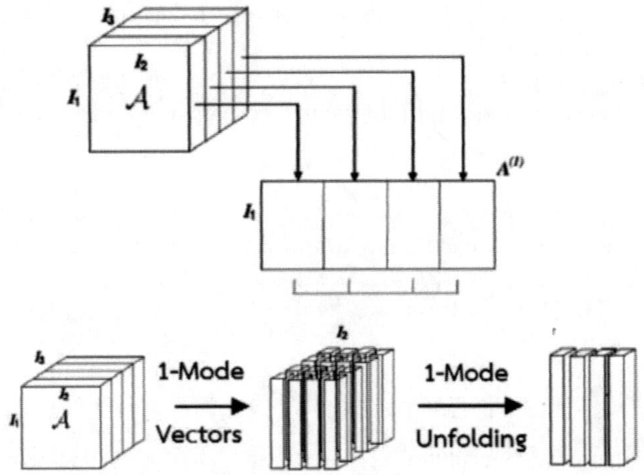

Figure 1. Block diagram of the basic idea in multi-linear principal component analysis (MPCA)

4. Color Face Hallucination with linear regression model in SO-MPCA method

In this section, we explain the Tensor-to-vector projection which is the building blocks of a **TVP**. In (4), we denote an elementary multi-linear projections (EMPs) as $\{\mathbf{u}^{(1)}, \mathbf{u}^{(2)}, ..., \mathbf{u}^{(N)}\}$, which consist of one unit projection vector in each mode and $\|\mathbf{u}^{(n)}\| = 1$ for $n = 1, ..., N$, where $\| \bullet \|$ is the Euclidean norm for vectors. Then, we can get the projection of a tensor $\mathcal{X}$ to a scalar y through the N unit projection vectors as:

$$y = \mathcal{X} \times_1 \mathbf{u}^{(1)^T} \times_2 \mathbf{u}^{(1)^T} ... \times_N \mathbf{u}^{(N)^T}. \tag{6}$$

The TVP of a tensor $\mathcal{X}$to a vector $\mathbf{y} \in \mathrm{R}^P$ consists of P **EMPs** $\{\mathbf{u}_p^{(1)}, ..., \mathbf{u}_p^{(N)}\}$, $p = 1, ..., P$, which can be written concisely as $\{\mathbf{u}_p^{(N)}\}_{p=1}^P$:

$$\mathbf{y}(p) = \mathcal{X} \times_1 \mathbf{u}_p^{(1)^T} ... \times_N \mathbf{u}_p^{(N)^T} = \mathcal{X} \times_{n=1}^N \{\mathbf{u}_p^{(N)}\}. \tag{7}$$

The purpose of SO-MPCA method is to find a TVP to maximize the variance of the projected samples in each projection direction, subject to the orthogonality constraint in only one mode which is defined as α-mode. The variance is measured by the total scatter S_p defined as

$$C_p = \sum_{m=1}^M \left(y_{m_p} - \bar{y}_p\right)^2, \tag{8}$$

Where $y_{m_p} = \mathcal{X} \times_{n=1}^N \left\{\mathbf{u}_p^{(N)}\right\}$, and $\bar{y}_p = \frac{1}{M}\sum_m y_{m_p}$, that is to say that the objective of SO-MPCA is to achieve the P **EMPs**. In addition, the pth **EMP** which maximized the variance in (8) can be determined as:

$$\{\mathbf{u}_p^{(n)}, n = 1, ..., N\} = \operatorname{argmax} \sum_{m=1}^M \left(y_{m_p} - \bar{y}_p\right)^2. \tag{9}$$

For orthogonality constraint in only one mode, each the pth **EMP** can be described as:

$$\mathbf{u}_p^{(n)^T} \mathbf{u}_p^{(n)} = 1 \text{ for } n = 1, ..., N \tag{10}$$

And

$$\mathbf{u}_p^{(\alpha)^T}\mathbf{u}_q^{(\alpha)} = 0 \text{ for } p > 1 \text{ and } q = 1, \dots, P-1. \tag{11}$$

The normalization constraint in (10) is applied for all modes and the orthogonality constraint in (11) is applied only in α-mode and there is no such constraint for the other modes.

However, the selection of mode α is free to choose any mode n as α to impose the orthogonality constraint in (11). We choose the mode with the highest dimension as α :

$$\alpha = \operatorname{argmax} I_n \,, \tag{12}$$

Such that $= \max_n I_n = I_\propto$. In this paper, we only keep on SO-MPCA with α determined by (12).

Following standard multi-linear algebra, any tensor can be expressed as the products

$$y_i^h = x_i^h \times_1 \mathbf{u}^{\mathrm{h}(1)^T} \times_2 \mathbf{u}^{\mathrm{h}(2)^T} \times_3 \mathbf{u}^{\mathrm{h}(3)^T}. \tag{13}$$

And

$$y_i^l = x_i^l \times_1 \mathbf{u}^{\mathrm{l}(1)^T} \times_2 \mathbf{u}^{\mathrm{l}(2)^T} \times_3 \mathbf{u}^{\mathrm{l}(3)^T}. \tag{14}$$

As y_i^h is in the index of high-resolution (HR) training set and y_i^l is the low-resolution (LR) training set respectively. In addition, the sets of $\mathbf{u}^{(1)}, \mathbf{u}^{(2)}, \mathbf{u}^{(3)}$ are followed in (10)-(11) and the correlation between the decomposition coefficients can be suppressed.

From the model of a mapping function $y_i^h = f(y_i^l)$, f can be thought of as a probability distribution function. Thus, we can consider the following conditional probability $P(y_i^h|y_i^l)$. When a new test color face image (LR) x^l is provided, and the HR tensorPCA subspace projection is given by:

$$y^h = \underset{y}{\operatorname{argmax}} P(y|y^l). \tag{15}$$

From [13], we use the assumption of a low-correlation between the coefficients in y^l, we can simplify this probability in (15) as

$$\hat{y}_{r,s,t}^h = \underset{y_{r,s,t}}{\operatorname{argmax}}\, P(y_{r,s,t}|y^l), \tag{16}$$

Next step, The Maximum Likelihood estimation is applied in (16) and we can express $\hat{y}_{r,s,t}^h$ in a linear regression model as:

$$\hat{y}_r^h = \sum_{p=1}^{Q_1Q_2Q_3}\left(\frac{w_{r,p}}{Q_1Q_2Q_3}\right) y_p^l. \tag{17}$$

5. Experimental Results

This section evaluates the proposed methods on third-order tensor data in terms of recognition rate and ability to reconstruct the color face images. Our face hallucination with SO-MPCA in a color facial image algorithm is tested against the traditional PCA and the MPCA method. We used face images from a subset of FERET databases [14] to form two data sets for training and testing images in two color models: RGB and HSV. In our experiments, we randomly selected 300 normal expression images of different people under the same light conditions and another 30 images in Figure. 2 were used for testing. According to demand, we manually cropped the most interesting region of the faces, and we standardized the images to a size of (30×30 pixels). In the degradation process, each testing image (LR) was introduced with Gaussian blur with a variance of 1 and resized by down-sampling 2:1 (15×15 pixels). We then added Gaussian noise with a variance of 10^{-6}.

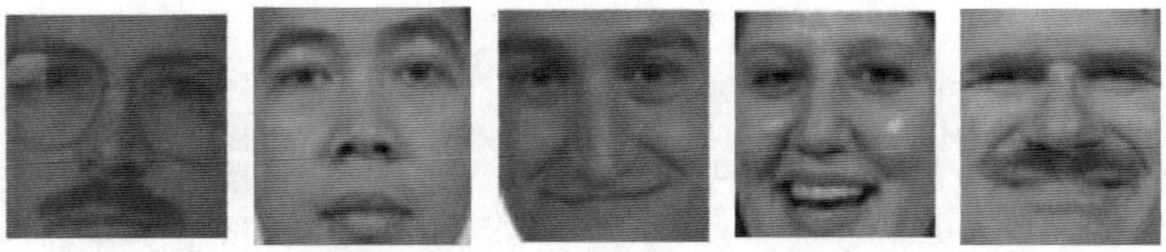

Figure 2. Some original high-resolution color face images from FERET (30 × 30) for testing.

Table 1. Face recognition rates in percentage (mean±std) by the nearest neighbor classifier on the FERET. subset.

L	P	PCA	MPCA	SO-MPCA
1	1	2.45±0.33	2.41±0.35	2.59±0.42
	5	13.12±1.48	15.24±1.56	16.07±1.61
	10	19.43±2.06	20.13±1.98	21.88±2.23
	20	26.85±2.31	28.17±2.39	30.72±2.51
	50	31.78±2.42	31.85±2.47	34.67±2.65
2	1	2.48±0.62	2.51±0.48	2.69±1.43
	5	18.94±1.55	20.03±1.27	23.65±1.89
	10	30.88±2.43	31.61±1.71	37.36±2.64
	20	41.71±2.52	40.71±2.07	45.28±2.71
	50	47.02±2.61	48.85±2.69	49.73±2.82
3	1	2.59±0.14	2.43±0.26	2.51±0.22
	5	24.07±1.62	25.63±1.57	28.92±1.85
	10	38.47±2.55	36.94±2.31	44.76±2.64
	20	49.63±2.67	47.28±2.55	52.89±2.53
	50	52.44±2.64	53.08±2.79	57.33±2.78
4	1	2.61±0.27	2.54±0.36	2.85±0.61
	5	26.87±1.79	27.95±1.81	28.93±1.56
	10	42.59±2.61	43.82±2.77	46.68±2.86
	20	54.97±2.81	57.23±2.81	58.07±2.91
	50	57.01±2.76	58.16±2.95	59.77±2.84
5	1	2.65±0.87	2.51±0.73	2.81±0.63
	5	28.93±1.91	30.67±1.81	31.65±2.34
	10	47.26±2.79	46.59±2.85	52.78±2.76
	20	57.72±2.96	60.93±3.08	61.47±2.95
	50	63.65±2.84	64.72±3.21	63.41±3.06

This section evaluates the proposed methods on third-order tensor data in terms of recognition rate and the outcome from reconstructed face images in RGB and HSV color models. In Table 1. show the color face recognition and we select L = 1,2,3,4,5 samples for each each subject as the training data and use the rest for testing. For SO-MPCA, we set both selected mode $\alpha = 1$ and number of iterations to 30. All features are sorted according to the scatters (captured variance) in descending order for classification. The Nearest Neighbor Classifier with the Euclidean distance is applied to classify the top P features which up to $P = 50$ features in face recognition. In Table 1. we shows the face recognition results for $P = 1, 5, 10, 20, 50$ and $L = 1, 2, 3, 4,$ 5, including both the mean and the standard deviation (std) over ten repetitions. We can see that the SO-MPCA can give more the recognition rates than other methods in all cases.

Table 2. PSNR of different methods for each facial image in RGB and HSV color space.

Case	RGB first row	RGB second row	RGB third row	HSV first row	HSV second row	HSV third row
1) 90 percent PCA	23.32 dB	29.56 dB	26.05 dB	22.76 dB	28.14 dB	25.23 dB

2) 95 percent PCA	23.35 dB	29.51 dB	26.41 dB	22.98 dB	28.56 dB	25.36 dB
3) 100 percent PCA	23.41 dB	29.62 dB	26.78 dB	23.15 dB	29.11 dB	25.87 dB
4) 90 percent MPCA	25.01 dB	31.46 dB	27.23 dB	24.69 dB	29.23 dB	26.88 dB
5) 95 percent MPCA	27.62 dB	32.87 dB	28.18 dB	26.13 dB	30.45 dB	27.26 dB
6) 100 percent MPCA	30.35 dB	33.74 dB	29.56 dB	29.84 dB	32.19 dB	28.59 dB
7) 90 percent SO-MPCA	25.03 dB	31.53 dB	27.29 dB	24.71 dB	29.28 dB	26.79 dB
8) 95 percent SO-MPCA	27.78 dB	32.82 dB	28.23 dB	26.32 dB	30.42 dB	27.23 dB
9) 100 percent SO-MPCA	30.32 dB	33.76 dB	29.54 dB	29.86 dB	32.26 dB	28.61 dB

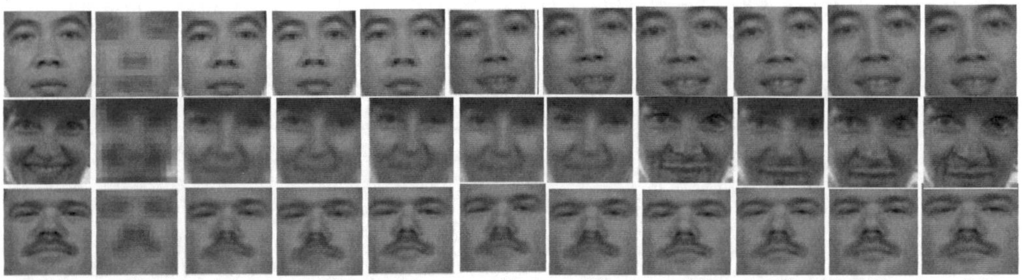

(a) (b) (c) (d) (e) (f) (g) (h) (i) (j) (k)

Figure 3. Color face hallucination results in RGB color space.

(a) original HR images (30 × 30 pixels); (b) input LR images (15 × 15 pixels) with noise, motion and blur in LR images; (c) face hallucination result with 90 percent traditional PCA; (d) face hallucination result with 95 percent traditional PCA; (e) face hallucination result with 100 percent traditional PCA; (f) face hallucination result with 90 percent MPCA; (g) face hallucination result with 95 percent MPCA; (h) face hallucination result with 100 percent MPCA; (i) face hallucination result with 90 percent SO-MPCA; (j) face hallucination result with 95 percent SO-MPCA; (k) face hallucination result with 100 percent SO-MPCA.

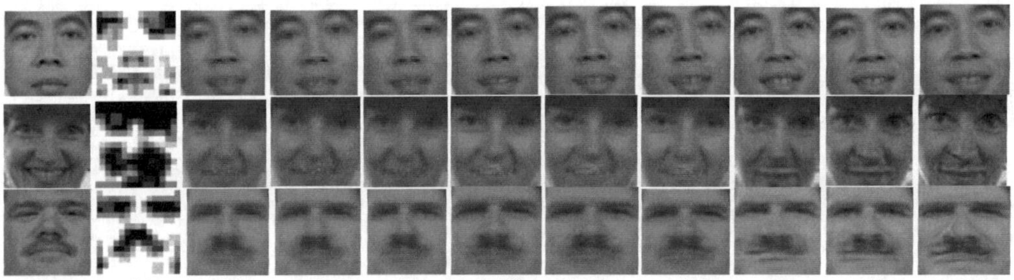

(a) (b) (c) (d) (e) (f) (g) (h) (i) (j) (k)

Figure 4. Color face hallucination results in HSV color space.

(a) original HR images (30 × 30 pixels); (b) input LR images (15 × 15 pixels) with noise, motion and blur in LR images; (c) face hallucination result with 90 percent traditional PCA; (d) face hallucination result with 95 percent traditional PCA; (e) face hallucination result with 100 percent traditional PCA; (f) face hallucination result with 90 percent MPCA; (g) face hallucination result with 95 percent MPCA; (h) face hallucination result with 100 percent MPCA; (i) face hallucination result with 90 percent SO-MPCA; (j) face hallucination result with 95 percent SO-MPCA; (k) face hallucination result with 100 percent SO-MPCA.

To provide a quantitative comparison of the color face reconstruction deviations from the ground truth data, we also computed the PSNR of each method. All the PSNR results of Figure 3-4 in the different color spaces and the number of PCA are shown in Table 2. The results shown that our proposed method has the highest PSNR values compared with the traditional PCA method on all test faces. In addition, the PSNR results form both MPCA and SO-MPCA methods are quite similar in the RGB and HSV space. The experimental results from Figure 3-4. clearly demonstrated that we can generate photorealistic color face images by using the SO-MPCA subspace with a regression model and that our hallucinated approach is suitable for the HSV color spaces.

6. Conclusion

This paper proposes a novel color face hallucination with linear regression model with algorithm under the TVP setting, named as semi-orthogonal multilinear PCA (SO-MPCA) method. This method can capture more variance and learn more features than full orthogonality. To test our proposed method, we used facial images from the FERET database to validate the algorithm. The experiments clearly demonstrated that we can generate photorealistic color face images by using the SO-MPCA subspace with a regression model and that our hallucinated approach is suitable for the RGB and HSV color spaces. The results showed that this proposed method can reconstruct reasonable HR color face images. This is the next step toward taking advantage of SO-MPCA for color image processing. Based on using the SO- MPCA in only one mode with the regression model, we will directly apply our technique to each mode across different views and under changing illumination conditions.

References

[1] Baker, S. and Kanade, T. Hallucinating faces. IEEE Int. Conf. Automatic Face and Gesture
[2] Elad, M. and Feuer, A. Restoration of a single superresolution image from several blurred, noisy, and undersampled measured images. IEEE Trans. Image Processing 6 , 12 (1997): 1646–1658.
[3] Kim, S. and Su, W. Recursive reconstruction of high-resolution reconstruction of blurred multiframes images. IEEE Trans. Image Processing 2 (June 1993): 534–539.
[4] Ma, X., Zhang, J., and Qi, C. An example-based two-step face hallucination method through coefficient learning. Image Analysis and Recognition :Lecture Notes in Computer Science. 5627 (2009): 471–480.
[5] Hardie, R., Barnard, K., and Armstrong, E. Joint map registration and high-resolution image estimation using a sequence of undersampled images. IEEE Trans. Image Processing 6 (June 1977): 1621–1633.
[6] Freeman, W. T. and Pasztor, E. C. Learning low-level vision. IEEE Int. Conf. Computer Vision.
[7] Wang, X. and Tang, X. Hallucinating face by eigentransformation. IEEE Trans. on Systems, Man and Cybernetics 35 , 3 (2005): 425–434.
[8] Tang, X. and Wang, X. Face sketch recognition. IEEE Trans. on Circuits, Systems, and Video Technology 14 , 1 (2004): 50–57.
[9] Vasilescu, M. and Terzopoulos, D. Multilinear subspace analysis of image ensembles. In Proc. of CVPR 2 (2003): 93–99.
[10] Wu, J. and Trivedi, M. M. A regression model in tensorpca subspace for face image super-resolution reconstruction. In Proc. IEEE Conf. Computer Vision and Pattern Recognition 3 (2006): 627–630.
[11] Lu, H., Plataniotis, K. N. K., and Venetsanopoulos, A. N. MPCA: Multilinear principal component analysis of tensor objects. IEEE Transactions on Neural Networks 19, 1 (2008): 18–39.
[12] Shashua and A. Levin. Linear image coding for regression and classification using the tensor-rank principle. IEICE Transactions Fundamentals, volume E94-A, No.8, pages 1724–1737, 2001
[13] K. Asavaskulkiet and S. Jitapunkul. Generalized color face hallucination with linear regression model in MPCA. In Proc. IEEE Int. Conf. on Computer Vision and Pattern Recognition, volume I, pages 42–49, 2001.
[14] Phillips, P. J., Moon, H., Rizvi, S. A., and Rauss, P. J. The FERET evaluation methodology for face recognition algorithms. IEEE Trans. PAMI 22.

IOP Conf. Series: Journal of Physics: Conf. Series **1004** (2018) 012017 doi:10.1088/1742-6596/1004/1/012017

Classification Influence of Features on Given Emotions and Its Application in Feature Selection

Yin Xing[1, a], Chuang Chen[2, b] and Li-Long Liu[1, c]
[1]College of Geomatics and Geoinformation, Guilin University of Technology, Guilin, China
[2]College of Electrical Engineering and Control Science, Nanjing Tech University, Nanjing, China

[a]xycc123@foxmail.com; [b]chenchuang@njtech.edu.cn; [c]hn_liulilong@163.com

Abstract. In order to solve the problem that there is a large amount of redundant data in high-dimensional speech emotion features, we analyze deeply the extracted speech emotion features and select better features. Firstly, a given emotion is classified by each feature. Secondly, the recognition rate is ranked in descending order. Then, the optimal threshold of features is determined by rate criterion. Finally, the better features are obtained. When applied in Berlin and Chinese emotional data set, the experimental results show that the feature selection method outperforms the other traditional methods.

1. Introduction

Speech emotion recognition is a research hotspot in the fields of artificial intelligence and signal processing. Its purpose is to enable the computer to perceive the surrounding environment, atmosphere and emotional state of the object as human beings, so as to provide the best conversation environment for dialogue objects [1]. The findings of Stanford University research by Reeves and Nass [2] show that the problems that need to be solved in human-computer interaction are in fact consistent with the important factors of human-human interaction. The most crucial is the ability of "emotional intelligence". Therefore, if the computer is able to more actively adapt to the needs of the operator, it must first identify the operator's emotion, and then adjust the way of interactive dialogue according to emotional judgment. The research has important application value in the fields of new human-computer interaction [3], call centre [4] and intelligent robot [5].

For speech emotion recognition system, it consists of three steps: acquisition of speech signal, extraction of emotion features and recognition of speech emotion [6]. At present, researchers have proposed many effective speech emotion features such as short-time energy, pitch frequency, formant, and Mel Frequency Cepstral Coefficient (MFCC). Although these features have achieved good results in speech emotion classification, the features often have higher dimensions. In particular, after the fusion of multiple types of features, the amount of feature data greatly increases. High-dimensional feature sets contain large amounts of noise and redundant data. On the one hand, it affects the recognition rate; on the other hand, it increases the computational complexity [7]. Therefore, it is necessary to perform feature reduction on the extracted features.

To sum up some literature in the field of speech emotion recognition, researchers mainly adopt the following methods of dimensionality reduction. Chiou *et al.* [8] used Principal Component Analysis (PCA) to process speech emotion features in Berlin emotional data set. After removing a lot of features, it still got 80% recognition rate. Xue *et al.* [9] took into account the nonlinear pop structure of speech signals and used Locally Linear Embedding (LLE) to realize feature reduction. Yuan *et al.* [10] used improved Linear Discriminant Analysis (LDA) to carry out experiments on Chinese emotional data set and achieved better rate. Zhang *et al.* [11] used Fisher Discriminant Ratio (FDR) to select emotion features and got 81% rate in computer-induced emotional corpus. Although the above methods have achieved a certain reduction effect, they all lack the classification influence analysis of the features on given emotions. Thus, there is a certain limit on recognition rate. Therefore, it is of great importance to develop a new feature selection method.

2. Feature analysis and selection (FAS) method

Considering difference contributions of a certain feature on the given emotion, each feature need to be deeply analyzed. It can give the rank of feature importance that affects the rate of given emotion, so as to find out the key features in high-dimensional emotion features. In order to facilitate the narrative, this paper will the feature analysis and selection method abbreviated as FAS method. The operation process of FAS method is described in detail as follows:

(1) Feature extraction. It is assumed that n -dimensional emotion features are extracted, forming the feature vector $\boldsymbol{F} = (f_1, f_2, \cdots, f_n)$.

(2) Given an emotion for classification. Based on the speech emotion classifier, the recognition rate of the $j\text{-}th$ $(1 \le j \le m)$ emotion using the $i\text{-}th$ feature f_i is r_{ij}. Then the established recognition rate matrix $\boldsymbol{R}$ is as follows:

$$\boldsymbol{R}_{n\times m} = \begin{bmatrix} r_{11} & r_{12} & \cdots & r_{1j} & \cdots & r_{1m} \\ r_{21} & r_{22} & \cdots & r_{2j} & \cdots & r_{2m} \\ \vdots & \vdots & \vdots & \vdots & \vdots & \vdots \\ r_{i1} & r_{i2} & \cdots & r_{ij} & \cdots & r_{im} \\ \vdots & \vdots & \vdots & \vdots & \vdots & \vdots \\ r_{n1} & r_{n2} & \cdots & r_{nj} & \cdots & r_{nm} \end{bmatrix} \tag{1}$$

(3) Sort of feature importance. Each column of the matrix $\boldsymbol{R}$ is descended to obtain a sorted recognition rate matrix $\boldsymbol{M}$.

(4) Determine the threshold k . The best k features of each emotion in matrix $\boldsymbol{M}$ are selected to form a new feature set $\boldsymbol{N}$ .The k value is chosen based on the actual recognition rate of speech emotion. That is, we think the k value is the best when maximum average recognition rate is found.

(5) Streamline operations. There may be repeated features in the feature set $\boldsymbol{N}$, and only one feature of them is taken as a representative.

This process of selecting emotion features using FAS method is simple and easy to implement. It takes the actual recognition rate as selection criteria and has a good applicability.

3. Feature extraction and classification algorithm

The experimental data in this paper are derived from Berlin emotional speech database [12] and Chinese emotional speech database [13]. The Berlin emotional speech database was recorded by from 10 (5 males and 5 females) non-professional actors. And a total of 800 emotional statements were obtained. These emotional statements include 7 emotions that are anger, happiness, neutral, sadness, fear, disgust and boredom. After 20 auditions, 535 sentences were retained. The Chinese emotional speech database was recorded by 4 professional speakers (2 males, 2 females). And a total of 1200 emotional statements were obtained. These emotional statements include 6 emotions that are anger, happiness, neutral, sadness, fear, and surprise. The speech samples of the two emotional speech databases are stored at 16000 sampling rate, 16bit quantization and wav format. In following

IOP Conf. Series: Journal of Physics: Conf. Series **1004** (2018) 012017 doi:10.1088/1742-6596/1004/1/012017

experiments, we selected 5 emotions as experimental data. The specific information is shown in Table 1.

Table 1. Introduction of experimental samples

Database	Number of samples						Number of features
	Anger	Happiness	Neutral	Sadness	Fear	Total	
Berlin emotional speech database	126	68	78	62	66	400	124
Chinese emotional speech database	200	200	200	200	200	1000	124

In this paper, 124 emotional features are extracted, which are maximum, minimum, mean and variance of short-term energy, pitch frequency, first-formant frequency, second-formant frequency, third formant frequency, MFCC and its first-order difference in 0 ~ 12th order. The calculation methods are detailed in reference [14]. Then, the K-Nearest Neighbor (KNN) algorithm is used for emotion classification. The basic idea of classification is that given a sample to be classified in feature space, if the majority of k nearest neighbors in a class belong to a certain class, then the sample to be classified currently also belongs to this class. In order to verify the reliability of FAS method, $K = 3$, $K = 5$ and $K = 7$ were used for classification experiment. Several experiments have found that the different values of K are only small changes in the recognition rate, but have no effect on the general trend of the recognition rate curve. Therefore, this paper selects $K = 5$ for the experiment.

4. Speech emotion classification experiment

First, the feature selection experiment was conducted on Berlin speech emotional database using FAS method. Given emotions are classified by each feature, shown in Table 2. Due to space limitations, only some of the results are given in Table 2. From Table 2, each feature has different contribution to different emotions. For example, the feature 1 has a greater contribution to anger, while feature 7 has a greater contribution to the fear.

Table 2. Analysis results of each feature

Feature	Anger	Happiness	Neutral	Sadness	Fear
1	0.4127	0.1176	0.1795	0.1290	0.3333
2	0.4921	0.0294	0.2308	0.0645	0.3636
3	0.3016	0.1471	0.1795	0.1935	0.2424
4	0.2857	0.0294	0.2564	0.2258	0.2121
5	0.5714	0.2353	0.2051	0.1935	0.3333
6	1.0000	0.7647	0.0000	0.0000	0.0000
7	0.2698	0.1176	0.1026	0.2258	0.3030
8	0.3492	0.1471	0.2821	0.1935	0.2424
9	0.4762	0.1765	0.2564	0.4194	0.5152
10	0.3333	0.1471	0.3590	0.1935	0.5152

The recognition rates of each emotion in Table 2 are arranged in descending order, and the feature importance ranking is obtained, shown in Table 3. It shows the best 10 features of each emotion. The experiment needs to select the best k features for the classification. For example, when k is 1, the feature set is $\{6, 6, 99, 107, 119\}$. Since the feature 6 has been repeated, we do not repeat the selection, then the new feature set is $N = \{6, 99, 107, 119\}$.

Table 3. Feature importance ranking of five emotions

Anger		Happiness		Neutral		Sadness		Fear	
Rate	Feature	Rate	Feature	Rate	Feature	Rate	Feature	Rate	Feature
1.0000	6	0.7647	6	0.4872	99	0.6774	107	0.6061	119
0.8889	79	0.7059	75	0.4615	75	0.5161	15	0.5758	115

0.7778	75	0.6765	87	0.4615	84	0.5161	92	0.5455	79
0.6825	95	0.4706	83	0.4103	19	0.4516	12	0.5455	111
0.6667	87	0.3824	14	0.3846	67	0.4194	9	0.5152	9
0.6190	103	0.3824	76	0.3590	10	0.3871	59	0.5152	10
0.6190	107	0.3824	103	0.3590	23	0.3871	96	0.4848	12
0.5714	5	0.3529	57	0.3590	79	0.3548	75	0.4848	103
0.5714	109	0.2941	13	0.3590	98	0.3548	82	0.4545	25
0.5556	110	0.2941	26	0.3590	120	0.3226	37	0.4545	66

Different k values are selected to correspond to different feature sets. Figure 1 shows the average recognition rate curve of speech emotion under different k values. From Figure 1, the average recognition rate of speech emotion first increases with the increase of k value. When k is 7, the average recognition rate reaches the maximum, which is 86.5%.When k continues to increase, the recognition rate decreases. Therefore, 7 is the optimal threshold, and the number of corresponding feature set $\boldsymbol{N}$ is 25.

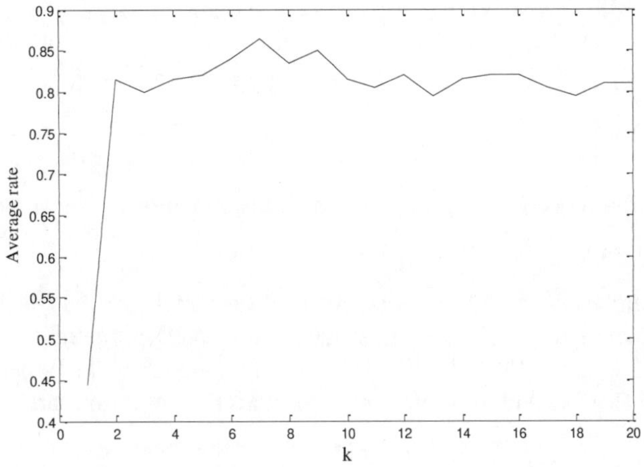

Figure 1. Average recognition rate curve of speech emotion under different k values

In order to verify the superiority of the FAS method proposed in this paper, we use PCA in [8], LLE in [9], LDA in [10] and FDR in [11] to reduce features. The experimental results are shown in Figure 2. From Figure 2, the recognition rate of FAS method proposed in this paper is obviously higher than that of the other 4 methods after reducing to 8 features, which shows the superiority.

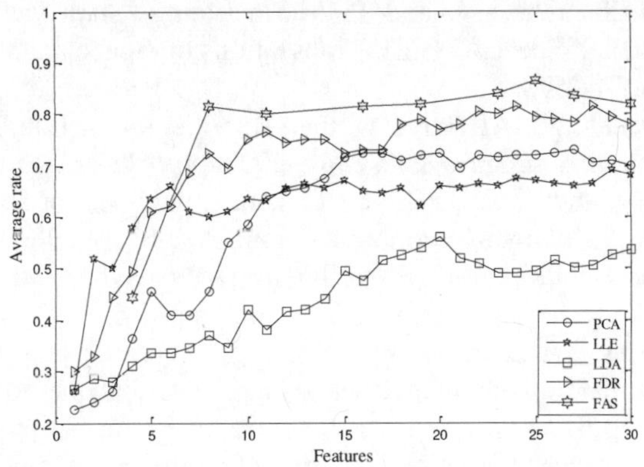

Figure 2. Recognition rate curves of different methods on Berlin speech emotional database

In order to verify the reliability of the FAS method, this paper also conducted emotion recognition experiments on the Chinese speech emotional database. The PCA, LLE, LDA, FDR and FAS methods are used to reduce speech emotion features respectively, and KNN algorithm is used to classify speech emotion. The experimental results are shown in Figure 3. From Figure 3, the recognition rate curve of the FAS method is also higher than that of the other 4 methods. In summary, it is demonstrated that the FAS method proposed in this paper is effective and superior to the PCA, LLE, LDA and FDR methods.

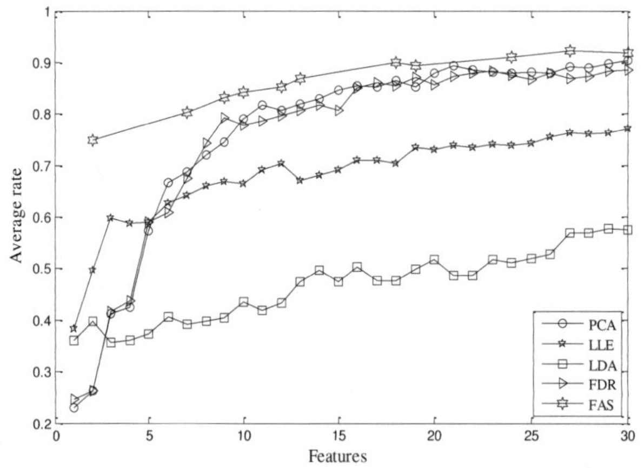

Figure 3. Recognition rate curves of different methods on Chinese speech emotional database

5. Conclusion

This paper extracts a number of speech emotion features and analyzes the importance of features on given emotions. And the feature selection is carried out by FAS method. In the comprehensive comparison and analysis with the PCA, LLE, LDA and FDR methods, the experimental results verify that the FAS method is effective and improves speech emotion recognition rate.

Acknowledgments

This work is supported by the National Natural Science Foundation of China (41664002) and Guangxi Provincial Key Fund Project (2017GXNSFDA198016).

References

[1] L. Zhao. "Speech signal processing (third edition)," Beijing: China Machine Press, 2016, pp. 213-220.

[2] Y. M. Cheng, D. O'Shaughnessy, and P. Mermelstein. "Statistical Recovery of Wideband Speech from Narrowband Speech," IEEE Transactions on Speech and Audio Processing, vol. 2, no. 4, 1994, 2(4), pp. 544-548.

[3] C. Jones and A. Deeming. "Affective Human-Robotic Interaction," Affect and Emotion in Human-Computer Interaction. Springer-Verlag, 2008, pp. 175-185.

[4] D. Morrison, R. Wang, and L. C. De Silva. "Ensemble methods for spoken emotion recognition in call-centres," Speech Communication, vol. 49, no. 2, 2007, pp. 98-112.

[5] R. W. Picard. "Robots with emotional intelligence," Proceedings of ACM/IEEE International Conference on Human-Robot Interaction. IEEE, 2009, pp. 5-6.

[6] C. C. Huang, W. Gong, W. L. Fu, and D. Y. Feng. "Research of speech emotion recognition based on DBNs," Journal of Computer Research and Development, no. s1, 2014, pp. 75-80.

[7] Y. S. Tao, K. X. Wang, J. Yang, and L. Li. "Hybridizing information gain and harmony search for feature selection on speech emotion," Journal of Chinese Computer Systems, vol. 38, no. 5, 2017, pp. 1164-1168.

[8] B. C. Chiou and C. P. Chen. "Feature space dimension reduction in speech emotion recognition using support vector machine," Proceedings of Signal and Information Processing Association Summit and Conference. IEEE, 2013, pp. 1-6.
[9] L. Xue and T. Qian. "Speech analysis based on locally linear embedding(LLE)," Proceedings of International Conference on Natural Computation. IEEE, 2010, pp. 2159-2162.
[10] J. J. Yuan, L. Chen, T. T. Fan, and J. Jia. "Dimension reduction of speech emotion feature based on weighted linear discriminant analysis," International Journal of Signal Processing, Image Processing and Pattern Recognition, vol. 8, no. 11, 2015, pp. 299-308.
[11] X. D. Zhang, F. Hu, and L. Zhao. "Recognition of practical speech emotion based on improved shuffled frog leaping algorithm and support vector machine," Signal Processing, vol. 27, no. 5, 2011, pp. 678-689.
[12] F. Burkhardt, M. Kienast, and A. Paeschke. "Berlin database of emotional speech (Technical University, Institute for Speech and Communication, Department of Communication Science, Berlin)," [2016-09-22]. http://pascal.kgw.tu-berlin.de/emodb/.
[13] Speech and human interaction research group, Institute of automation, Chinese Academy of Sciences. "Chinese database of emotional speech," [2016-11-17]. http://www.datatang.com/data/39277.
[14] R. Y. Liang, L. Zhao, and X. Wei. "Experimental course on speech signal processing," Beijing: China Machine Press, 2016, pp. 231-234.

Query Auto-Completion Based on Word2vec Semantic Similarity

Taihua Shao[1,]*, Honghui Chen[1] and Wanyu Chen[1]

[1]Key Laboratory of Information Systems Engineering, National University of Defense Technology, Changsha, Hunan, China

*E-mail: hunsth@outlook.com

Abstract. Query auto-completion (QAC) is the first step of information retrieval, which helps users formulate the entire query after inputting only a few prefixes. Regarding the models of QAC, the traditional method ignores the contribution from the semantic relevance between queries. However, similar queries always express extremely similar search intention. In this paper, we propose a hybrid model FS-QAC based on query semantic similarity as well as the query frequency. We choose word2vec method to measure the semantic similarity between intended queries and pre-submitted queries. By combining both features, our experiments show that FS-QAC model improves the performance when predicting the user's query intention and helping formulate the right query. Our experimental results show that the optimal hybrid model contributes to a 7.54% improvement in terms of MRR against a state-of-the-art baseline using the public AOL query logs.

1. Introduction

Query Auto Completion (QAC) has its task to help the user formulate her query while she is typing only the prefix, e.g. several characters [1]. Its main purpose is to predict the user's search intention by completing the query with fewer characters. The emergence of Google Instant is a major trigger for query auto-completion research. After that, how to build a better query completion becomes very important, which determines the time users wait for to get intended information [2].

With a QAC system, queries that match the query prefix are pre-generated and stored in an efficient data structure. Figure 1 provides a QAC instance. When one user inputs several characters, the search engine will provide a set of query completions matching user's input with a drop-down menu in real time [3].

A basic method in most QAC systems is to extract the query and its prefix from the search records for a period of time, and to count up the number of times that the query is submitted in the previous history, i.e. the frequency [4]. Then, queries can be sorted according to their frequencies. This method assumes that the future frequency of query is constant with the current frequency. The performance of a QAC system with this method is in good satisfaction but far less than the best. Because it fails to consider other important factors like time and semantic similarity relationship between queries, etc [5] [6]. However, these information always affect the query auto-completion ranking performance [7]. To smooth the deficiencies of QAC research, we study on how to consider these information in a QAC system.

IOP Conf. Series: Journal of Physics: Conf. Series **1004** (2018) 012018 doi:10.1088/1742-6596/1004/1/012018

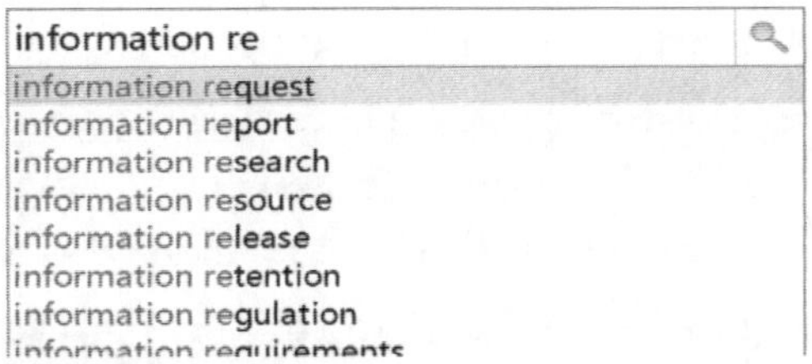

Figure 1. Query auto-completion instance.

We firstly propose a query auto-completion model S-QAC based on semantic similarity feature between queries, which is calculated by word2vec method. Next, by combining the frequency feature from traditional MPC model [2] and semantic similarity feature from our S-QAC model, a hybrid query auto-completion model FS-QAC is proposed.

We obtain the optimal hybrid model and analyze its performance by adjusting the weight of each feature. We also discuss the performance of our query auto-completion model under different prefix length. The results show that combining the frequency and semantic similarity does help to improve the performance of query auto-completion.

Our contributions can be summarized as follows:

- A new method considering semantic similarity feature in QAC system with the help of word2vec method, which indicates that semantic similarity is efficient for improving the QAC performance in terms of Mean Reciprocal Rank (MRR);
- A hybrid model combining both frequency and semantic similarity features for QAC, whose experiment results show that our optimal hybrid model significantly outperforms the state-of-the-art MPC method.

2. Approach

In this section, we describe our semantic-similarity-based model S-QAC, as well as a hybrid model FS-QAC, which considers both frequency and semantic similarity features. We introduce the related concepts of our models, and give concrete steps of modeling.

2.1. S-QAC model

Many methods have been proposed to measure the semantic similarity between two words. After comparing various semantic similarity measuring methods, we build the S-QAC model, which considers semantic similarity between the candidate queries and their previous queries submitted in the same session, on the basis of word2vec method.

Word2vec method [8] would firstly train the input corpus, namely, the sentences consisted by word collection: $term_1$, $term_2$... $term_n$. Its purpose is to predict the terms around through the term vector representation outputted by the model [9]. This goal can be achieved by maximizing the average log probability, i.e.:

$$L = \max\Big(\sum_{term_i \in C} \log P(term_i \mid context(term_i))\Big) \quad (1)$$

Where C is the size of the context used for training. It can be seen that the word2vec method highly depends on the training text dataset [10]. That's to say, the queries to be represented must appear in the training corpus before.

The pre-trained Google News model provided by Google includes hundreds of millions of text statements. However, in real world information retrieval system, queries submitted by users could sometimes be rarely used in the pre-trained model. In order to solve this problem, we first train our experimental AOL dataset [11]. After generating AOL dataset based model, we combine it with the pre-trained Google News model to ensure each query in test process could be represented as vector.

With the two word2vec models above, we could measure the semantic similarity between the candidate queries and the previous queries submitted in the same session. We could obtain two

IOP Conf. Series: Journal of Physics: Conf. Series **1004** (2018) 012018 doi:10.1088/1742-6596/1004/1/012018

semantic similarity scores of a certain candidate query: $Score_A(q_c,q_x)$ and $Score_G(q_c,q_x)$. Among them, $Score_A(q_c,q_x)$ represents the score between the candidate query q_c and the query q_x in current session calculated by AOL word2vec model while $Score_G(q_c,q_x)$ is the semantic similarity score calculated by Google News word2vec model. They were synthesized as follows:

$$ScoSim(q_c,q_x) = \omega \cdot Score_A(q_c,q_x) + (1-\omega) \cdot Score_G(q_c,q_x) \quad (2)$$

Where ω is the weight parameter controlling the contributions between the similarity calculated by the two word2vec models.

As the prefix p is also a role reflecting user's query intention. We synthesize it into our model to improve the accuracy. The formula is as follows:

$$Score_S(q_c) = \sum_{x=1}^{N} \gamma_x \cdot ScoSim(q_c,q_x) + \delta \cdot ScoSim(q_c,q_x) \quad (3)$$

Where N is the number of queries that had already been submitted in current session. γ and δ are weight parameters.

2.2. FS-QAC model

Besides the semantic similarity score of each candidate query q_c obtained in § 2.1, it is reasonable to assume that the query with higher frequency in the search history is more likely to be submitted again.

Thus in this study, we propose to combine the semantic similarity feature and the frequency feature by using the linear fitting formula in our hybrid model FS-QAC as follows:

$$Score_{FS}(q_c) = \alpha \cdot Score_S(q_c) + (1-\alpha) \cdot Score_F(q_c) \quad (4)$$

Where α denotes the tradeoff between query semantic similarity and query frequency. By adjusting α, we can linearly integrate two features. After groups of experiments, we could discover the optimal tradeoff for our FS-QAS model. Algorithm 1 details the major steps of our hybrid model FS-QAC.

Algorithm 1 Frequency and Semantic Similarity Based Query Auto-Completion (FS-QAC).

Input: AOL dataset: **Q;** Completion List: **L;** Current Session Queries List: **C;**
Output: Frequency Score List: **F;** Semantic Similarity Score List: **S;** Re-ranked List: **R;**

```
for each a ∈ L do
    ScoreF (a) = count(a in Q);  F.append(ScoreF (a))
end for
AOL.model = models.save.Word2Vec(Q);
model.load(AOL.model + Google News.model)
for each a ∈ L do
    for each b ∈ C do
        ScoreS(a)+ = model.similarity(b, a)
    end for
    S.append(ScoreS(a));
end for
ScoreFS(a) = α·ScoreS(a) + (1 - α)·ScoreF (a)
dic ={ a ∈ L : ScoreFS(a)}
R = L.sorted(dic; key = dic.getitem; reserve = True)
return Re-ranked List: R;
```

3. Experiments

We begin with the model summary and detail the research questions that we aim to answer. We then describe the experiment setup for our study.

IOP Conf. Series: Journal of Physics: Conf. Series **1004** (2018) 012018 doi:10.1088/1742-6596/1004/1/012018

3.1. Model summary

The baseline to be compared is the Most Popular Completion (MPC) model, which is simple and direct, but widely used. This model predicts users' submission probability according to the frequency of queries in long-term search logs based on the maximum likelihood estimation method.

The models to be discussed in our paper are:

- **MPC:** a traditional query auto-completion model based on the frequency of queries, which is the baseline to be compared in our study;
- **S-QAC:** a query auto-completion model based on semantic similarity feature between candidate queries and the previous queries submitted in the same session;
- **FS-QAC:** a hybrid query auto-completion model proposed by us which combining both frequency and semantic similarity features;

The research questions guiding our experiments are:

- **RQ1:** Could semantic similarity feature improve the performance of QAC system? (See § 4.1)
- **RQ2:** How could semantic similarity feature improve the performance of QAC system together with frequency feature? (See § 4.2)
- **RQ3:** How does prefix length influence the performance of models? (See § 4.3)

3.2. Dataset and parameters

Our experiments was carried out on the AOL dataset [11]. AOL is one of few datasets that contain real query data. The number of query records could fully guarantee statistical significance. Queries from AOL were obtained in real search engine from March 1, 2006 to May 31st, 2006. We process queries submitted within 30 minutes as a session. For the sake of temporal coherence, we divide the AOL dataset into two parts. 75% of the data is used as the train set, and the remaining 25% as the test set. According to the prefix length, we divide test dataset into 5 different data blocks, their prefix lengths are 1, 2, 3, 4 and 5 respectively. Table 1 details the statistics of the dataset used.

Table 1. Dataset statistics.

Variables	Train	Test
Queries	2,908,666	1,121,126
Unique queries	452,980	452,980
All Prefixes	2,430,110	867,344
Prefix-1	216,925	75,455
Prefix-2	384,909	132,786
Prefix-3	574,980	203,830
Prefix-4	644,897	232,613
Prefix-5	608,399	222,660

To train our word2vec model, we set the vector space dimension size as 200, the least number of occurrences min_count as 1, the number of parallel training process workers as 4. To simplify our models, we set coefficients γ and δ as $1/(N+1)$ to make $\mathrm{ScoSim}(q_c, Q_L)$ the arithmetic mean of all relevant semantic similarity scores of q_c, and the linear integration parameter ω as 0.5. α is also a linear integration parameter which control the proportion of semantic similarity in FS-QAC model. In this study, we adjust the α from 0 to 1 with the step increase 0.1 to conduct 11 groups of experiments.

In order to evaluate the effectiveness of query auto-completion method, we choose Mean Reciprocal Rank (MRR) to evaluate the performance of query auto-completion. MRR performed well in many other studies that allow returning multiple results. It is often used to measure the effectiveness of query auto-completion experiments.

IOP Conf. Series: Journal of Physics: Conf. Series **1004** (2018) 012018 doi:10.1088/1742-6596/1004/1/012018

In the drop list of query auto-completion system, if query is in the position $rank_i$, then it corresponds to the score $RR=1/rank_i$. Finally, the Mean Reciprocal Rank score is the mean of all RRs for all prefixes. The formula is as follows:

$$MRR = \frac{1}{Q}\sum_{i=1}^{|Q|}\frac{1}{rank_i} \tag{5}$$

Where Q is the sample dataset, |Q| represents the number of samples in Q, and $rank_i$ represents the position of the query in the query completion list.

4. Results and Discussion

4.1. *Performance of S-QAC Model*

To answer **RQ1,** we examine the query auto-completion performance (MRR) of the baseline as well as our S-QAC model under various prefix lengths. See Figure 2 for the results.

As it illustrates in Figure 2, when prefix length is 1, 2 or 5, the MRR of S-QAC model is higher than MPC model. Especially when prefix length is short, S-QAC model performs significantly better than MPC model. As for the whole MRR of test dataset, we have calculated that the performance of S-QAC model is 3.58% higher than that of MPC model, where $MRR_{S\text{-}QAC}=0.6712$ and $MRR_{MPC}=0.6480$. The result shows that our S-QAC model based on semantic similarity is helpful to improve the performance of QAC system especially when there is not enough input prefix for the QAS task.

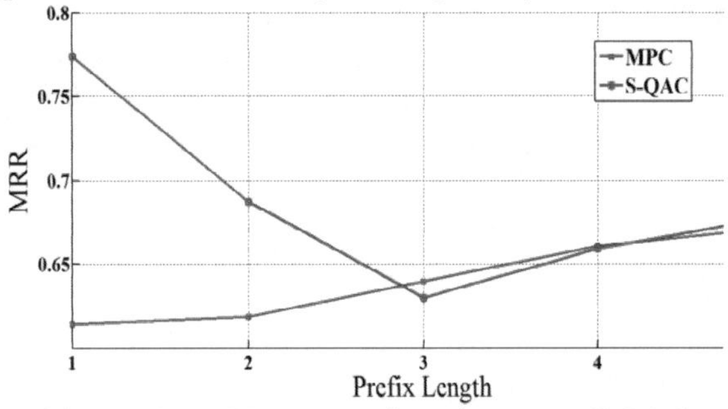

Figure 2. MRR comparison between S-QAC and MPC.

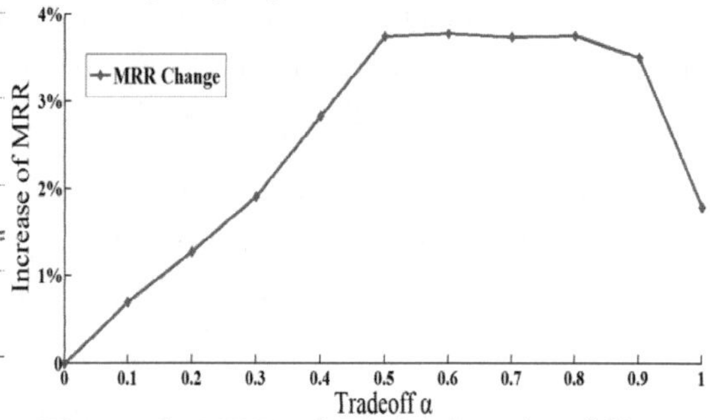

Figure 3. MRR of FS-QAC under different tradeoff

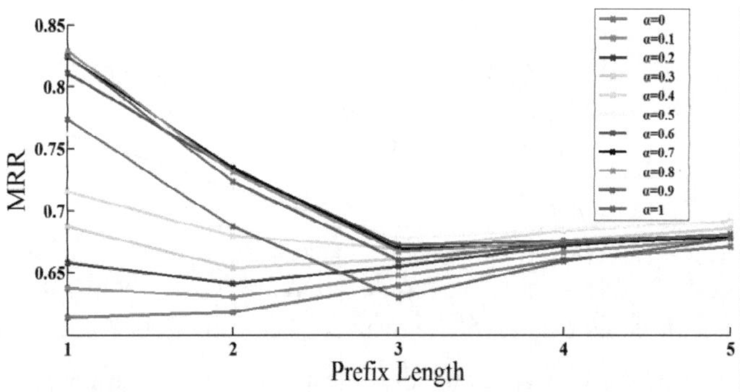

Figure 4. Performances under all prefix lengths.

4.2. *Performance of FS-QAC Model*

In order to verify the performance improvement of our proposed hybrid model FS-QAC in **RQ2**, we conduct 11 groups of experiments and calculate the MRR of FS-QAC model under different trade off parameter α. The results is showed in Figure 3.

As shown in Figure 3, our proposed hybrid model FS-QAC improves the performance of QAC system, and outperforms the plain model MPC (α=0) and S-QAC (α=1). When the tradeoff varies from 0 to 0.6 with the step increase 0.1, growth rate of MRR increases with α. When the tradeoff varies from 0.6 to 1, growth rate of MRR decreases with α. With different tradeoff, the performance improvement rate is different, among which the best performance is when α=0.6, where $MRR_{FS\text{-}QAC}$=0.6969, the growth rate against baseline is 7.54%. The result shows that our proposed hybrid model FS-QAC could significantly improve the performance of QAC system by well synthesizing frequency feature and semantic similarity feature.

4.3. Discussion on Different Prefix Length

To answer **RQ3**, we examine the MRR of our FS-QAC model with different tradeoff parameter under various prefix lengths. See Figure 4 for the results.

As it illustrates in Figure 4, when the prefix length is short (≤3), the MRR values of the models under different tradeoff are higher than the results when the prefix length is long (>3). With the growth of prefix length, the performance tends to be stable. It indicates that when prefix length is short, the proportion of semantic similarity feature could obviously increase the MRR value of QAC system, as the performance of our FS-QAC models (α>0.5) outperforms those models with α<0.5 especially when the prefix length is short (≤3). However, when the prefix length is long (>3), there is no too much differences among those models with various values of parameter α in terms of MRR.

5. Conclusions and Future Work

We propose a model FS-QAC based on hybrid ranking scores by combining two features, i.e. query semantic similarity and the query frequency. We calculate the semantic similarity score by integrating Google News model and AOL dataset model. Our experiments show that the semantic similarity together with query frequency can provide a satisfactory idea to predict the user's query intention and help formulate the right query. We verify the effectiveness of our best performer FS-QAC, showing significant improvement over the baseline MPC model in terms of MRR. The results show that the optimal hybrid model contributes to a 7.54% improvement in MRR against well-known baseline.

As future work, we plan to evaluate our models on other dataset so as to verify the effectiveness of our proposal. In addition, we also want to investigate the sensitivity of involved parameters, e.g. the personalized indicators of query [12].

References

[1] Cai F and Rijke M D 2016 A survey of query auto completion in information retrieval. *Foun. Tre. in Inf. Retr.* **10(4)** 273-363
[2] Bar-Yossef Z and Kraus N 2011 *Proc. Int. Conf. on WWW* (Hyderabad: ACM) pp 107-116
[3] Shokouhi M and Radinsky K 2012 *Proc. of Int. Acm Sigir Conf. on Res. Dev. in Inf. Retr.* (Portland: ACM) pp 601-610..
[4] Santo G D, Mccreadie R, Macdonald C and Ounis I 2015 *Proc. of Int. Acm Sigir Conf. on Res. Dev. in Inf. Retr.* (Santiago: ACM) pp 775-778
[5] Cai F and Rijke M D 2016 *Proc. of Int. Acm Sigir Conf. on Res. Dev. in Inf. Retr.* (Pisa: ACM) pp 993-996
[6] Cai F, Liang S and Rijke M D 2014 *Proc. of 23rd ACM Int. Conf. on Inf. Kno.Mana.* (Shanghai: ACM) pp 1599-1608
[7] Jiang J Y, Ke Y Y, Chien P Y and Cheng P J 2014 *Proc. of 37th Int. Acm Sigir Conf. on Res. Dev. in Inf. Retr.* (Gold Coast: ACM) pp 445-454
[8] Mikolov T, Sutskever I, Chen K, Corrado G and Dean J 2013 *Proc. of 26th Int. Conf. on Neur. Inf. Proc. Sys. (vol 2)* (Lake Tahoe: Curran Associates Inc) pp 3111-3119
[9] Mikolov T, Chen K, Corrado G and Dean J 2013 Effcient estimation of word representations in vector space. *DBLP: J. CoRR* **abs/1301.3781**
[10] Cai F and Rijke M D 2016 *J. Inf. Proc. and Mana.: an Int. J. arch.* **52(4)** 628-643
[11] Pass G, Chowdhury A and Torgeson C 2006 *Proc. of 1st Int. Conf. on Sca. Inf. Sys.* (Hong Kong: ACM) p 1
[12] Shokouhi M 2013 *Proc. of 36th Int. Acm Sigir Conf. on Res. Dev. in Inf. Retr.* (Dublin: ACM) pp 103-112

IOP Conf. Series: Journal of Physics: Conf. Series **1004** (2018) 012019 doi:10.1088/1742-6596/1004/1/012019

Location Estimation of Urban Images Based on Geographical Neighborhoods

Jie Huang[1, a] and Sio-Long Lo[1, b]

[1] Faculty of Information Technology, Macau University of Science and Technology, Macau S.A.R., China
[a]must.jhuang@gmail.com; [b]sllo@must.edu.mo

Abstract. Estimating the location of an image is a challenging computer vision problem, and the recent decade has witnessed increasing research efforts towards the solution of this problem. In this paper, we propose a new approach to the location estimation of images taken in urban environments. Experiments are conducted to quantitatively compare the estimation accuracy of our approach, against three representative approaches in the existing literature, using a recently published dataset of over 150 thousand Google Street View images and 259 user uploaded images as queries. According to the experimental results, our approach outperforms three baseline approaches and shows its robustness across different distance thresholds.

1. Introduction

Location estimation of images is a challenging computer vision task, with the aim of estimating the geographical location where a query image was captured. The increasing amount of images available on the Internet provide an opportunity for research in location estimation of images. Attention from the research community has been drawn to the location estimation of images [1, 2, 3, 4, 5, 6] in the recent decade, due to the prospect of various applications in landmark recognition [7], city reconstruction [8], and robot vision [9].

One of the pioneering approaches was proposed by Hays and Efros in [3], where they adopted a data-driven approach to estimating the location of a query image, when given a large reference database of various images captured from all across the world. Succeeding research efforts have been devoted to addressing the problem of localizing social network images [10, 11], street view images [12, 13, 14], etc.

There are two related tasks for image geo-localization. One is finding a matching scheme, with the purpose of finding correspondences between image descriptors of the query image and the reference images in the database. Sattler *et al.* formulated the problem of image geo-localization, as a descriptor matching problem [15]. The other is location estimation of the query image, based on the matching results. However, we are more interested in the latter, i.e., how to make a location estimate of the query based on the matches returned by a matching scheme, for example, image retrieval.

Our goal in this paper is to explore the possibilities of improving the location estimation accuracy of images captured in urban environments. We note that the distribution of reference images in the database has an impact on the estimation accuracy, i.e., if there are sufficient reference images taken at, or near the query location in the database, it is more likely for the query image to be geo-localized correctly. Moreover, the visual content of query images also has an influence on the estimation accuracy. Location estimation of generic images (for example, images of the sky, plants, and indoor scenes) are beyond the scope of this paper. We confine our discussion to the problem of location estimation of urban images captured outdoors.

To pursue the goal mentioned above, we propose an approach that consists of three major steps. In the first step, *visual candidates retrieval*, we retrieve multiple most similar images (or visual candidates) from the reference database, for each query image. Then we build a neighborhood for every candidate, which includes images of the shortest geographical distance to the candidate, in the second step, *geo-neighborhood construction*. Finally, for each neighborhood, we compute the average visual similarity to the query, based on which we rank all the visual candidates. The location of the highest ranked candidate will be propagated to the query image, as the location estimate. The novelty of our approach lies in addressing the shortcomings of existing approaches including [3, 11], and improvement in accuracy of location estimation compared to these approaches.

The remainder of this paper is structured as follows. Section 2 reviews related work on location estimation of images. Section 3 introduces our approach to location estimation of urban images in more detail. The setup of our experiments is presented in Section 4. Section 5 demonstrates the performance of our approach, against three baseline approaches. We conclude the paper in Section 6.

2. Related Work

In this section, we will review a number of retrieval-based approaches related to our research work. Conventionally, image retrieval returns one or multiple images with matching descriptors to the query image. And retrieval-based approaches estimate the query location based on the retrieved images.

Hays and Efros first proposed adopting first-nearest neighbor (1-NN) in estimating the location of a single image [3]. Essentially, when there is only one candidate image for the query, its location will be directly propagated to the query image. This approach assumes that if two images are visually similar, then they are likely to be in the proximity of each other geographically. One weakness of directly propagating the location of the 1-NN to the query is that it is likely to fail in cases when two images are visually similar, but continents away from each other.

The idea of extracting multiple visual nearest neighbors (NNs) of the query, then clustering the location of the NNs using mean-shift clustering was proposed in [3]. Eventually, the mode location of the largest cluster will be assigned to the query image. The assumption behind the clustering approach is that if a location cluster has more images than the others, the mode of which is more likely to be the closest to the query location. However, the assumption does not hold for cases when the location closest to the query location does not have sufficient reference images in the database, to become the largest cluster.

Li *et al.* proposed and matured the idea of "geo-visual ranking" (GVR) [10, 11] as an alternative approach to location estimation, overcoming the weakness of both approaches proposed in [3]. The GVR approach first extracts k nearest neighbors of the query, then assigns k images into different clusters based on their geo-location. Finally, the top m clusters, where $m < k$, are ranked based on the aggregate visual similarity to the query. The location estimate of the query, will be the centroid of GPS coordinates of all the images in the highest ranked cluster [11]. The shortcoming of the ranking approach [11] is that it requires pre-defining a number of parameters during the location estimation process.

3. Location Estimation Based on Geo-Neighborhoods

Our approach assumes that, images in the proximity of one another geographically, have visual elements in common thus making them appear visually similar. Hence, we propose to first retrieve multiple visual candidates of a query, then evaluate each candidate based on the visual similarity of its neighborhood to the query. Geo-neighbors in a candidate's neighborhood are defined as the images of the shortest geographical distance to a candidate. The rationale is the neighbors of a candidate provide additional visual information of a neighborhood where a candidate is located. If the visual similarity of a neighborhood to the query is high, this can serve as supportive evidence that a candidate is visually similar to the query. There are three steps in our approach, namely, 1) *visual candidates retrieval*, 2) *geo-neighborhood construction*, and 3) *visual candidates ranking*. In the succeeding subsections, we will elaborate on each of the steps in detail.

IOP Conf. Series: Journal of Physics: Conf. Series **1004** (2018) 012019 doi:10.1088/1742-6596/1004/1/012019

3.1. Visual candidates retrieval

When given a query image q, the purpose of this step is to retrieve images that are most likely to have been captured at the same location as the query, solely based on the visual content. Hence, we compute the visual similarity between the query q and each image in the reference database $D = \{I_i\}_{i=1}^N$, and further retrieve the images of the highest similarity. The k-nearest neighbor (kNN) classifier will be applied to retrieve visually similar images from the databases D, given that there exist multiple views of a single location in the database adopted in our experiments. As a result, K visual nearest neighbors of the query will be retrieved from the database as $N_K(q)$

$$N_K(q) = \{(I_i^{(q)}, d_i^{(q)})\}_{i=1}^K \tag{1}$$

Where $I_i^{(q)} \in D$, and $d_i^{(q)} \in \mathbb{R}_{\geq 0}$ is the visual distance, or dissimilarity between a database image and the query. In addition, 2-norm (L2) distance is applied to computing the dissimilarity between two images when using the kNN classifier

$$\text{dist}_{\text{vis}}(p, q) = \|f_p - f_q\|_2 \tag{2}$$

Where f_p and f_q are the feature representation of image p and q. To represent an image with a feature vector, gist descriptors [16, 17] are adopted to describe an image on the whole, due to its usefulness in semantic image retrieval [18]. In addition, we accommodate the gist descriptors of an image in a 600-dimensional feature vector.

3.2. Geo-neighborhood construction

After retrieving K visual candidates of the query image, the aim of the second step is to create a geographical neighborhood for each candidate $r \in N_K(q)$, by extracting from the reference database D, images of the shortest geographical distance to each candidate. For a visual candidate $r \in N_K(q)$, its ground-truth geo-location can be found in the corresponding GPS database $G : D \to L \in R^2$, $l_i = G(I_i)$, and $l_i = (\text{lat}_i, \text{lon}_i)$ for $i = 1, 2, \ldots, N$ are the latitude and longitude of image I_i in the reference database D. The M-geographical neighbors of a candidate r are defined as images of the shortest geographical distance to r as

$$R_M(r) = \{(I_i^{(r)}, h_i^{(r)})\}_{i=1}^M \tag{3}$$

Where $I_i \in D$, $h_i^{(r)} \in \mathbb{R}_{\geq 0}$ is the geographical distance to the location of candidate r, and the value of M varies among visual candidates. The Haversine formula [19], denoted $H(p, l_i)$, is applied to computing the geographical distance between GPS coordinates of a candidate $p = G(r)$ and those of images in the reference database $l_i = G(I_i)$ where $I_i \in D$. Images $I_i \in D$ of the shortest Haversine distance to the candidate r are extracted and form a neighborhood of the candidate r.

The rationale of this step is, the neighbors $R_M(r)$ of a visual candidate r provide additional visual information regarding the geo-location of r. If the visual similarity of the neighborhood of a candidate r is high, this can serve as additional evidence that r is visually similar to the query. The approach by Li *et al.* [10], where they also consider extracting all the images within a certain radius (for example, 1 kilometer), in order to build the neighborhood of a visual candidates, is prone to an unbalanced number of neighbors for each candidate when constructing a neighborhood. Instead, we consider extracting only images of the shortest geographical distance to the candidate, and compute the average similarity of all the images in the neighborhood, including the candidate itself, in order to overcome the bias caused by the unbalanced number of neighbors for each candidate as proposed in [10].

3.3. Visual candidates ranking

To obtain a reasonable geo-location estimate, we first calculate the visual distance between the neighborhood of every candidate r, $R^{(r)} = R_M(r) \cup \{(r,0)\}$, $r \in N_K(q)$, and the query image q. We define the average visual distance $\text{dist}_{\text{rank}}(R^{(r)}, q)$ between the neighborhood $R^{(r)}$ of a candidate r, and the query image q as

$$\text{dist}_{\text{rank}}(R^{(r)}, q) = \frac{1}{|R^{(r)}|} \cdot \sum_{\alpha \in R^{(r)}} \text{dist}_{\text{vis}}(\alpha, q) \tag{4}$$

Where $|R^{(r)}|$ is the number of images in $R^{(r)}$, and $\alpha \in R^{(r)}$ represents every image belonging to the neighborhood $R^{(r)}$. We rank all the visual candidates $r \in N_K(q)$ based on the average visual distance of their corresponding neighborhood $R^{(r)}$.

$$r^* = \underset{r \in N_K(q)}{\arg\min}\ \text{dist}_{\text{rank}}(R^{(r)}, q) \tag{5}$$

Finally, GPS coordinates l_{r^*} of the candidate r^*, which has the highest neighborhood similarity, will be propagated to the query q, as the location estimate, $l_{q^*} = l_{r^*} = G(r^*)$.

4. Experimental Setup

4.1. Dataset

To perform a fair comparison, we evaluate the location estimation accuracy of our approach against three baseline approaches on a Google Street View Dataset contributed by Zemene *et al.* in their recent work [14], comprising of 152,536 images taken in 14 cities from different parts of the world, along with a separate query dataset of 259 images, both after elimination of faulty images. The distribution of reference and query images in the UCF Google Street View Dataset [14] is shown below in Table 1.

Table 1. Distribution of reference and query images from the UCF Google Street View Dataset [14].

Continent/country	City	# of reference images	# of query images
Europe (5 cities)	Amsterdam	15,608	17
	Frankfurt	8,788	0
	Rome	15,729	13
	Milan	15,986	2
	Paris	16,588	28
Australia (2 cities)	Sydney	15,986	16
	Melbourne	14,190	10
USA (7 cities)	Las Vegas	5,950	54
	Los Angeles	12,610	13
	Phoenix	4,233	13
	Houston	10,361	11
	San Diego	3,079	15
	Dallas	4,740	18
	Chicago	8,688	49
Total number		152,536	259

IOP Conf. Series: Journal of Physics: Conf. Series **1004** (2018) 012019 doi:10.1088/1742-6596/1004/1/012019

Figure 1 shows sample reference images and queries from the UCF Google Street View Dataset [14]. Each location is represented with at most five reference images in the database. As demonstrated in Figure 1a, the reference images depict each location from different perspectives, and are complementary to one another. In Figure 1b, twelve sample query images are demonstrated. The visual content of query images varies, however, most queries are ground-level images.

(a) Twelve sample reference images.

(b) Twelve sample queries.

Figure 1. Sample reference images and queries from the UCF Google Street View Dataset [14].

4.2. Experimental design and evaluation metric

To compute the visual similarity between images, we represent each image with a feature vector, then compute the distance between two feature vectors. The shorter the distance, the more similar two images are. In our experiments, we build a feature vector of gist descriptors[16, 17] for every image, with the spatial resolution of 5 by 5 , where every bin stores the average response of the image region to filters at 6 orientations and 4 scales. We arrive at a 600-dimensional vector for every image and apply L2 distance to computing the distance between feature vectors. The settings are identical to those used in [3].

In order to evaluate the accuracy of location estimation, we define a distance threshold, θ_d, which controls the evaluation accuracy and tolerance to noise in ground-truth GPS coordinates [11]. An image is considered to be geo-localized correctly, if its estimated coordinates, denoted l_{q^*}, fall within the distance θ_d, of the ground-truth location l_q. As formally expressed in Eq.6, the correctness of geo-localization of an image can be obtained using the function f_{θ_d}, with respect to the distance threshold θ_d, where $\text{dist}(l_q, l_{q^*})$ refers to the geographical distance between l_q and l_{q^*}, calculated using the Haversine formula [19].

$$f_{\theta_d}(l_q, l_{q^*}) = \begin{cases} \text{correct}, & \text{if dist}(l_q, l_{q^*}) \leqslant \theta_d \\ \text{incorrect}, & \text{otherwise} \end{cases} \tag{6}$$

4.3. Baseline approaches

The three baseline approaches, against which we compare our approach, in terms of geo-localization accuracy, are all retrieval-based approaches, i.e., approaches that first retrieve visually similar images(or visual nearest neighbors) of a query, then make an estimate of the query location based on the location of retrieved images.

- 1-NN: Following [3], we retrieve from the reference database, the most similar image to the query using the k-nearest neighbor classifier, then assign the location of the retrieved image to the query, as the location estimate.
- Clustering: Instead of retrieving the most similar image, k images of the highest visual similarity are retrieved [3]. These retrieved images will be assigned to the corresponding cluster using mean-shift clustering algorithm. Eventually, the cluster with the most images will be selected, and GPS coordinates of the cluster centroid will be propagated to the query.

• Ranking: The ranking approach [11], used as a baseline in our experiments, is an extension of the one proposed in [10]. k nearest neighbors will first be retrieved for the query, then the neighbors will be assigned into different clusters, based on their ground-truth location. However, only the top m cluster, where $m < k$, will enter the location estimation process. Each cluster will be given a ranking score, i.e., the sum visual similarity of images in the cluster to the query. The cluster with the highest ranking score will be selected, and the centroid of GPS coordinates of all the images will be the location estimate of the query.

In our experiments, when applying the k-nearest neighbor classifier, we set the number of nearest neighbors to 20, i.e., when given a query image, we retrieve 20 images of the highest visual similarity to the query. Moreover, the bandwidth is set to 0.2 for mean-shift clustering. Regarding the ranking approach [11], we also set the number of nearest neighbors to 20 during image retrieval, while the number of clusters eligible for the location estimation process to 10, following the optimal parameter combination in [11].

5. Experimental Results

5.1. Comparison with the baseline approaches

As reported in Table 2, our approach outperforms three baseline approaches in terms of location estimation accuracy, across all distance thresholds. Moreover, one of Hays and Efros' findings about the performance gap between 1-NN and the clustering approach [3] is also validated in our experiments: 1-NN performs better when it comes to precise location estimation. In our experiments, 1-NN yields better accuracy when the distance threshold is set to 25 or 200 kilometers. While the clustering approach outperforms the 1-NN approach, in terms of accuracy, as the distances threshold grows towards 2,500 kilometers.

Table 2. Comparison with three baseline approaches regarding location estimation accuracy.

Distance threshold (km)	25	200	750	2,500
1-NN	8.49%	9.27%	15.06%	27.03%
Clustering	6.18%	6.95%	15.44%	27.41%
Ranking	8.88%	8.88%	20.08%	35.91%
Ours	**10.42%**	**10.42%**	**20.85%**	**36.68%**

5.2. Discussion

Compared to 1-NN [3] and the clustering approach [3], the advantage of our approach is its robustness. Hays and Efros concluded in [3] that 1-NN performs better at precise location estimation, while the clustering approach is more suitable for global-scale localization. Our approach, however, results in better performance than both baselines when measured with different distance thresholds. The estimation accuracy of our approach also exceeds that of the ranking approach [11], where we have witnessed an encouraging increase of 1.54% in estimation accuracy, when the distance threshold is 25 or 200 kilometers, and a slight increase in accuracy as well, when the threshold is set to 750 or 2,500 kilometers.

In addition, our approach does not require tuning any additional parameters after retrieving visual candidates of a query. Moreover, we overcome the problem of the unbalanced number of images in the neighborhood among visual candidates, introduced by the ranking approach in [10]. The problem is tackled by ranking the candidates on the basis of the average similarity of images in the neighborhood, where it is guaranteed that each neighborhood will have at least one neighbor for the candidate.

6. Conclusion

We have proposed a new approach to location estimation of urban images, based on geographical neighborhoods. Our approach consists of three steps, namely, visual candidates retrieval, geo-neighborhood construction, and visual candidates ranking. Experiments have been carried out to compare the performance of our approach against three baseline approaches. The results demonstrate the robustness of our approach across different distance thresholds. In addition, our approach has outperformed all three baseline approaches, in terms of estimation accuracy. In the future work, we plan to retrieve visual candidates of a query using additional geo-informative features, and improve on how we build the geo-neighborhood for a visual candidate.

Acknowledgment

This work was supported partially by the Macao Science and Technology Development Fund grants (No. 019/2014/A1 and No. 112/2014/A3).

References

[1] Zhang W and Kosecka J 2006 *3D Data Processing, Visualization, and Transmission, Third International Symposium on* (IEEE) pp 33–40
[2] Philbin J, Chum O, Isard M, Sivic J and Zisserman A 2007 *Computer Vision and Pattern Recognition, 2007. CVPR 2007. IEEE Conference on* (IEEE) pp 1–8
[3] Hays J and Efros A A 2008 *Computer Vision and Pattern Recognition, 2008. CVPR 2008. IEEE Conference on* (IEEE) pp 1–8
[4] Serdyukov P, Murdock V and Van Zwol R 2009 *Proceedings of the 32nd international ACM SIGIR conference on Research and development in information retrieval* (ACM) pp 484–491
[5] Crandall D J, Backstrom L, Huttenlocher D and Kleinberg J 2009 *Proceedings of the 18th international conference on World wide web* (ACM) pp 761–770
[6] Irschara A, Zach C, Frahm J M and Bischof H 2009 *Computer Vision and Pattern Recognition, 2009. CVPR 2009. IEEE Conference on* (IEEE) pp 2599–2606
[7] Chen D M, Baatz G, Köser K, Tsai S S, Vedantham R, Pylvänäinen T, Roimela K, Chen X, Bach J, Pollefeys M *et al.* 2011 *Computer Vision and Pattern Recognition, 2011. CVPR 2011. IEEE Conference on* (IEEE) pp 737–744
[8] Tuite K, Snavely N, Hsiao D, Tabing N and Popovic Z 2011 *Proceedings of the SIGCHI Conference on Human Factors in Computing Systems* (ACM) pp 1383–1392
[9] Meier L, Tanskanen P, Fraundorfer F and Pollefeys M 2011 *Robotics and automation (ICRA), 2011 IEEE international conference on* (IEEE) pp 2992–2997
[10] Li X, Larson M and Hanjalic A 2013 *Proceedings of the 3rd ACM conference on International conference on multimedia retrieval* (ACM) pp 81–88
[11] Li X, Larson M and Hanjalic A 2015 *IEEE Transactions on Multimedia* **17** 674–686
[12] Zamir A R and Shah M 2010 *European Conference on Computer Vision* (Springer) pp 255–268
[13] Zamir A R and Shah M 2014 *IEEE Transactions on Pattern Analysis and Machine Intelligence* **36** 1546–1558
[14] Zemene E, Tariku Y, Idrees H, Prati A, Pelillo M and Shah M 2017 *arXiv preprint arXiv:1702.01238*
[15] Sattler T, Weyand T, Leibe B and Kobbelt L 2012 *BMVC* vol 1 p 4
[16] Oliva A and Torralba A 2001 *International journal of computer vision* **42** 145–175
[17] Oliva A and Torralba A 2006 *Progress in brain research* **155** 23–36
[18] Hays J and Efros A A 2007 *ACM Transactions on Graphics (TOG)* vol 26 (ACM) p 4
[19] Robusto C C 1957 *The American Mathematical Monthly* **64** 38–40

An Effective Hybrid Evolutionary Algorithm for Solving the Numerical Optimization Problems

Xiaohong Qian[1, a], Xumei Wang[1], Yonghong Su[1, b] and Liu He[1]
[1]College of Information Engineering, Wuhan Huaxia University of Technology, Wuhan 430223, China

*Corresponding author: [a]21424691@qq.com; [b]461782640@qq.com

Abstract: There are many different algorithms for solving complex optimization problems. Each algorithm has been applied successfully in solving some optimization problems, but not efficiently in other problems. In this paper the Cauchy mutation and the multi-parent hybrid operator are combined to propose a hybrid evolutionary algorithm based on the communication (Mixed Evolutionary Algorithm based on Communication), hereinafter referred to as CMEA. The basic idea of the CMEA algorithm is that the initial population is divided into two subpopulations. Cauchy mutation operators and multiple paternal crossover operators are used to perform two subpopulations parallelly to evolve recursively until the downtime conditions are met. While subpopulation is reorganized, the individual is exchanged together with information. The algorithm flow is given and the performance of the algorithm is compared using a number of standard test functions. Simulation results have shown that this algorithm converges significantly faster than FEP (Fast Evolutionary Programming) algorithm, has good performance in global convergence and stability and is superior to other compared algorithms.

1. Introduction

Evolutionary Programming (EP), Evolution Strategies (ES) and Genetic Algorithm (GA) are three typical evolutionary algorithms, all of which are based on simulating the evolution of a biological population composed of individuals to solve complex optimization problems, especially those that are difficult to solve using other methods or not at all.

Although EP was initially considered an effective solution to artificial intelligence, in recent years, it has been successfully applied to solve numerical optimization problems [1]. In order to solve the problems that the traditional EP algorithm is prone to premature convergence, the Cauchy mutation is introduced to EP, and then FEP (Fast EP) is formed [2]. FEP is more effective than traditional EP in solving multi-peak and multi-local optimization problems. However there is not much improvement for the single peak and multi-peak less local extremum problems and the convergence speed is not fast.

Recently Guo Tao and so on [3] proposed a multi-parent crossover operator for solving function optimization problems. Practice has proved that Guo Tao algorithm has many advantages in solving optimization problems [4]. Its algorithm is simple and computational efficiency is high where you can find more than one optimal solution convergence speed is particularly fast and so on. After experiments, we find that Guo Tao algorithm can converge to the global optimal solution very quickly for the optimization of unimodal fu9nctions and multi-peaks with few local extremums.

However it is difficult to find the optimal solution for some multi-peak and multi-local optimization problems. For example, Generalized Rastrigin's Function has the minimum of zero, but Guo Tao algorithm has an average result of only about 12.

In this paper the Cauchy mutation in FEP and the multi-parent hybrid operator in Guo Tao algorithm are combined to propose a hybrid evolutionary algorithm based on the exchange model (Mixed Evolutionary Algorithm based on Communication), hereinafter referred to as CMEA. In this algorithm, evolution takes place between two parallel populations which uses two different operators separately. Evolution is not only the process of natural selection by environmental individuals, but also the result of exchange among different populations. Subpopulation reorganization leads to the flow of individuals among subgroups and this flow is the manifestation of information exchange between subgroups. Finally 23 test function benchmark are used as an example to verify that the CMEA algorithm has good global convergence and robustness to all kinds of numerical optimization problems.

2. FEP algorithm and introduction of Guo Tao algorithm

This paper discusses the following form of numerical optimization problems: Known n-dimensional real-valued function f: S→R, S∈Rn is a closed set in space Rn, please find a point Xmin∈S which makes f (X)the smallest, that is ∀X∈S, f (Xmin) ≤ f (X).

2.1 Fast Evolution Planning (FEP)

FEP is described as follows:

Step 1: An initial population (Xi,μ_i) containing (i = 1, 2,⋯,μ) is generated. The current evolution algebra k = 1.

Step 2: The function value of all the individuals in the current population is calculated.

Step 3: Each individual generates a descendant according to equation (1).

$$xi'(j) = xi(j) + \eta'_i(j)\,\delta j \quad (1)$$

$$\eta i'(j) = \eta i(j) \exp(\tau' N(0, 1) + \tau Nj(0, 1)) \quad (2)$$

xi '(j) and ηi'(j) is the jth component of the ith child (Xi ' η') .N (0 1) represents the Gauss random number whose mean is 0 and the standard deviation is 1. Nj (0 1) means to generate Gauss random numbers for each component respectively. δj means that the average value of each component is 0 and Cauchy random number with standard deviation 1. Generally $\tau = (\sqrt{2\sqrt{n}}) - 1$,$\tau' = (\sqrt{2n}) - 1$.

Step 4: The function value of the subbody (Xi ' "i') (i = 1 2,⋯,μ) is calculated.

Step 5: The tournament is held in the set consisting of the parent (Xi , ηi) and the child (Xi ', ηi") (i = 1, 2,⋯,μ).For each of the individuals , q individuals are chosen from 2μ randomly to compare with it, and the number of the q individuals inferior to the evaluated individuals is taken as the score of the individual.

Step 6: μ individuals with the highest score selected from the set consisting of (Xi , ηi) and (Xi ', ηi") (i = 1, 2,⋯,μ) are formed to the next generation population.

Step 7: If the shutdown condition is reached, evolution is discontinued, otherwise evolving algebra k = k+ 1 and go to step 3.

2.2 Guo Tao algorithm (GT)

Guo Tao algorithm introduced the subspace concept. The subspace occupied by the M points Xj '= (x1j' ,x2j ',⋯, xnj'), j = 1, 2 ,⋯,M in S recorded as follows:

$$V = \{X \in D | X = \textstyle\sum_{j-1}^{M} \alpha_j X_j\} \quad (3)$$

Where α_j satisfies the following conditions: $\sum_{j-1}^{M} \alpha_j = 1$,- 0.5 $\leqq \alpha_j \leqq$ 1.5.

The algorithm is described as follows.

Step 1: N initial individuals are randomly generated. P = {X1, X2,⋯,Xn} .Xi∈S, the current evolution algebra k = 1.

Step 2: The best individual Xbest in the population and the worst individual Xworst are found.

Step 3: If | f (Xbest) - f (Xworst) | ≤ε then go to step 8.

Step 4: M points X1, X2 ,···, XM are randomly selected from P form the subspaces V.
Step 5: A point X is chosen from V randomly.
Step 9: If f (X ') <f (Xworst), then Xworst = X'.
Step 7: k = k= 1, go to step 2.
Step 8: Output Xbest.

3. CMEA algorithm flow

The basic idea of the CMEA algorithm is that the initial population is divided into two subpopulations. The first N individuals are classified as the first population, Cauchy mutation [6] is used to generate N descendants and the last $\mu - N$ individuals are classified as the second population using multiple paternal crossover operators to produce $\mu - N$ offspring. The random competition method is then used to rank all 2μ individuals, taking the first μ individuals as the next generation population and then continuing to evolve recursively until the downtime conditions are met. The entire algorithm flow is as follows:

Step 1: Parameter initialization. Setting the number of individuals in the population to μ, if the first population contains N individuals, then the second population contains μ-N individuals. The biggest evolutionary algebra is MaxGeneration, the number of randomly competing individuals is q with initial η0, parent number of multi-parent crosses (That is Zhang Cheng subspace size) is M and evolution algebra is k.

Step 2: Population Initialization. An initial population (Xi , ηi), (i = 1, 2,···,μ) with k = 1 is generated in the feasible solution space of the problem.

Step 3: Evolution is stopped to judge. If k == MaxGeneration is satisfied, evolution aborted and the result is outputted, otherwise go to step 4.

Step 4: Breeding. In the first population N, each individual uses the formula (1) to generate a progeny. The second population uses equation (3) to generate μ- N offsprings. A total ofμ offspings were generated. In order to prevent loss of search ability when the step size is very small, the method in literature [5] is used: if ηi <10-4 , then ηi = 10-4.

Step 5: Population communication. The integration of the parent (Xi, ηi) and the child (Xi ', ηi") (i = 1, 2,···,μ) is sorted using the random q competition method. The first μ individuals are the next generation. In this way, the first population and the second population communicate through the flow of individuals.

Step 6: k = k+ 1. Go to step 3.

4. CMEA optimization simulation examples and result comparison

In order to make the experiment representative, this paper chooses 23 different types of standard test functions. It is necessary to choose a large number of standard test functions so that we can analyze the performance of CMEA algorithm for different functions. For example, for some optimization problems it has good performance while for others the effect is not good.

4.1 Standard test function description

In the 23 test functions, f1-f13 is a high-dimensional optimization problem, f1-f5 is a unimodal function, f6, is a single peak discontinuous step length function ,f7 is a function of noise; f8- f13 is a multi-peak function and the local extremum points grow geometrically as the dimension increases. These six functions are the most difficult optimization problem type. The function f14-f23 is a multimodal function with low-dimensional and local extrema.

4.2 Experimental Procedures and Results

If the number of individuals in the first population N = 80, the number of individuals is 20 in the second population. The value of parameter M for multi-parental crossover is slightly different for the optimization problem. Table 1 shows the specific parameter settings. For all the experiments in this paper, the remaining parameters are consistent with literature [2], that is q = 10, η0 = 3.μ = 100.

IOP Conf. Series: Journal of Physics: Conf. Series **1004** (2018) 012020 doi:10.1088/1742-6596/1004/1/012020

Table 1. The setting of parameter M

Function	f_1-f_4	f_5	f_6	f_7	f_8	f_9	f_{10}	f11	f_{12}-f_{13}	f_{14}-f_{23}
M	4	6	4	3	5	7	4	5	4	5

In order to reflect the stability of the algorithm, we run the algorithm 50 times independently for each instance and then calculate the average and standard deviation to compare with the FEP algorithm and the SPMEP algorithm[7]. As can be seen from Table 2, the CMEA algorithm is obviously superior to the other two algorithms for the unimodal function f1-f6 and the multi-peak less local exponential function f15-f23. In this case the multi-parent hybrid operator shows a superior performance. With the introduction of multiple patrons in a crossover operation it reduces the possibility that super individuals copy themselves into children, which means that brings a more diverse solution space search results so that the evolutionary algorithm converges to the global optimum. Fig. 1 and Fig. 2 show the evolution curves of the optimal individual function values in the population with evolutionary algebra when f3 and f21 are solved (Fig. 1 shows the ordinate in logarithmic coordinates). As can be seen from the figure, the CMEA algorithm converges significantly faster than the FEP algorithm. The CMEA algorithm is superior both in the early stage of evolution and in the late stage of evolution and CMEA algorithm has a higher rate of evolution, which is the result of combined action between Cauchy mutation operator and multi-parent hybrid operator.

Table 2. CMEA algorithm compared with FEP and SPMEP, "Mean best" indicates the average value found by the algorithm, and "Std dev" represents the standard variance

Function	Number of Generations	CMEA		SPME[7]		FEP[2]		CME	CMEA-F
		Mean best	Std dev	Mean best	Std dev	Mean best	Std dev	*t*-test	*t*-test
f_1	1500	9.5×10^{-9}	5.1×10	1.3×10^{-4}	1.0×1	5.7×1	1.3×10	-9.19	-31.00
f_2	2000	1.27×10	2.9×10	5.1×10^{-4}	1.5×1	8.1×1	7.7×10	-230.	-74.26
f_3	5000	8.0×10^{-4}	3.7×10	---	---	1.6×1	1.4×10	---	-7.67
f_4	5000	1.2×10^{-3}	7.8×10	6.5×10^{-3}	2.0×1	0.3	0.5	-17.4	-4.23
f_5	20000	2.399	2.75	---	---	5.06	5.87	---	-2.90
f_6	1500	0	0	0	0	0	0	0	0
f_7	3000	9.3×10^{-3}	4.2×10	1.0×10^{-2}	1.5×1	7.6×1	2.6×10	-1.11	2.43
f_8	9000	-12543.4	59.4	-12569.5	9.1×1	-1255	52.6	3.10	0.98
f_9	5000	0.15	0.4	2.9×10^{-7}	6.0×1	4.6×1	1.2×10	2.65	1.83
f_{10}	1500	1.0×10^{-4}	1.2×10	1.9×10^{-3}	4.4×1	1.8×1	2.1×10	-27.9	-60.17
f_{11}	2000	6.1×10^{-3}	7.9×10	5.6×10^{-3}	1.7×1	1.6×1	2.2×10	0.43	-2.99
f_{12}	1500	6.68×10	4.1×10	8.5×10^{-7}	9.7×1	9.2×1	3.6×10	-13.5	-17.82
f_{13}	1500	9.2×10^{-6}	2.4×10	1.4×10^{-5}	9.2×1	1.6×1	7.3×10	-1.32	-13.87
f_{14}	100	1.04	0.19	1.00	1.6×1	1.22	0.56	1.48	-2.15
f_{15}	4000	$3.0756\times$	3.65×1	4.5×10^{-4}	1.5×1	5.0×1	3.2×10	-6.71	-4.25
f_{16}	100	-1.03	0	-1.03	4.3×1	-1.03	4.9×10	0.0	0.0
f_{17}	100	0.398	0	0.398	5.7×1	0.398	1.5×10	0.0	0.0
f_{18}	100	3.00	0	3.00	5.6×1	3.02	0.11	0.0	-1.28
f_{19}	100	-3.86	0	-3.86	1.5×1	-3.86	1.4×10	0.0	0.0
f_{20}	200	-3.3	4.5×10	-3.25	5.4×1	-3.27	5.9×10	-5.03	-2.86
f_{21}	100	-9.0	2.6	-6.63	3.5	-5.52	1.59	-3.84	-8.07
f_{22}	100	-9.97	1.4	-7.4	3.00	-5.52	2.12	-5.49	-12.38

f_{23}	100	-10.19	1.38	-6.53	4.00	-6.57	3.14	-6.12	-7.46

Cauchy's mutation dominates in the multi-peak multi-local exponential function f8-f13, parent hybrid operator only plays fine-tuning role. For function f9, CMEA algorithm has poor showing. Figure 3 is an image of a function in two dimensions. As can be seen from the image, this is a complex function with multiple peaks, multiple local maxima and close-spaced peaks. This function grows geometrically with the increase of the number of local extremum points. At this time the Cauchy variation in CMEA algorithm dominates. The domain of function f9 is [- 5.12,5.12]. As shown in Figure 4, the Cauchy mutation has more steps in the small multimodal neighborhood with large back and forth in the peak and peak disturbance but don't have the ability to climb, so it is difficult to find the optimal solution.

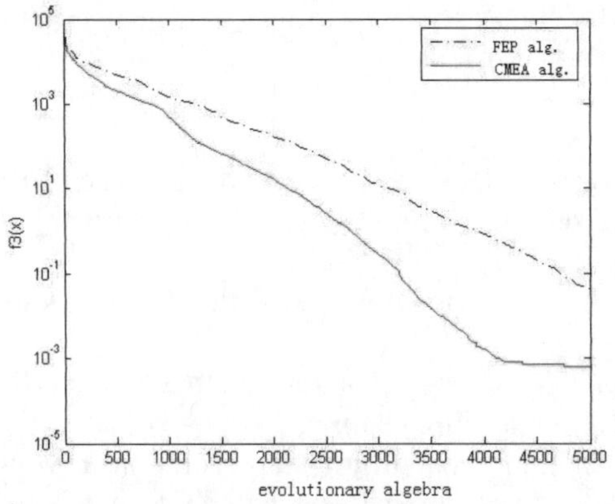

Figure 1. The image of a function f_9 in two dimensions

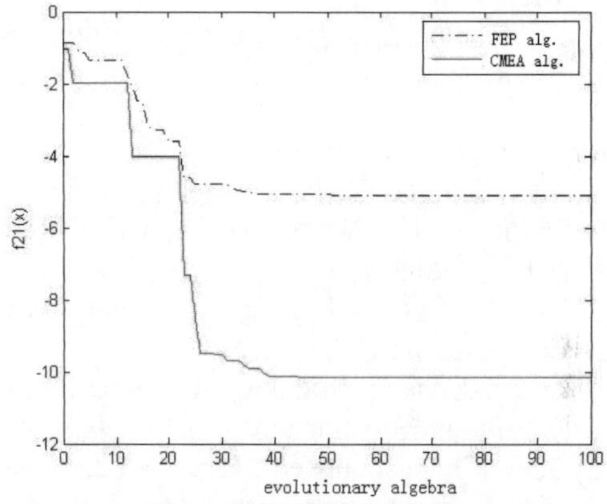

Figure 2. The optimal individual curve of function f_{21}

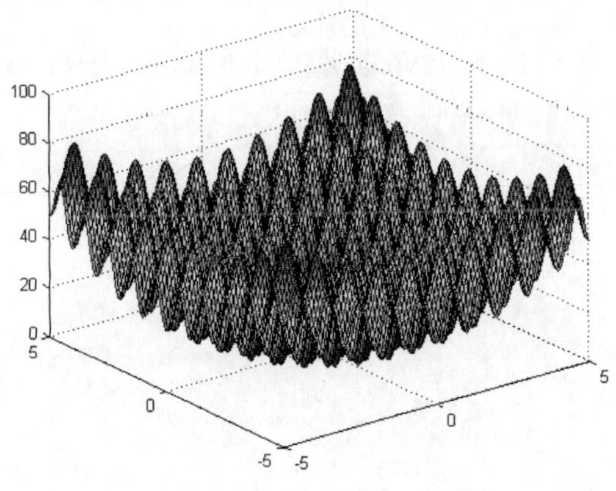

Figure 3. The optimal individual curve of function f_3

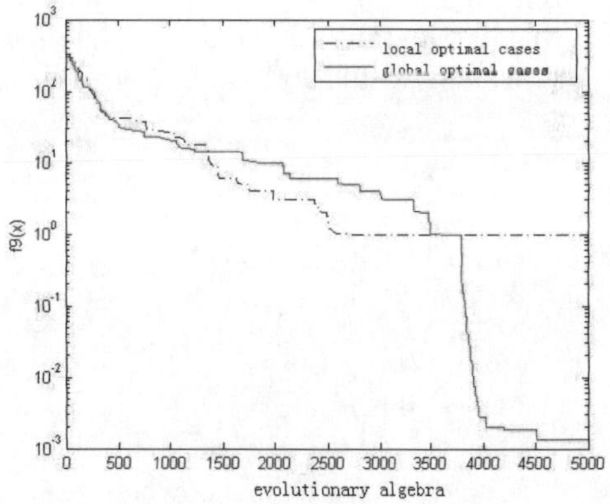

Figure 4. The optimal individual curve of CMEA for function f_9

5 Conclusion and the next step of research work

According to the respective advantages of fast evolution programming and Guo Tao algorithm, this paper proposes the hybrid evolutionary algorithm based on AC by combining Cauchy mutation operator and multi-parent hybrid operator. In this algorithm, evolution proceeds between two populations. Communication between the two populations is achieved by the reorganization of the populations. The algorithm flow is given and the performance of the algorithm is compared using a number of standard test functions. The experimental results show that the CMEA algorithm has good performance in global convergence and stability.

The CMEA algorithm exhibits excellent performance for both unimodal functions and multi-peak less local extremum functions. However there is not much improvement on the multi-peak local extremal function over the FEP algorithm. The next steps of work are as follows: (1) Replacing the Cauchy mutation with Lévy mutation and T mutation to improve the shortcoming that Cauchy mutation step is too large. (2) Improving the parameter M of multi-parent hybrid operator in CMEA algorithm so that has its own adapt to the adjustment function for different types of functions and improving the robustness of the parameters. (3) The CMEA algorithm is used in optimization problems of various linear and nonlinear constraints to make the algorithm more general.

Acknowledgements
This work was partly supported by the Educational Science Planning Project of Hubei Province (No 2017GB088).

References
[1] Guo-sheng HAO,Na GUO,Gai-ge Wang,Zhao-jun ZHANG,De-xuan ZOU. Scheme of Big-Data Supported Interactive Evolutionary Computation, Science and Engine erring Reaearch Center, 2017,(1),26-31.
[2] Yao X, Liu Y, Lin G.Evolutionary programming made faster[J].IEEE Trans Evolutionary Computation, 1999(3) : 82- 102.
[3] Li Xia, Zhang Jingmin, Li Ruihua, Liu Kunqi, A cultural algorithm based on differential evolution algorithm and Guo Tao algorithm,Computer Engineering, 2012, (13):178-181.
[4] Deng Changshou,Ren Hongwei,Peng Hu. An improved differential evolution algorithm for mixed integer nonlinear programming problems,Application Research of Computers, 2012, 2: 445-448.
[5] Yousef Alipouri,Javad Poshtan,Yagub Alipouri. A modification to classical evolutionary programming by shifting strategy parameters,Applied Ingelligence, 38(2) : 175-192, 2013.
[6] Quan Ji,Wang Xianjia, Evolutionary games in a generalized Moran process with arbitrary selection strength and mutation,Chinese Physics B, pages 25-30, 2011.
[7] M. Ji, H. Tang, J. Guo. A single-point mutation evolutionary programming. Information Processing Letts, 90(2): 293-299, 2004.

Ingenious Snake: An Adaptive Multi-Class Contours Extraction

Baolin Li[1,*] and Shoujun Zhou[1]

[1]Shenzhen Institutes of Advanced Technology, Chinese Academy of Sciences, Shenzhen, China

*E-mail: sj.zhou@siat.ac.cn

Abstract. Active contour model (ACM) plays an important role in computer vision and medical image application. The traditional ACMs were used to extract single-class of object contours. While, simultaneous extraction of multi-class of interesting contours (*i.e.*, various contours with closed- or open-ended) have not been solved so far. Therefore, a novel ACM model named "Ingenious Snake" is proposed to adaptively extract these interesting contours. In the first place, the ridge-points are extracted based on the local phase measurement of gradient vector flow field; the consequential ridgelines initialization are automated with high speed. Secondly, the contours' deformation and evolvement are implemented with the ingenious snake. In the experiments, the result from initialization, deformation and evolvement are compared with the existing methods. The quantitative evaluation of the structure extraction is satisfying with respect of effectiveness and accuracy.

1. Introduction

ACM, namely Snake for the parametric model, has been studied for a long time and is widely used in the closed contour extraction [1-4]. The vast majority of studies involved the construction and optimization of its external energy conditions [2-4]. For example, improving external dynamic range of the gradient vector flow (GVF) field [2, 4], constructing an external force condition based on the image area content [5], and initializing Snake contours [6,7,8]. Another few studies relate to open snake, namely, open curve contour model [9-11], which is early proposed by Wong et al. [9] and used by Li et al. [10] to enable the dynamic contour extracting the linear or linear objects. Different from closed snake, the internal forces at the ports of open snake are cancelled, and the pull forces at the ports are used as the external force condition at both ends. Considering that open snake can perform axial elongating and radial deforming, we use it to describe the interesting boundaries, or linear objects. Another, existing open snake is faced with the external force defect as well as the initialization problems (*e.g.*, a manual and time-consuming process), which severely limited in their application.

In summary, the existing ACMs only apply for single-class contour with finite level of initialization. The proposed ingenious snake applies to fast-adaptively extract multi-class contours, it can adaptively select the internal force and external force condition for the closed contour deformation and open curve evolution respectively, and finally approaches the interesting contours.

2. Method

2.1. Ingenious Snake

As a special application of ACM, ingenious snake model is still expressed as $\mathrm{r}(s) = (x(s), y(s)), s \in [0,1]$. The difference is that it can express both the open curve and the closed contour. The total energy is: $E_{snake} = E_{int} + E_{ext}$. The internal energy term of the model definition:

$$E_{int} = \int_0^1 (\alpha(s)|\mathrm{r}_s(s)|^2 + \beta(s)|\mathrm{r}_{ss}(s)|^2) ds \tag{1}$$

Traditionally, the α and β are set as constant for the closed Snake contour [1], while be set as zero at the both tips for the open Snake curves [11]. The external energy term of ingenious snake E_{ext} is composed of image term E_{img} and stretch force term E_{str}:

$$E_{ext} = \int_0^1 k_{img} \cdot E_{img}(\mathrm{r}(s)) + \zeta \cdot k_{str} \cdot E_{str}(\mathrm{r}(s)) ds \tag{2}$$

The appearance of the logical variable ζ, namely $\zeta = 0$ or 1, in Eq. (2) differentiates it from the traditional ACM. The model automatically identifies the two port types (namely, open contour and closed contour). For Eq. (2), the port stretch force is generated when $\zeta = 1$. The stretching forces are applied to both ends of the open curve, and point at the tangent direction of the front-points. The size of the stretching force is referred to in [10].

If $\zeta = 0$, Eq. (2) is reduced to describe the evolution of closed contours. The form is same to its traditional one, where the image energy conditions $k_{img} \cdot E_{img}(\mathrm{r}(s))$ act as the external energy conditions of the contour. The corresponding external force structure can utilize GVF field, that is, $-\nabla I_{GVF}$, as in [2].

When the automated initial curves are deforming and evolving using the above model Eq. (1) ~ (2), the ingenious snake activates an adaptive coefficient setting of α, β and the logical variable ζ in terms of the two port types. Let R_i be the i-th initial curve, $R_i = \{X, \zeta\}_i$, where X is a point sequence composed of N points $\mathrm{x}_n, n = 1, \dots, N$; and ζ represents the on/off logic variable of this curve. For the internal energy condition of the model, as shown in equation (1), we modulate the "elasticity" and "stiffness" coefficients α and β of the curve front-points by introducing the window functions $\omega(n), n = 1, \dots, N$ of length N:

$$\omega(n) = \begin{cases} 0, & n = 1,2, N-1, N \\ 1, & others \end{cases} \tag{3}$$

Thus, the α, β in the internal force term E_{int} of the model are expressed as follows:

$$\alpha(n) = \alpha_0 \cdot \omega(n) \cdot \zeta + \alpha_0 \cdot (1 - \zeta), \quad n = 1, \dots, N$$
$$\beta(n) = \beta_0 \cdot \omega(n) \cdot \zeta + \beta_0 \cdot (1 - \zeta), \quad n = 1, \dots, N \tag{4}$$

For the external energy condition of the model, as expressed in Eq. (2), its discrete external energy E_{ext} is denoted by:

$$E_{ext}(X) = k_{img} \cdot E_{img}(X) + \zeta \cdot (k_{str} \cdot E_{str}(X)) \tag{5}$$

Let $X_k = (\mathrm{x}_{0,k}, \dots, \mathrm{x}_{1,k}, \dots, \mathrm{x}_{N,k})^{\mathrm{T}}$, k=0,1 be the vector containing all the kth-dimension coordinates of the point sequence, X_k^t at iteration t can be computed iteratively after deriving the Euler-Lagrangian equation:

$$X_k^t = (A + \gamma I)^{-1} \cdot (\gamma X_k^{t-1} - \frac{k_{img} \partial E_{img}(X_k^{t-1})}{\partial \mathrm{x}_k} - \zeta \cdot \frac{k_{str}\, \partial E_{str}(X_k^{t-1})}{\partial \mathrm{x}_k}) \tag{6}$$

Where A is composed of coefficients α and β, γ is used to control the step size, and I is the unit matrix. The parameters $\zeta = 0$ and $\zeta = 1$ are respectively used to control the updating mode of the two kinds of curves. For the closed contour ($\zeta = 0$), it can approach to the object and make the Eq. (6) convergence with no more than three iteration-steps (n_{max}). For one open curve ($\zeta = 1$), the curve keep evolving along the linear structure and stops evolving until it meets some termination terms. For example, stretch force disappearing, or the front-points' intensities are less than a low threshold.

2.2. Preprocessing

Ingenious snake firstly need to enhance the feature (*e.g.*, ridge characteristics of boundaries or linear structures) of interesting contours. In the preprocessing, K-means classification and morphological operators are used to separate the region shape and a linear structure. Then, the ridges of region edge and linear structure are enhanced respectively. The enhancement of edge ridges is done using the Canny operator following with Gaussian kernel G_σ filter. While the ridges of the linear-objects are enhanced by using the multi-scale filter [12]. The resultant ridge feature map is used to generate the GVF fields.

2.3. Automated initialization

2.3.1. Ridge-points Extraction. The GVF field [2] is normalized and denoted by $F(x, y)$, which can be obtained by solving a Euler iteration diffusion equation. Firstly, the GVF field is acquired through computing the ridge feature map. Then, the ridge-points are obtained by the direction estimation of the GVF field. Given the two-dimensional vector $\mathrm{F}_x^{(i,j)}$ and $\mathrm{F}_y^{(i,j)}$ of a GVF at each point $p(i, j)$, it will be defined as a ridge-point if they satisfy one of the following conditions

$$\begin{cases} F_x^{(i-1,j)} F_x^{(i+1,j)} + F_y^{(i-1,j)} F_y^{(i+1,j)} \leq 0 & if\ F_x^{(i,j)} > 0 \\ F_x^{(i,j-1)} F_x^{(i,j+1)} + F_y^{(i,j-1)} F_y^{(i,j+1)} \leq 0 & if\ F_y^{(i,j)} > 0 \end{cases} \tag{7}$$

Restricting $F_x^{(i,j)}$ and $F_y^{(i,j)}$ to be greater than 0 can reduce image-marginal noise point. For example, in figure 1, the 2D coronary angiography X-ray image is shown in figure 1(a), figure 1(b) shows the GVF field overlapping with the ridge feature image, wherein the local amplification shows that GVF points to the vessel central gradient vectors. The result of ridge-points extraction via Eq. (7) is shown in figure 1(c).

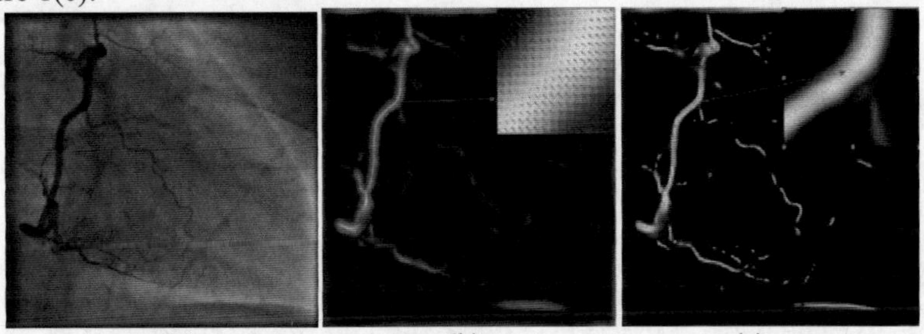

(a) (b) (c)

Figure 1. An example of ridge-point extraction: (a) Original coronary X ray radiography image; (b) multi-scale enhancement image GVF field; (c) the results of ridge-point extraction.

2.3.2. Generation of ridgelines. We provide a fast algorithm to automate a tracking process.

- **Starting a tracking process using seed point.** See figure 2, a seed point p_0 is randomly generated from the ridge-point set P, when the neighbouring two points p_{-1} and p_1 are detected with the angle of the two direction vectors being no less than a big obtuse angle (e.g., $2\pi/3$). Otherwise, the seed point is removed from the ridge-point set.
- **Generation of ridgelines based on tracking.** Given seed point p_0, the tracking process continues to search the candidates along the path of opposite directions. Suppose that v_t and v_{-t} represent two reverse tracking directions, p_t and p_{-t} represent two front-points of the tracking. The tracking process is to sequentially update $p_{\pm t \pm 1}$, $v_{\pm t \pm 1}$, so as to form a ridgeline in X. In t-th step of tracking, the ridge-points on the corresponding path of the ridgeline obtained in step $t - 1$ should be deleted, thereby updating the ridge-points set P.

IOP Conf. Series: Journal of Physics: Conf. Series **1004** (2018) 012021 doi:10.1088/1742-6596/1004/1/012021

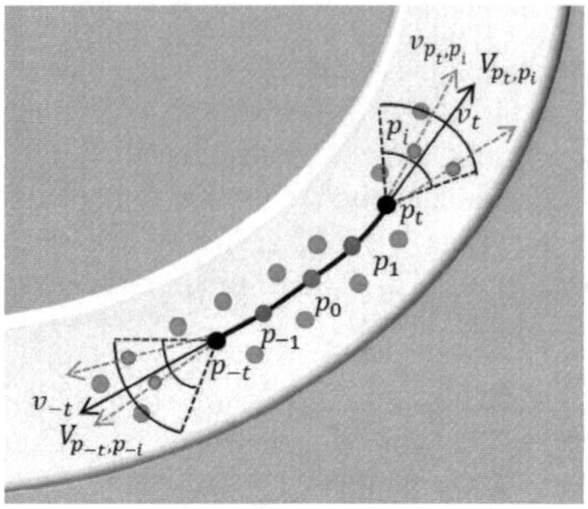

Figure 2. Generation a ridgeline using ridge-point tracking. All the points in the graph are ridge-points, in which the red point is a seed point to start the tracking process, the blue point is the initial detection point, the black point is the front-point of tracking path, and the candidates locate in the fan shaped region.

- **Termination criteria.** A tracking process should be stop with the suitable termination criteria. We devise the terms considering the characteristic of both open curve and closed contour and detail them as follows: (1) No candidates can be used to generate the front-points p_t and p_{-t}. (2) After a length of path tracking, the distance between the left and right front-points is less than a low-limit ε_0 (usually $\varepsilon_0 = r_{min}$).(3) When the whole ridge space P is empty, there is not seed points to start a new tracking, thus the tracking process is finished.

The resulting ridgelines are used as the initial contours for Ingenious snake in the aforementioned Sections 2.1. Each ridgeline $R_i = \{X, \zeta\}_i$ contains the initial coordinate sequence $X = [\mathrm{x}_1, \mathrm{x}_2, \mathrm{x}_3, \dots]$ and ridgeline's port type parameter ζ. When the termination term meets the second criteria, a closed contour is determined and labelled with ζ=0; Otherwise, if the termination term meets the first criteria, the curve will be determined as an open curve labelled with ζ=1.

3. Experiments

Existing contour extraction methods involve the center of divergence (CoD) method [6], force field segmentation (FFS) method [7], poisson inverse gradient (PIG) method [8] (website: http://cn.mathworks.com/matlabcentral/fileexchange/22871-parametric-active-model-toolbox) and stretching open active contours (SOACs) [11] (website: http://www.cse.lehigh.edu/~idea-lab/soax/). To extract multi-class contours, the proposed method acquired obvious verification on their visual comparisons. Parameters are manually assigned with $\alpha = 0.1$, $\beta = 0.5$, $\gamma = 2$, $k_{img} = 2$, $k_{str} = 3$, $r_{min} = 2$ and $n_{max} = 3$. In the evolvement, curves are resampled to maintain point spacing strictly with single pixel. The workstation configuration is as follows, Intel (R) Core (TM) i7-6700 CPU @ 3.41GHz, RAM 32G, operating system Win10, and programming language MATLAB r2014a.

3.1. Experimental of Phantom

We constructed the Phantom image to quantify the precision of Ingenious snake. As shown in figure 3, figure 3 (a) is the Phantom with Gaussian mixing noise, in which the linear object with label (1)-(6) is manually generated as the central line gold standard. The initialization ridgelines are shown in figure 3 (b). figure 3(c) shows the result of multi-class contours extraction, which could describe the structure pattern. The segmentation accuracy of the boundary obtained by Dice coefficient is 0.9882. To evaluate the precision of the extracted central lines, we refer to the document [11] and use the Vertex error and Hausdorff distance to evaluate. The results are shown in Table 1, the Vertex error of the 6 lines is on average 0.593 pixels, and the Hausdorff distance is on average 3.011 pixels. The results show that the method can extract multi-class contours with good precision.

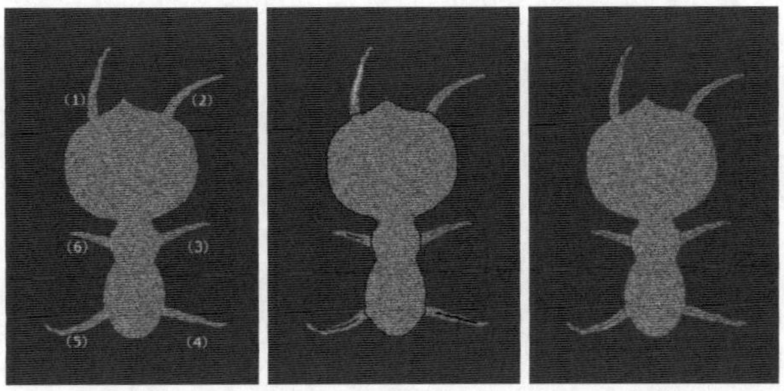

(a) (b) (c)

Figure 3. Verification of multi-class contours extraction from phantom: (a) Original image, (b) initialization, and (c) extraction results by ingenious Snake. In the subfigure (c), the red lines denote the final result.

Table 1. Evaluation of central lines extraction of the Phantom

line No.	Vertex error (pixel)	Hausdorff distance (pixel)
(1)	0.584	4.120
(2)	0.611	3.609
(3)	0.568	3.104
(4)	0.489	2.690
(5)	0.716	3.254
(6)	0.591	1.290
mean	0.593	3.011

3.2. Experimental of medical image

We use different methods to extracted multi-class contours in figure 4. In the MR image, the organ's boundary and vessels' central lines are both the interesting contours. The CoD method initializes 42 active contours, most of which are inside the vessels. And its convergent result could not express the vessels. The FFS method initializes outside the objects, and converges at the vessels' boundary or the organ's contour, thus it couldn't distinguish between the vessels and organ. The PIG method only initializes within the organ, while lack completeness for multi-class contours. The SOACs method generates 212 initial ridgelines and eventually converged to 42 contours, as a result, it obtains discontinuous and inaccurate organ boundaries and vessels. The proposed method obtains the multi-class contours at the same time.

(a) Initialization

(b) Convergence

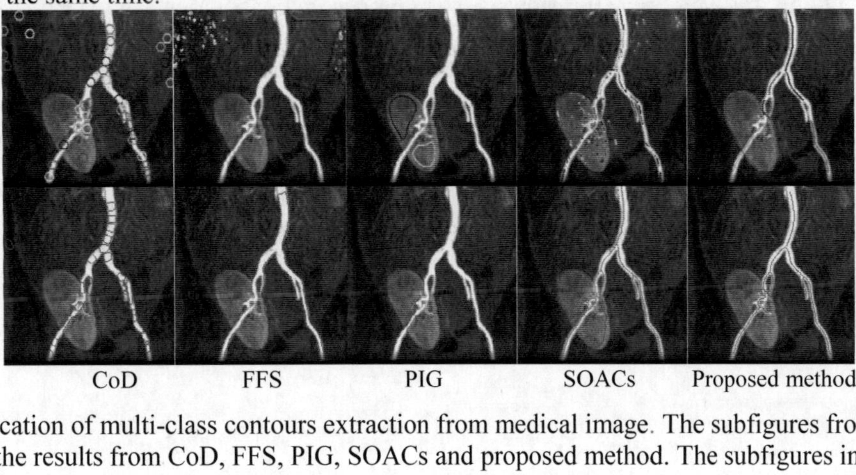

CoD FFS PIG SOACs Proposed method

Figure 4. Verification of multi-class contours extraction from medical image. The subfigures from left to right show the results from CoD, FFS, PIG, SOACs and proposed method. The subfigures in the row (a) and (b) are the results of the initialization and the corresponding convergent results.

4. Conclusion

Considering that the traditional ACM model cannot solve simultaneously open contour and closed contour, we proposed the ingenious snake to effectively initialize the closed/open contour, and it has a good performance in the multi-class contours extraction. For the follow-up works, we should develop the method with better evolution scheme to describe complex scene, in companied with deep learning.

Acknowledgments

The study has been financially supported by the following grants: National High Technology Research and Development Program of China (863 Program) (Nos. 2015AA043203), National Science Foundation Program of China (Nos. 61471349), Basic Discipline Layout Project of Shenzhen City (Nos. JCYJ20150731154850923), Technology innovation program of Shenzhen City (Nos. JSGG20160229203812944), and Shenzhen Engineering Laboratory for Key Technologies on Intervention Diagnosis and Treatment Integration.

References

[1] Kass M, Witkin A and Terzopoulos D 1988 Snakes: active contour models *Int. J. Comput. Vision* **1**(4) 321–31.

[2] Xu C and Prince J L 1998 Snakes, shapes, and gradient vector flow *IEEE T. Image Process.* **7**(3) 359–69.

[3] Li B and Acton S T 2007 Active contour external force using vector field convolution for image segmentation *IEEE T. Image Process.* **16**(8) 2096-106.

[4] Zhu G, Zhang S, Zeng Q and Wang C 2010 Gradient vector flow active contours with prior directional information *Pattern Recogn. Lett.* **31**(9) 845–56.

[5] Shih F Y and Zhang K 2006 Locating object contours in complex background using improved snakes *Comput. Vis. Image Und.* **102** (2) 93–8.

[6] Ge X and Tian J 2002 An automatic active contour model for multiple objects *Proc. Int. Conf. Pattern Recognition* pp 881-4.

[7] Li C, Liu J and Fox MD 2005 Segmentation of external force field for automatic initialization and splitting of snakes *Pattern Recogn.* **38** 1947–60.

[8] Li B and Acton S T 2008 Automatic Active Model Initialization via Poisson Inverse Gradient *IEEE T. Image Process.* **17**(8).

[9] Wong Y Y, Yuen P C and Tong C S 1998 Segmented snake for contour detection *Pattern Recogn.* **31**(11) 1669–79.

[10] Li H, Tian S, Smith M B and Fujiwara I 2009 Automated actin filament segmentation, tracking and tip elongation measurements based on open active contour models *Proc. IEEE Int. Symp. on Biomedical Imaging* p 1302.

[11] Xu T, Vavylonis D and Huang X 2014 3D actin network centerline extraction with multiple active contours *Med. Image Anal.* **18** 272–84.

[12] Frangi A F, Niessen W J, Vincken K L and Viergever M A 1998 Multiscale vessel enhancement filtering *Proc. Med. Image Comput. Assist. Interv.* **1496** pp 130–7.

IOP Conf. Series: Journal of Physics: Conf. Series **1004** (2018) 012022 doi:10.1088/1742-6596/1004/1/012022

Chinese License Plates Recognition Method Based on A Robust and Efficient Feature Extraction and BPNN Algorithm

Ming Zhang[1,2]**, Fei Xie**[2,4*]**, Jing Zhao**[3,4]**, Rui Sun**[5,6]**, Lei Zhang**[2] **and Yue Zhang**[2,4]

[1]Department of Electronic Engineering, City University of Hong Kong, Hong Kong, 999077, China
[2]School of Electrical and Automation Engineering, Nanjing Normal University, Nanjing 210042, China
[3]School of Automation, Nanjing University of Posts and Telecommunications, Nanjing 210023, China
[4]Jiangsu Key Laboratory of 3D Printing Equipment and Manufacturing, Nanjing 210042, China
[5]College of Civil Aviation, Nanjing University of Aeronautics & Astronautics, Nanjing 211106, China
[6]Centre for Transport Studies, Imperial College London, London SW7 2AZ, UK

* xiefei@njnu.edu.cn

Abstract. The prosperity of license plate recognition technology has made great contribution to the development of Intelligent Transport System (ITS). In this paper, a robust and efficient license plate recognition method is proposed which is based on a combined feature extraction model and BPNN (Back Propagation Neural Network) algorithm. Firstly, the candidate region of the license plate detection and segmentation method is developed. Secondly, a new feature extraction model is designed considering three sets of features combination. Thirdly, the license plates classification and recognition method using the combined feature model and BPNN algorithm is presented. Finally, the experimental results indicate that the license plate segmentation and recognition both can be achieved effectively by the proposed algorithm. Compared with three traditional methods, the recognition accuracy of the proposed method has increased to 95.7% and the consuming time has decreased to 51.4ms.

1. Introduction

Vehicle plate license recognition (VPLR) plays a significant role in the fields of traffic management, vehicle monitoring and suspect vehicle tracking these days. For example, in some cities in China, a new VPLR technology which enables drives to pay parking fee using electronic wallet in a short time automatically without leaving cars has received widely favourable reception. Due to the increasing demand of human society, it's of great significance to enhance the result of license plate location and recognition in terms of improving vehicle management and alleviating traffic congestion. However, because the license plates can be easily affected by external factors such as lightning conditions, weather and backgrounds, locating and detecting the license plate from original images accurately and efficiently is still the main obstruction for successful license plate recognition.

In the past several years, many researchers have developed various methods to extract features from specific characters [1-4]. In [5], the researchers find that a vertical traverse density (VTD) vector and horizontal traverse density (HTD) vector can be used to describe each character object. However, due to the similar structures in different characters, the presented algorithm has difficulty distinguishing between Z and E, T and L, and some other groups.

In this paper, a robust license plate detection and character recognition method based on a novel combined feature extraction model and BPNN algorithm is presented. The combined feature extraction method uses VTD, HTD features and edge distance features as training samples of the VPLR classifier. It means that the method which contains more useful information for network training has a good recognition accuracy.

2. License plate location and segmentation of characters before recognition

2.1. Pre-processing of original image and license plate detection

Considering practical conditions, there exists much interference in original images such as lighting condition and image quality, which cause damaging influence to the recognition performance. So a pre-processing of the original image is important to locate the license plate quickly and accurately. Figure 1 shows the original vehicle image and the gray-scale image respectively.

(a) (b)

Figure 1. Original image and the gray-scale image of the vehicle

The whole steps of the original image pre-processing are described as follows: Firstly, a grey level stretch processing is applied to enhance the contrast between the license plate area and the other parts of image. Then, detecting edge using Roberts operator to highlight the difference between the license plate frontier and the background. After this, a set of candidate regions are extracted after image erosion and morphological closed operation, excluding the small parts which are certainly not parts of the license plate region. According to empirical experience [6], the true region of the license plate parts is verified while the other candidate regions are all removed from the image.

2.2. Accurate positioning of the license plate image and character segmentation

The horizontal integral projection which is used to locate the license plate accurately in vertical direction can be depicted by the following equations:

$$r(i,j)=\left|f(i,j)-f(i,j-1)\right|,\ i=2,3,...,m,\ \ j=1,2,3,4,...,n \tag{1}$$

Where, $r(i,j)$ is the pixel value of the image r, $f(i,j)$ is the pixel value of the image f, m and n are the height and the width of the image f respectively. Then the projection value of the row i named $T_1(i)$ can be obtained by accumulating the pixel value of image r per row, which can be depicted as

$$T_1(i)=\sum_{j=2}^{n} r(i,j) \tag{2}$$

Similarly, applying the vertical integral projection to locate the license plate accurately in horizontal direction. The difference in coarse image, vertical direction processed image and accurately poisoning image of the license plate is shown in figure 2 respectively.

IOP Conf. Series: Journal of Physics: Conf. Series **1004** (2018) 012022 doi:10.1088/1742-6596/1004/1/012022

Figure 2. Coarse image, vertical direction processed image and accurately poisoning image

The license plate image consists of seven characters, a dot and the space between them. To get the images only contains each character, it's natural to seek for the starting and ending points of each character. Thus, an algorithm based on calculating the times of white and black transformations in each column is presented. The histogram of transformation projection is shown in figure 3.

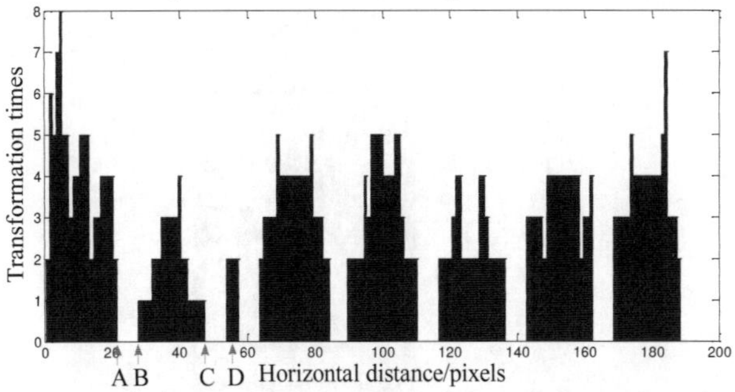

Figure 3. Histogram of transformation projection

Figure 4. Character segmentation results

Thus, the first character can be divided ranging from column 1 to column A from the image. The second character begins with point B and ends with point C. Due to the fact that the point D represents for the dot which need to be removed, the remaining image should start at the ending point of the dot. In this way, the rest five characters can be segmented one after another. The final segmentation result is shown in the figure 4.

3. Character recognition based on a novel combined feature extraction

3.1. Features extraction algorithm based on VTD and HTD methods

The traditional traverse dense features consist of HTD and VTD features. Choosing a character object "N" to explain the extraction algorithm clearly. As is shown in figure 5, the character N is selected after the binary operation. Scanning the column and row one by one of the image pixels and recording the number of changes in black and white pixels. The algorithm can be described as follows:

$$\textit{feature}\ \text{value}=\begin{cases}1, & \text{while the first pixel of the row or column is white}\\ \textit{feture value}+1, & \text{while } r(\text{i,j})\text{=0\&\&}r(\text{i,j+1})\text{=1in row}\|\ r(\text{i,j})\text{=0\&\&}r(\text{i+1,j})\text{=1 in column}\end{cases} \tag{3}$$

Where, the initial feature value is 0 and $r(\text{i,j})$ is the pixel value of the image. In the column scan of the character "N", the most number of alternating black and white pixels is one. These characteristic values are remarked as small black squares shown in figure 6, the number of squares means the feature value of the column. A series of feature values are obtained from left to right, recorded as VTD features. In the same way, scanning the object from top to bottom, the image has altered form black to white twice in side parts and three times in the intermediate part, describing the HTD features clearly.

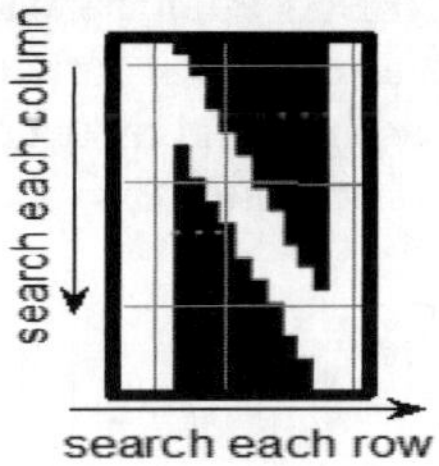

Figure 5. The processing of character N

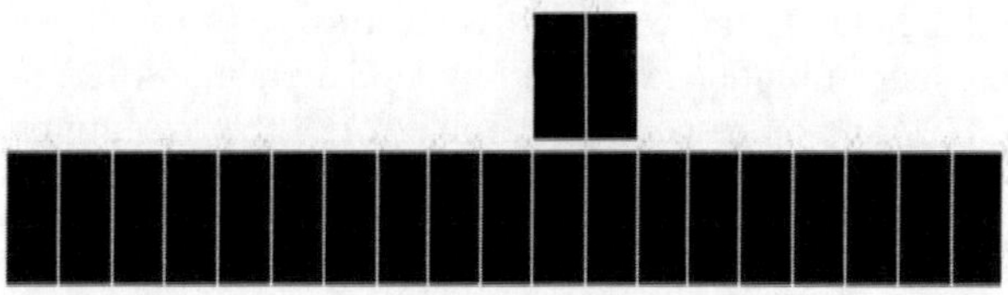

Figure 6. The VTD features of character "N"
VTD (N) =1111111112211111111

The dimension of the VTD and HTD features are the width and height of the processed character image. According to the experimental results, the final optimal license plate size is 35×20, that is to say, the dimension of the two feature vectors are 35 and 20 respectively.

3.2. Improved features extraction algorithm using edge distance features

The algorithm mentioned above uses VTD and HTD features as two sets of characteristic vectors. However, it may cause incorrect results in recognition when handling with some words which have similar traverse density structure like the letter "E" and "Z", the letter "T" and "L". As shown in figure 7, the character "E" and "Z" have the almost identical VTD and HTD feature vectors which is indistinguishable.

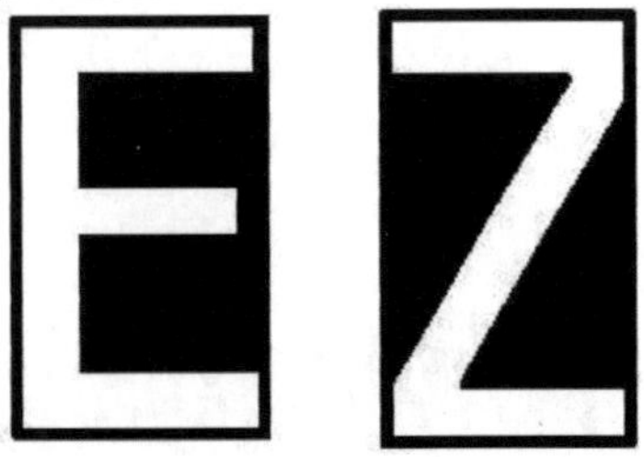

Figure 7. The character "E" and "Z"

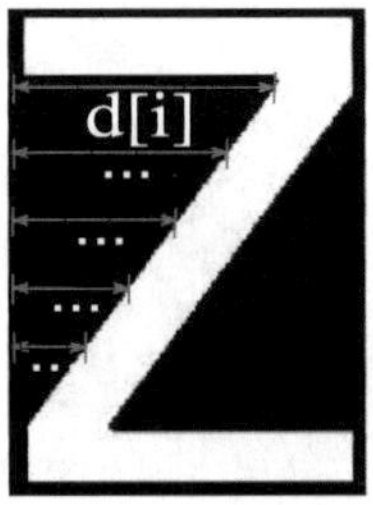

Figure 8. Extraction processing of edge distance feature

In order to solve this problem, another set of features based on edge distance is presented to compensate for the deficiency in traditional feature extraction algorithm. The proposed method of extracting the edge distance features can be described as followings:

From the up to the bottom, searching the each pixel of row from the left to the right of the character image. Then, finding the first pixel of the row i whose value is 1 (white), recording the index of the pixel, setting as $d[i]$, where, $i = 1, 2, ..., n$, n is the height of the character image. Finally, we obtain the array $d[i]$, the elements in which consist of the feature vector. As shown in figure 8, the edge distance feature values of the character "Z" are changing from the row 1 to row n, while the most edge distance feature values of the character "E" are identical. The feature values of these two characters is obvious different when adding this set of features.

Thus, three different sets of features have been defined in the improved features extraction algorithm: the vertical traverse density, the horizontal traverse density and the edge distance features.

3.3. Characters classification and recognition using Back-propagation Neural Network

In order to recognize the numbers 0-9 and the English letters A-Z of the license plate, choosing BPNN as a training model to utilize the specific learning process of the neural network [7-8].

Since the dimension of the HTD features, VTD features and the edge distance features of a specific character are 35, 20 and 35 respectively, connecting three sets of features into one feature vector, so the number of input layer nodes is 90. There are 26 English letters, 10 digital numbers, 36 alphanumeric characters waited to be recognized in the system, so the output layers has 26 nodes, 10 nodes and 36 nodes corresponding to the English letters classifiers, digital numbers classifiers and mixed classifiers respectively. According to the experimental results and empirical formula, the node number of hidden layers in English letters classifier, digital numbers classifier and mixed classifier are 48, 27 and 64 respectively. The error of the training processing of letters and numbers mixed classifier is plotted in the figure 9.

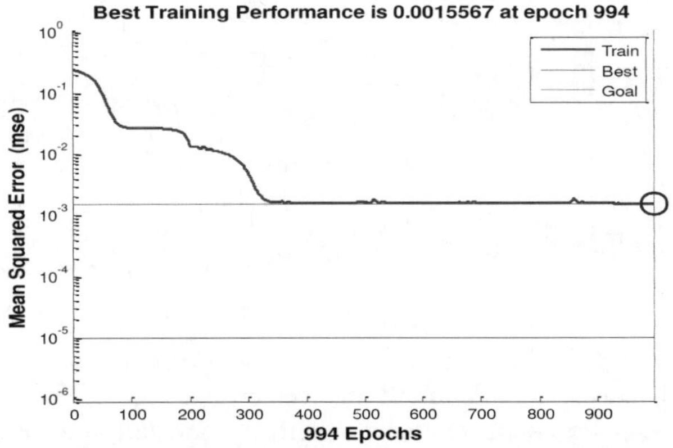

Figure 9. Error curve of the neural network training performance

4. Experiment and analysis

The experimental results of image pre-processing and license plate detection are shown on figure 10.

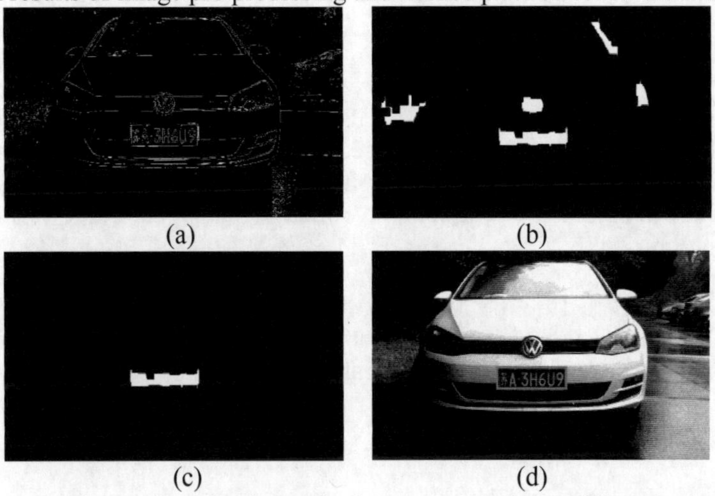

Figure 10. The results of each step in the pre-processing and license plate detection

As seen from the above figure, the edge is detected by Roberts operator in figure 10(a), then after erosion operation and morphological closed operation, there are five candidate regions of license plate waiting to be checked shown in figure 10(b). Then the license plate region is determined which is marked white colour shown in figure 10(c) while the other candidate regions are all removed. Finally, the license plate region is successfully located in the original grey-scale image shown in figure 10(d). The figure 11 shows the entire recognition processing of the original image applying the VPLR system.

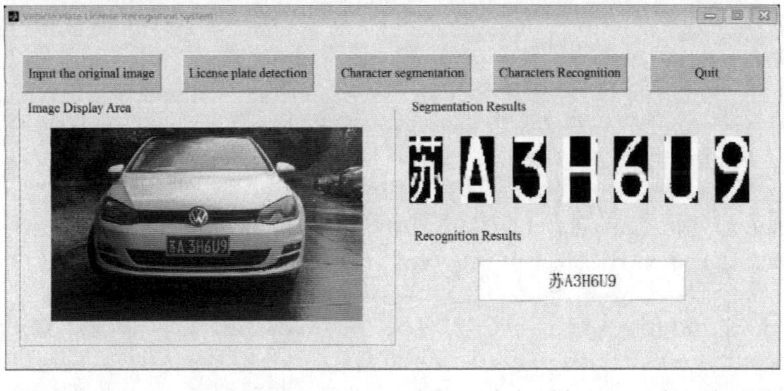

Figure 11. The entire recognition processing of the original image using the VPLR system

Table 1. Identification function of each classifier

Classifier	Parameter	Value
Letters	Recognition Rate (%)	94.8
	Recognition Times (ms)	55.2
Numbers	Recognition Rate (%)	96.3
	Recognition Times (ms)	46.3
Mixed	Recognition Rate (%)	95.7
	Recognition Times (ms)	51.4

Table 2. Comparison of the proposed method and conventional methods of two factors

Method	Precision (%)	Computational time (ms)
Proposed method	95.7	51.4
Line processing[1]	94.3	70.4
Peak position[2]	95.2	66.7
Pixel density[3]	94.9	59.9

Finally, 35 license plates collected in different backgrounds and illumination condition are used. As a result, 28 license plates are completely correct identified and the license plate recognition accuracy is 80 percent. There are total 201 of 210 valid characters are correctly recognized and the character recognition accuracy is 95.7 percent. The recognition rate of each classifier and the consuming time of single character are shown in table 1. Compared with the traditional methods, the proposed algorithm occupies a dominant position in the precision and the computation time. Experimental results are summarized in table 2.

5. Conclusion

A novel combined feature extraction model with BPNN algorithm is proposed to realize license plate recognition in this paper. The advantages of proposed method are higher accuracy and can be applied efficiently under complicated backgrounds. In the future, the algorithm would be further studied to implement on hardware in order to fit more complex conditions and reduce time-cost.

Acknowledgments

This work is partially supported by the National Natural Science Foundation of China (Grant No. 61601228, 61403210), and Natural Science Foundation of Jiangsu (BK20161021, BK20170780), the Natural Science Foundation of Jiangsu Higher Education Institution (17KJB510031) and sponsored by NUPTSF (NY214173) and the Open Program of Jiangsu Key Laboratory of 3D Printing Equipment and Manufacturing, Item number 3DL201607.

References

[1] Obeid H, Zantout R 2007 Line processing: An approach to ALPR character recognition *IEEE Int. Conf. on Computer Systems and Applications,* Amman, Jordan.

[2] Almustafa K, Zantout R and Obeid H 2011 Peak position recognizing characters in Saudi license plates *GCC Conf. and Exhibition* pp 186-189.

[3] Khaled M, Rached Z and Hasan O 2010 Pixel density: Recognizing characters in Saudi license plates *10th Int. Conf. on Intelligent Systems Design and Applications (ISDA)* pp 308-313.

[4] Alshami S, Elzaart A and Zantout R, et al 2015 A new feature extraction method for license plate recognition *5th Int. Conf. on Digital Information and Communication Technology and its Applications* pp 64-69.

[5] Tan, Chew Lim, et al 2002 Imaged document text retrieval without OCR *IEEE Transactions on Pattern Analysis and Machine Intelligence* **24** 838-844.

[6] Bai H, Zhu J and Liu C 2003 A fast license plate extraction method on complex background *Proc. of Intelligent Transportation Systems* **2** 985-987.

[7] Qu Z, Chang Q and Chen C, et al 2014 An Improved Character Recognition Algorithm for License Plate Based on BP Neural Network *Open Electrical & Electronic Engineering Journal* **8** 202-207.

[8] Dong J, Sun M and Liang G, et al 2015 The Improved Neural Network Algorithm of License Plate Recognition *Int. Journal of Signal Processing, Image Processing and Pattern Recognition* **8** pp 49-54.

IOP Conf. Series: Journal of Physics: Conf. Series **1004** (2018) 012023 doi:10.1088/1742-6596/1004/1/012023

Research on Image Encryption Based on DNA Sequence and Chaos Theory

Tian Tian Zhang[1,2,*], Shan Jun Yan[2,3], Cheng Yan Gu[1], Ran Ren[1] and Kai Xin Liao[4]

[1]City University of Hong Kong, Hong Kong, China
[2]School of Information and Electrical Engineering, Xuzhou University of Technology, Xuzhou Jiangsu, China
[3] China University of Mining and Technology, Xuzhou Jiangsu, China
[4] Nanjing Institute of Technology, Nanjing Jiangsu, China

*e-mail: ttzhang9-c@my.cityu.edu.hk

Abstract. Nowadays encryption is a common technique to protect image data from unauthorized access. In recent years, many scientists have proposed various encryption algorithms based on DNA sequence to provide a new idea for the design of image encryption algorithm. Therefore, a new method of image encryption based on DNA computing technology is proposed in this paper, whose original image is encrypted by DNA coding and 1-D logistic chaotic mapping. First, the algorithm uses two modules as the encryption key. The first module uses the real DNA sequence, and the second module is made by one-dimensional logistic chaos mapping. Secondly, the algorithm uses DNA complementary rules to encode original image, and uses the key and DNA computing technology to compute each pixel value of the original image, so as to realize the encryption of the whole image. Simulation results show that the algorithm has good encryption effect and security.

1. Introduction

In recent years, contributed by the rapid development of information technology and data processing, the problem concerning image secrecy has become more and more important. Encryption is an efficient way to keep image data free from attackers. In these encryption techniques, the chaotic image encryption method is more and more concerned because of the unpredictability of chaotic signals, the sensitivity to control parameters, the presence of the initial value, etc[1]. Many scholars are engaged in the research on the application of the combination of chaos theory and cryptography in encrypting image and improving the encryption level of cipher. Chaos theory has been widely applied to many subjects and become an important frontier science. However, the security of some existed algorithms has to be improved. At present, many proposed encryption methods simply utilize chaotic encryption, which have some drawbacks[2]-[6].

Today, researchers have proposed a number of encryption schemes based on DNA. DNA encryption is a combination of DNA technology and cryptography, thereby generating a new cryptographic technology and providing secure and efficient encryption service. The interesting feature lies in DNA structure is the complementary rules proposed by Watson and Crick. This principle has been widely used in different domains[7]-[9].

2. The basic concept and the results of the study

2.1. Chaos theory
Chaotic system is a kind of peculiar motion form, which has a very good randomness. We often use this property to generate random sequences, thus disrupting the original image and getting satisfactory results. In this paper, we will choose one-dimensional logistic mapping as the chaotic system.
One-dimensional logistic mapping is a typical chaotic mapping, which is demonstrated below:

$$x_{n+1} = \mu x_n (1 - x_n), \mu \in [0,4], x_n \in (0,1), n = 0,1,2\cdots. \quad (1)$$

The results of study show that the system reaches chaos under the following criteria:

$$3.569945 < \mu \le 4 \quad (2)$$

2.2. DNA
DNA is like the human body's password, its code word is A, T, C, G appearing with pairs. In image encryption, we can apply this rule to the encoding. Figure 1 shows the coding depending on DNA. Figure 2,3 propose DNA addition and subtraction rules.

0	1	2	3	4	5	6	7
A–00	A–00	C–00	C–00	G–00	G–00	T–00	T–00
C–01	G–01	A–01	T–01	A–01	T–01	C–01	G–01
G–10	C–10	T–10	A–10	T–10	A–10	G–10	C–10
T–11	T–11	G–11	G–11	C–11	C–11	A–11	A–11

Figure 1. Eight encoding rules for DNA sequences

+	A	C	G	T
A	T	A	C	G
C	A	C	G	T
G	C	G	T	A
T	G	T	A	C

Figure 2. DNA addition rule

-	A	C	G	T
A	C	G	T	A
C	A	C	G	T
G	T	A	C	G
T	G	T	A	C

Figure 3. DNA subtraction rule

3. The proposed algorithm program

3.1. Flow chart
This flow chart of this algorithm is shown as figure 4, which uses DNA encoding and DNA sequence:

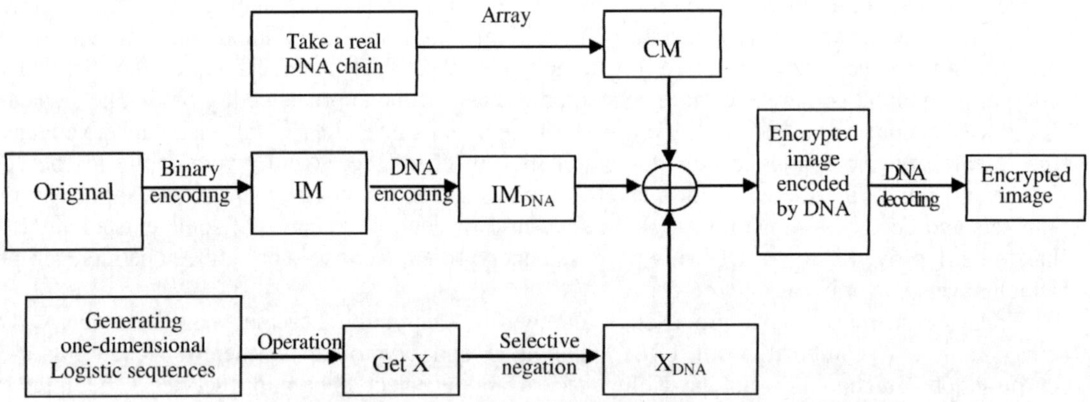

Figure 4. Flow chart of algorithm

3.2. Key generation
Input: Original image, DNA, value of Logistic mapping
Output: Encrypted image

3.2.1. Pretreatment

Step1: Choose a real DNA sequence.

Step2: Arrange the sequence into a matrix(512*4*512), which is named CM.

Step3: Using a programmer to produce a sequence of a one-dimensional logistic chaotic mapping, which is given initial value:

$$\mu = 4, X_0 = 0.53.$$

Step4: Choose the eighth bit of each number, named as X. If X is even, change the value into 0, or into 1.

Step5: Using the former DNA method to encode X, then we get X_{DNA}.

3.2.2. Encryption

Step6: Convert the image into a matrix, whose grey value is between 0-255, named as IM.(two-dimensional)

Step7: Using the DNA encoding method again to encode the matrix IM. The result is named as IM_{DNA}.

Step8: Adding the previous three matrices, which is from step2, step5 and step7, the result is called matrix P.

Step9: Applying the selective transformation into P. Define the formula of supplemental transformation: (it is the key step in the algorithm)

$$\text{Complement}_z(Y) = \begin{cases} Y, \text{if } X_i = 0, \\ \text{Complement}(Y), \text{if } X_i = 1, \end{cases} \tag{3}$$

End: Obtain the encrypted image, which is shown in figure 5.

(a) (b)

Figure 5. The result using the MATLAB software (a) original image (b) encrypted image

This algorithm can also reverse the above operations and obtain the original image.

4. Results analysis

We know that if the image encryption algorithm is security, it should be very exquisite and reasonable. In this part, the algorithm is allowed to face many attacks to test its security.

4.1. Resist brute force attack

If we want to resist brute force attacks, enlarging the key space should be an effective method. This requires designer to plan long key space or the keys will be destroyed for an exhaustive search within a limited time. In the encryption scheme, $\{x_0, \mu\}$ is the key, $\mu \in (3.56995, 4)$. 64-bit double precision 64 is 2^{64}, so the key space is about 2^{64}. In the DNA sequence portion, a 2559 DNA sequence was taken, the key space is 4^{2559}, The two key spaces added equal to the key space of this algorithm, much larger

IOP Conf. Series: Journal of Physics: Conf. Series **1004** (2018) 012023 doi:10.1088/1742-6596/1004/1/012023

than the AES's practical symmetric encrypted maximum key space 2^{256}. After testing, it is big enough to stand up to brute force attacks.

4.2. Anti-statistical attack

4.2.1. Information entropy. We often use information entropy to evaluating the gray pixel values distribution in the image. If the image is more chaotic, it's information entropy will be higher. Therefore, information entropy can be regarded as a measure of the ordering of the system. In image encryption, if the image encryption effect is very good, then the pixel is chaotic, it is equivalent to the system is disorder. Under ideal conditions, if the entropy of the image is 8, means that the encrypted image is chaotic. The information entropy of the encrypted image we calculate is H(M)= 7.9979.

$$H = \sum_{i\in\{A,T,G,C\}} P(i)\log_2 \frac{1}{p(i)}. \tag{4}$$

This result is very close to 8, so we can roughly think this algorithm can achieve the goal of encryption well.

4.2.2. Histogram analysis. In image encryption, the distribution of the gray pixel values is often displayed on histogram. The histogram of the encrypted image should change into a flat pattern. Figure 6 shows the histogram of two images. As they vividly show below, in the original image, the pixels concentrate in some values. In contrast, pattern of the encrypted image histogram is very flat, and almost every pixel is divided into each value. This way can resist statistical attack effectively.

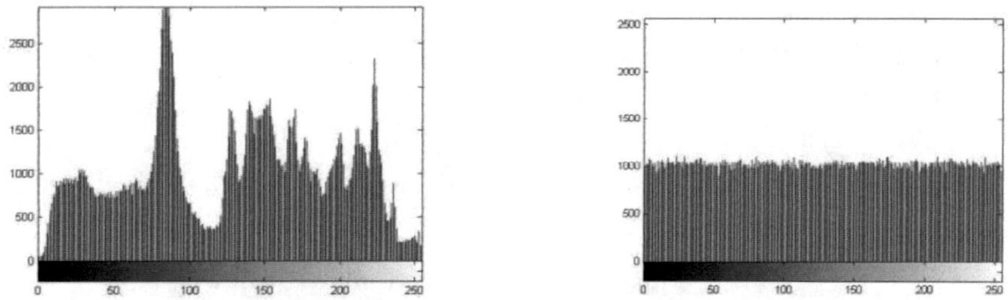

(a) (b)

Figure 6. histogram (a) original image (b) encrypted image

4.2.3. Correlation analysis. Correlation analysis refers to the analysis of two or more variables, so as to measure the correlation degree of two variables. In the original image, the correlation between two adjacent pixels is very tight.

$$E(x) = \frac{1}{N}\sum_{i=1}^{N} x_i, D(x) = \frac{1}{N\sum_{i=1}^{N}(x_i - E(x))^2}, \tag{5}$$

$$\mathrm{cov}(x,y) = E(x - E(x))E(y - E(y)), r_{xy} = \frac{\mathrm{cov}(x,y)}{\sqrt{D(x)}\sqrt{D(y)}}, \tag{6}$$

The encryption to do is to break or reduce the correlation, so that the attacker can't adopt the pixel analysis to restore the original picture, in order to achieve the effect of encryption. The smaller the correlation between two adjoining pixels, the more secure the cryptographic image. We randomly selected 1000 pairs of adjacent pixels from the original image comparing with encrypted images. The correlation through the MATLAB software to run the original two adjacent pixels is 0.8214, 0.0060 after encryption. We can see the correlation was significantly reduced.

4.3. Resist the chosen plaintext attacks

In the chosen plaintext attack, attacker can arbitrarily select a certain number of plaintext, let the algorithm to encrypt it, and get the corresponding ciphertext. In the worst case, the attacker can get the key for decryption directly. We analyse a chosen plaintext attack named differential attack.

The difference analysis involves the comparison of ciphertext pairs and plaintext with some characteristic. NPCR(Numbers of Pixel Change Rate), UACI(Unified Average Changing Intensity) are used to judge the effect of resisting differential attack.

$$\mathrm{NPCR} = \frac{\sum_{ij} D(i,j)}{m \times n} \times 100\%, \quad \mathrm{UACI} = \frac{1}{m \times n}\left[\sum_{ij} \frac{\left|C_1(i,j) - C_2(i,j)\right|}{255}\right] \times 100\%, \tag{7}$$

$C_1 \sim C_2$ represent the encrypted image after changing the gray value of a pixel respectively. m represents the size of the row, n represents the size of the column.

For the figure of 512*512, We select 5 representative points, which are located in the four corners and centres of the image, slightly change pixel values to test.

We can see the results in the table 1, the values of NPCR are all close to 100%, the values of UACI are about 33%.

Table 1. Encryption scheme anti-differential attack capability

Location	Primitive pixels	Changed pixels	NPCR	UACI
(1,1)	172	171	99.57%	33.41%
(1,512)	84	85	99.61%	33.44%
(256,256)	224	225	99.66%	33.35%
(512,1)	131	132	99.73%	33.47%
(512,512)	48	49	99.24%	33.46%

4.4. Comparison of algorithms

The algorithm compilation of references 8 and 9 are compiled in figure 7. The correlation of reference 8 and 9 is 0.0254 and -0.0106, while the correlation of this paper is 0.0060.

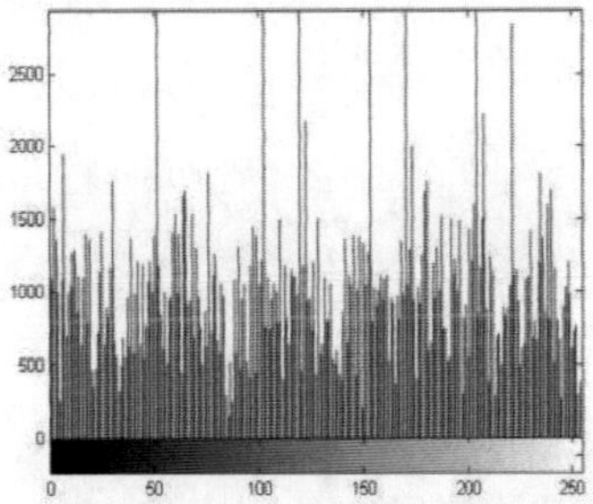

(a)

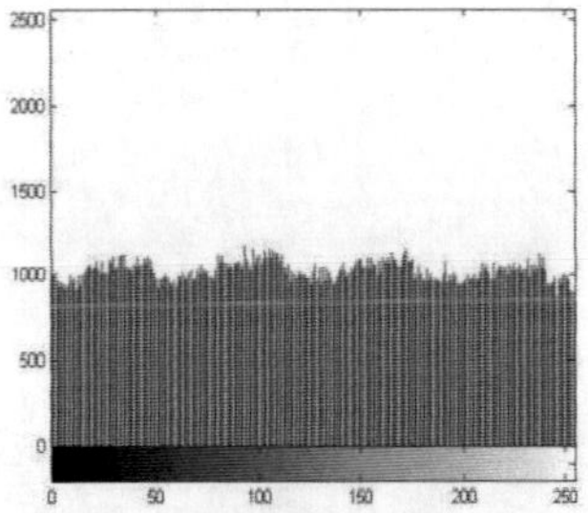

(b)

Figure 7. Histogram of encrypted image from the reference (a) from reference 8(b) from reference 9

5. Conclusion

This paper proposed an algorithm combining DNA and Logistic chaotic mapping. The paper first shows that how to put the real DNA chain into the image and introduces logistic chaotic mapping, which is used for image encryption. Then a new algorithm is proposed which combines the logistic method and DNA sequence to encrypt the grayscale image. And a random matrix is generated on the basis of logistic chaotic mapping and DNA sequence, so that the addition and complement operations are performed on it. This algorithm successfully solves irreversible problems and expands the key space. It can resist attacks, such as the chosen plaintext attack, brute force attack and statistical attack.

References

[1] Peng J, Jin S J, Lei L and Han Q. Research on a Novel Image Encryption Algorithm Based on the Hybrid of Chaotic Maps and DNA Encoding 2013 *J. IEEE International Conference on Cognitive Informatics & Cognitive Computing*, p 403-8.

[2] Wang Q, Zhang Q and Zhou C J. A Multilevel Image Encryption Algorithm Based on Chaos and DNA Coding 2009 *J. International Conference on Bio-inspired Computing*, p 1-5.

[3] Liu H J, Wang X Y, Abdurahman K. Image Encryption Using DNA Complementary Rule and Chaotic Maps 2012 *J. Applied Soft Computing*, **12**(5) p 1457-66.

[4] Zhang Q, Xue X L, Wei X P. A Novel Image Encryption Algorithm Based on DNA Subsequence Operation 2012 *J. The Scientific World Journal*, **6736** p 286.

[5] Kar N, Majumder A, Saha A, Jamatia A, Chakma K and Pal M C. An Improved Data Security using DNA Sequencing 2013 *J. ACM Mobihoc Workshop on Pervasive Wireless Healthcare*, p13-8.

[6] Zhang Q, Guo L, Wei X P. Image Encryption Using DNA Addition Combining with Chaotic Maps 2010 *J. Mathematical and Computer Modeling*, **52**(11-12) p 2028-35.

[7] Hermassi H, Belazi A, Rhouma R, Wei X. Security Analysis of An Image Encryption Algorithm Based on A DNA Addition Combining with Chaotic Maps 2014 *J. Multimedia Tools Application*, **72**(3) p 2211 -24.

[8] Ritu G, Anchal J. A New Image Encryption Algorithm based on DNA Approach 2013 *J. International Journal of Computer Applications*(0975-8887), **85**(18) p 27-31.

[9] Jin X., Tian Y L, Song C G, Wei G Z, Li X D, Zhao G and Wang H C. An invertible and anti-chosen plaintext attack image encryption method based on DNA encoding and chaotic mapping 2016 *J. Chinese Automation Congress,* p 1159-64.

IOP Conf. Series: Journal of Physics: Conf. Series **1004** (2018) 012024 doi:10.1088/1742-6596/1004/1/012024

A Novel Visual Interface to Foster Innovation in Mechanical Engineering and Protect from Patent Infringement

Salvatore Sorce[1,]*, Alessio Malizia[2], Pingfei Jiang[1], Mark Atherton[1] and David Harrison[1]

[1] Brunel University London, College of Engineering, Design and Physical Sciences
[2] University of Hertfordshire, School of Creative Arts

*salvatore.sorce@brunel.ac.uk

Abstract. One of the main time and money consuming tasks in the design of industrial devices and parts is the checking of possible patent infringements. Indeed, the great number of documents to be mined and the wide variety of technical language used to describe inventions are reasons why considerable amounts of time may be needed. On the other hand, the early detection of a possible patent conflict, in addition to reducing the risk of legal disputes, could stimulate a designers' creativity to overcome similarities in overlapping patents. For this reason, there are a lot of existing patent analysis systems, each with its own features and access modes. We have designed a visual interface providing an intuitive access to such systems, freeing the designers from the specific knowledge of querying languages and providing them with visual clues. We tested the interface on a framework aimed at representing mechanical engineering patents; the framework is based on a semantic database and provides patent conflict analysis for early-stage designs. The interface supports a visual query composition to obtain a list of potentially overlapping designs.

1. Introduction

Intellectual Property (IP) disputes can be resource intensive and time-consuming. A 2010 study indicated at least 24% of UK companies had experienced an IP dispute in the previous five years, damages averaging £75k-£115k were agreed in 30% of cases [1]. A more recent study found that there is a downward trend in the number of actual litigations because several disputes end in out-of-court settlement and roughly half that reach court find the patent invalid. Despite this trend, the time needed to resolve the disputes, both by litigation or settlement, remains quite long [2]. In addition, the final agreement often has a cost for both the patent incumbent and the challenger, being the result of a mediation between the two parties. On the other hand, patent applications have grown year-on-year 8.3% worldwide and 5% in Europe, with mechanical engineering accounting for 22.1% of worldwide patent applications in 2015 and maintaining an average growth of 6.4% over the preceding decade [3].

These two facts pave the way to an increasing number of disputes, due to the increasing number of patents and to the consequent increasing difficulty in data mining. Indeed, the great number of documents to be mined and the wide variety of technical language used to describe inventions, along with the different formats used to digitally represent them, represent a big obstacle for a cost and time-effective search.

In this context, an increased awareness of prior art at the beginning of a design, or during an emerging design, would help to limit these disputes. There are many available systems and tools for automated patent analysis, based on different principles and approaches: visualization, citation analysis and text mining, each using different techniques, such as Natural Language Processing, semantic analysis, property-function analysis, and so on [4]. Regardless of the approach, all patent analysis systems follow the same workflow, sketched in Figure 1. First, patents data — in form of structured representation — is stored in a database. Then, a suitable patent analysis is carried out, by means of the available querying/mining capabilities of the database.

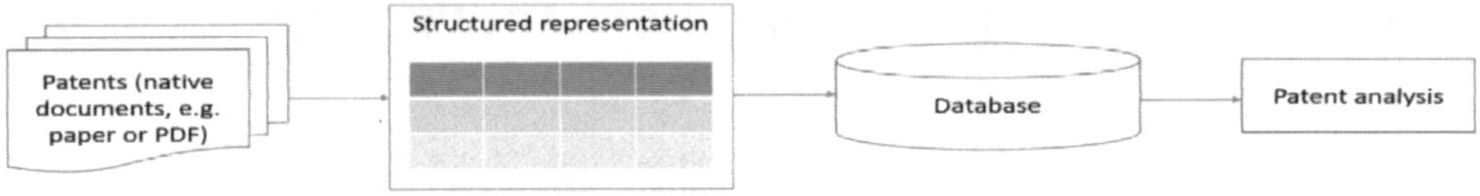

Figure 1. The patent representation and analysis workflow

Focusing on the mechanical engineering field, designers have to deal with two specific issues. One is that, while patents in such field mainly rely on graphical/visual descriptions (images, sketches), available patent analysis systems are text-based [4][5]. In these cases, the designer will typically need to enter keywords in a patent retrieval system to identify relevant prior art. A single search is rarely sufficient to capture the entire prior art. Natural Language Processing (NLP) with machine learning (e.g. IBM Watson SIIP platform [6]) has been applied to patent text search, often using statistical inference and weightings to enable text search beyond keywords. However, statistical NLP is semantically weak and is only able to predict with acceptable accuracy when trained with large datasets [7]. Commercial patent retrieval systems only employ text-based search methods and the need for advanced approaches is becoming more important as text-based techniques are increasingly challenging [8]. However, some possible approaches, such as content-based image retrieval techniques, are not well-suited to patent-images, because they mainly exploit colour images whereas patent images are mostly monochrome [9]. The requirements of a generic Patent Image Retrieval (PIR) system have been defined in the literature [10] and include a semantic-level interpretation of patents still to be developed in patent search systems [4]. Hitherto, they have been limited to a text description of images present in a patent [8]. Other PIR research has focused on image segmentation and feature-extraction but without properly capturing the semantics.

The second issue relates to detailed knowledge on query languages or patterns that normally is not part of a mechanical engineering background.

In this paper, we present a novel visual interface for helping designers to access prior art via a semantic database without requiring any specific background on query languages. The interface is designed to give visual aids for composing queries and visualising results from a patent database. It can be easily programmed to provide access to any underlying database, keeping the same layout and the corresponding affordable interaction paradigm.

In order to test the interface effectiveness, we developed our interface on top of an existing patent analysis framework in mechanical engineering. The framework is mainly composed of a domain-specific ontology stored in a semantic database that can be queried by means of the SPARQL Protocol and RDF Query Language (SPARQL) [11]. Such language, although very effective and being a de-facto standard in semantic databases, is fairly technical and not suitable for non-experts (such as mechanical designers, architects, etc.). The proposed interface allows the designers to visually compose their queries intuitively without requiring any SPARQL knowledge. It also allows for an immediate and interactive visualisation of queries results, effectively allowing designers to quickly check overlapping prior art and facilitating innovative and patentable design solutions. The interface is web-based and therefore highly interoperable including intuitive visual elements for patent analysis.

2. Background

The framework we used as a case study is composed of a patent functional representation, a domain-specific ontology and a semantic database. Figure 2 shows the relations among functional components of the framework and our interface representing the front-end.

The *functional representation* aims at expressing patents in terms of geometric features and their functional interactions. The *domain-specific ontology* enables knowledge sharing and conceptualisation, providing a standardised vocabulary for describing patented designs. The vocabulary and the relationships among geometric features and their functional interactions are encoded in a *semantic database*; this structured representation models similar working principles between an emerging design and prior art. The whole framework allows for early identification of potential conflicts and thereby can help designers steer their emerging designs away from overlapping patents.

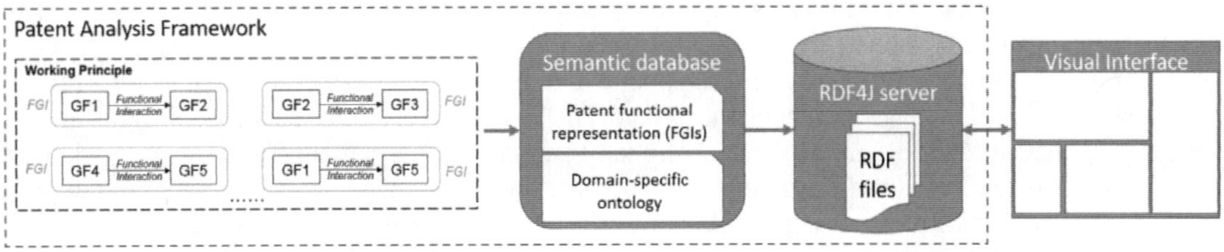

Figure 2. The patent representation and semantic querying framework

2.1. The Semantic Database

In mechanical engineering, functions are mainly realised by combinations of interrelationships between physical effects, geometric and material characteristics, known as the *working principles* [12]. Explaining the working principle in many mechanical design patents relies heavily on illustrating the Functional Interactions (FI) among the involved Geometric Features (GF). Here the structured representation of patents is obtained by means of Functional-Geometry Interactions (FGIs). They represent interacting geometric features (embodying physical effects and material characteristics) that carry a functional significance in a working principle. Both the FGIs and the semantic relations defined in the domain-specific ontology are expressed using a triple-store approach, which is a widely adopted solution for the storage and retrieval of data through semantic queries [13]. The basic form of a triple is Subject-Predicate-Object, which can be suitably used to describe GF1, FI and GF2 (Figure 2) respectively. In the framework we used, one FGI (GF1-FI-GF2) corresponds to one triple. Patent working principles, represented as FGIs, along with the domain-specific ontology are finally encoded into Resource Description Framework (RDF) format, which is a standard model for data interchange [14]. The generated RDF files are then uploaded to an RDF4J server [15], which is an open-source framework for querying and analysing RDF data. This provides access to the semantic database both through a Web interface (for browser-based access) and from an URI (Uniform Resource Identifier – for programmatic access). The server supports SPARQL queries, thus allowing the description of working principles of an emerging design in form of simple or complex queries about any existing overlap in prior art.

3. The Interface Layout

The need for different inputs other than text-based and NLP is widely discussed in literature [5]. For instance, in order to provide designers with an intuitive interface to compose queries, a visual metaphor can be used. Following a recent trend in the end-user programming community [16], block-oriented programming presents program logic as compositions of visual blocks. Tools such as Scratch, Blockly, Code.org's lessons, and App Inventor have introduced programming and computational thinking to a huge audience, reaching people of all ages and backgrounds. Supporting a block-oriented metaphor, we aim at easing the designer's cognitive load in formulating queries and freeing him from the burden of learning SPARQL language. Query results are visualised including some interactive feature (such as a thumbnail-preview to allow a quick navigation of the patents ranking).

Figure 3 shows the interface layout organised right-left and top-down, following the standard F-shaped pattern of reading [17]. With reference to Figure 3, the title bar (area #1) shows the title "D4i" (Design for Invention), and three option buttons. The "Viewer" button activates the design viewer that

IOP Conf. Series: Journal of Physics: Conf. Series **1004** (2018) 012024 doi:10.1088/1742-6596/1004/1/012024

in our case is a 3D rendering of the design in area #2. The "Simple" button hides the design view, thus making more room for the other areas and items in our interface. Indeed, a designer may need to check for patent infringements before starting any new design. For such reason, the 3D view can be closed, thus making our interface also suitable in cases where a 3D view is not applicable/available. The "Guide" button shows a short video guide on how to use the tool to improve usability.

The area #2, when activated, shows the design viewer. It is worth noting that, since we are considering the specific domain of mechanical engineering, the design viewer shows a 3D design, along with common features for view manipulation (rotate, pan, zoom and explode to show the design components). The visual interface is also able to import any 3D design in OBJ format (a standard file format for 3D objects to support interoperability with existing CAD software). The area #3 shows the query result — a list of existing patents overlapping with the proposed design; the visualisation of the list shows the number of patents matching the design and can be reordered; a single click on an item activates a preview of the patent and a double-click opens the corresponding file. The area #4 shows query block-oriented commands available for querying the database. Each block can be drag-and-dropped in the area #5 to compose a query. Finally, area #6 shows a preview of the selected document.

3.1. Interface Features and Functions

Our use case focus on an example of a typical workflow for searching a given FGI (Functional-Geometry Interaction), and in particular for searching products having a hole in a plate.

The designer starts composing a query by picking the appropriate functional blocks from the "Available blocks" area (#4 in Figure 3) and drags those to the "Query composition" area (#5 in Figure 3). Each query starts with the "Search for products" block, and all other blocks must be wedged after it. The block shapes suggest possible ways to compose a query and prevent from wedging meaningless or wrong blocks to queries, following the poka-yoke principle [18]. According to the triple-based structure of our semantic database, each FGI is composed of a Geometric Feature GF, then a Functional Interaction FI, and then another GF. In this case, a query to search for a single FGI is composed by drag-and-drop of a GF (orange) block under the "Search for products" (green) block. Then a FI (purple) block is wedged to the first GF, and last another GF block is placed at the end of the line. This block arrangement semantically composes the statement: "Search for products with the geometric feature GF1 functionally interacting by means of FI with the geometric feature GF2", i.e. a triplet.

Once a block is correctly placed, a dropdown list allows the user to select a specific item (a GF or a FI, depending on block type). In our example, we selected "Hole", "Locate at", and "Plate" respectively. The dropdown lists are populated at the start, by querying the database for the available GFs and FIs.

In order to run a query, the user must click on the green disc with a triangular symbol, the "Play" button. This will start the background process to translate the visual query to its textual version in SPARQL and send it to the server that stores the semantic database, and returns the results. This translation step (from blocks to SPARQL) can be adapted to different target output languages, thus making our interface adaptable to different patent analysis systems. The matching results are listed in the list box on the right side of the interface. Figure 3 shows the interface status at the end of this process. The results list is interactive, and in particular, it allows users to see a preview of a given product within a patent by clicking on the corresponding name in the list. The query may be saved by clicking on the blue disc with a downward arrow, the "download" button below the "Query composition" area.

It is worth noting that more complex or detailed queries can be composed by adding block lines below the first one, following the example above. Furthermore, simpler queries can be composed by adding fewer blocks, for instance, supporting a single GF. In this case, the tool will search for products having that GF, leading to a (possibly) longer list of results. Last, it is not mandatory to choose an item from the dropdown lists. In this case, the search will be carried out as if there is a "*" wildcard in the corresponding SPARQL field, meaning "search for all". Finally, the two available blocks "Function action" and "Function object" allow for database-specific search. Indeed, the databased used in this case study included additional features coded to describe a product or a patent itself. These features allow for generic search, which can turn out to be useful especially at the beginning of a design process.

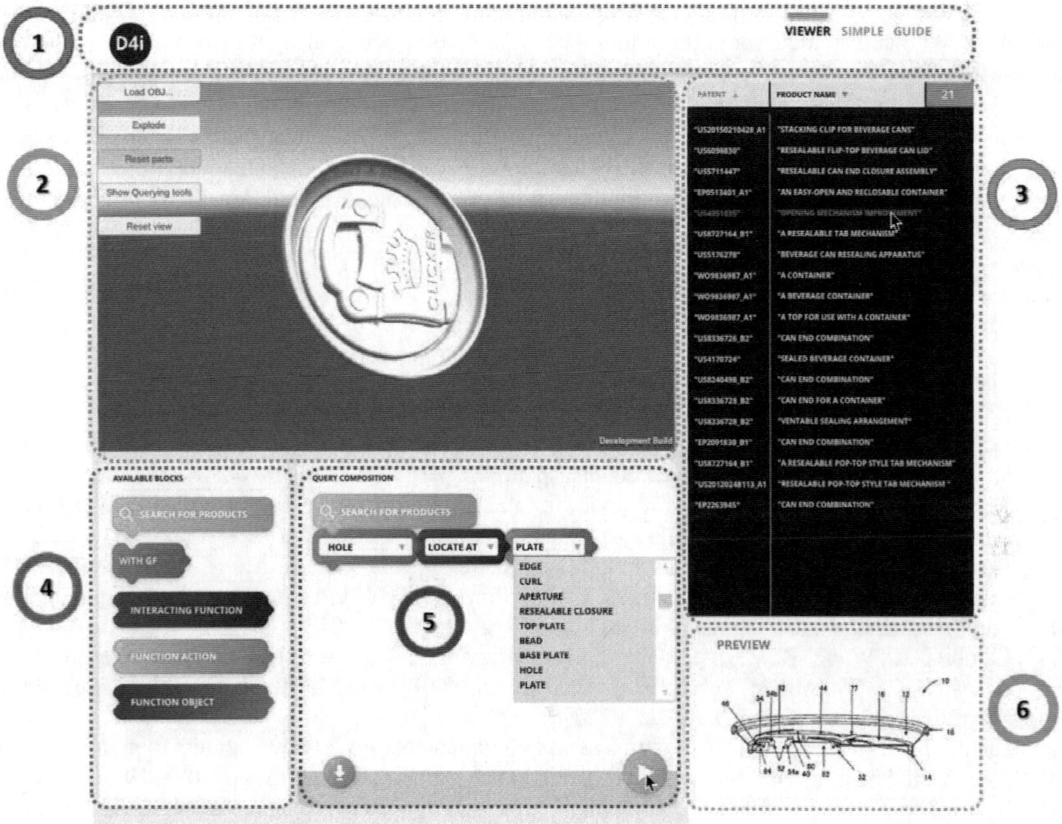

Figure 3. The interface layout

4. Discussion & Future Work

As a first preliminary evaluation, here we briefly report an informal and qualitative testing, conducted on three colleagues skilled in mechanical design. The test was conducted individually, each session took 10 minutes at most, and none of the participants had seen the interface before. The current layout and shape of the visual items have been perceived as suitable and intuitive, including the position and size of the interactive 3D viewport. All three participants reported that the interface affords the drag-and-drop action for the blocks, and the drop-down lists suggest to choose one item among the available ones. Next, the meaning of interface symbols, terms and acronyms need to be clearly defined. This is an expected result, since we used our interface to access a specific database but it does not affect the interface effectiveness. Lastly, we observed some uncertainty about how and when to visualise the results of the query in terms of the role of the green "Play" button in the "Query composition" area was not so clear; and there was no clear understanding of when a query could be executed depending on the number and type of blocks added. This suggests us to implement a "continuous" view of the results, so that every time a change occurs in the "Query composition" area, the "Results" area should show the corresponding results. This will require one less step of interaction (the click on the "Play" button), and should make it more evident that adding blocks will lead to a refined list of results.

We are currently working on the definition of a possible model of the "state" reached in the patent checking process. This would allow designers to save snapshots of their work at given steps, and to compare them in order to check the effectiveness of their design choices. Furthermore, we are currently planning a thorough evaluation, involving people from the design engineering profession.

In this paper, we presented a visual interface supporting an easy formulation of semantic queries for early detection of possible patent infringements. The interface is based on the well-established principle of "block-oriented programming". The visual interface highlights the potential prior art conflicts and areas of innovation of the emerging design in the patent space. As a desirable side-effect, the early detection of existing relevant prior art should foster design creativity, suggesting how to achieve the same functions with different operating principles or geometric features. Despite our case study being mechanical design, the proposed interface may be adapted to other functional representations in different fields where intellectual property protection and development is a relevant issue.

References

[1] Greenhalgh, C., et al (2010), "Intellectual Property Enforcement in Smaller UK Firms: a report for the Strategic Advisory Board for Intellectual Property Policy (SABIP)."
[2] Jeon, H., (2015), Patent infringement, litigation, and settlement, In Economic Modelling, Volume 51, 2015, Pages 99-111, ISSN 0264-9993, https://doi.org/10.1016/j.econmod.2015.07.019.
[3] World Intellectual Property Organization (WIPO), "World Intellectual Property Indicators 2017", http://www.wipo.int/publications/en/details.jsp?id=4234
[4] Abbas, A., et al (2014), "A literature review on the state-of-the-art in patent analysis", World Patent Information, vol. 37 issue 4, Jun. 2014, pp. 3–13.
[5] Bonino, D., Ciaramella, A., Corno, F. (2010). "Review of the state-of-the-art in patent information and forthcoming evolutions in intelligent patent informatics", World Patent Information, 32(1), 30-3
[6] http://researcher.watson.ibm.com/researcher/view_group.php?id=2134
[7] Cambria, E., White, B., (2014), "Jumping NLP Curves: A Review of Natural Language Processing Research [Review Article]," in IEEE Computational Intelligence Magazine, vol. 9, no. 2, pp. 48-57, May 2014, doi: 10.1109/MCI.2014.2307227
[8] Bhatti, N., Hanbury, A. (2013), "Image search in patents: A review", International Journal of Document Analysis and Recognition (IJDAR), vol. 16 issue 4, Dec. 2013, pp. 309–329.
[9] Vrochidis, S., et al (2010), "Towards content-based patent image retrieval: A framework perspective", World Patent Information, vol. 32 issue 2, Jun. 2010, pp. 94–106.
[10] Vrochidis, S., et al (2012), "Concept-based patent image retrieval", World Patent Information, vol. 34 issue 4, Dec. 2012, pp. 292–303.
[11] Jiang, P., Atherton, M., Harrison, D. Malizia, A., (2017) "Framework of mechanical design knowledge representations for avoiding patent infringement", In: Proceedings of the 21st International Conference on Engineering Design (ICED17), Vol. 6: Design Information and Knowledge, Vancouver, Canada, 21.-25.08.2017.
[12] Pahl, G., Beitz, W., (2006), Engineering Design: A Systematic Approach. Third. Springer-Verlag London. doi:10.1007/978-1-84628-319-2.
[13] Rusher, J., (2003), "TripleStore." In Semantic Web Advanced Development for Europe (SWAD-Europe), Workshop on Semantic Web Storage and Retrieval.
[14] Atherton, M., Jiang, P., Harrison, D. and Malizia, A. (2017), "Design for invention: annotation of Functional Geometry Interaction for representing novel working principles". Research in Engineering Design. ISSN: 0934-9839.
[15] RDF4J "Eclipse RDF4J – a Java Framework for RDF", http://rdf4j.org/
[16] "Foreword," presented at the Blocks and Beyond Workshop (Blocks and Beyond), IEEE, 2015
[17] Nielsen, J., (2006), "F-Shaped Pattern For Reading Web Content", https://www.nngroup.com/articles/f-shaped-pattern-reading-web-content-discovered/
[18] Nikkan Kogyo Shimbun (1988). Poka-yoke: improving product quality by preventing defects. Productivity Press. p. 111. ISBN 978-0-915299-31-7

IOP Conf. Series: Journal of Physics: Conf. Series **1004** (2018) 012025 doi:10.1088/1742-6596/1004/1/012025

Unreal Interactive Puppet Game Development Using Leap Motion

An-Pin Huang[1], Fay Huang[1,*] and Jing-Siang Jhu[1]
[1]CSIE, National Ilan University, Taiwan

*fay@niu.edu.tw

Abstract. This paper proposed a novel puppet play method utilizing recent technology. An interactive puppet game has been developed based on the theme of a famous Chinese classical novel. This project was implemented using Unreal Engine, which is a leading software of integrated tools for developers to design and build games. On the other hand, Leap Motion Controller is a sensor device for recognizing hand movements and gestures. It is commonly used in systems which require close-range finger-based user interaction. In order to manipulate puppets' movements, the developed program employs the Leap Motion SDK, which provides a friendly way to add motion-controlled 3D hands to an Unreal game. The novelty of our project is to replace 3D model of rigged hands by two 3D humanoid rigged characters. The challenges of this task are two folds. First, the skeleton structure of a human hand and a humanoid character (i.e., puppets) are totally different. Making the puppets to follow the hand poses of the user and yet ensuring reasonable puppets' movements has not been discussed in the literatures nor in the developer forums. Second, there are only a limited number of built-in recognizable hand gestures. More recognizable hand gestures need to be created for the interactive game. This paper reports the proposed solutions to these challenges.

1. Introduction

Puppetry has an ancient history in various cultures with different forms and manipulation methods. In traditional point of view, puppetry is an art performance to deliver stories and/or communicate the ideas among human societies. They are performed in the theaters, special stages, TV shows, or even in the movies. Puppets are controlled by the puppeteers using their hands, arms, or control devices such as rods or strings.

In this project, we proposed a new way to control and manipulate puppets using recent technology. Puppetry nowadays can be performed in virtual world such as using VR gears, or it can be shown by a 3D displaying device or holographic projection. Moreover, puppetry is no longer just a performance by artists, but can also be interactive games accessible to all range of users.

Our interactive game has equipped one Leap Motion (LM) Controller as input device. It is a small USB peripheral device supporting the understanding of hand gestures, analogous to a mouse, but requires no hand contact or touching [1]. It has been widely used for interactive game developments [2, 3]. It is consist of two monochromatic IR cameras and three infrared LEDs. The sensing range is about one meter in distant within a roughly 45-degree pyramid shaped region. The LEDs generate pattern-less IR light [4] and the cameras generate almost 200 frames per second of reflected data. The 3D position data is then synthesized by comparing the 2D frames generated by the two cameras [5 ,6]. The most essential spirit of puppetry is the fact that it has to be manipulated by hand gestures, not just by pressing some buttons on controllers or keyboards. Thus, Leap Motion device has make this possible due to its capability of sensing fingers movements in real time.

Figure 1. The snapshot of the developed digital puppet game using Leap Motion Controllers as input device.

There are many game engines available in the market. Among them, Unreal Engine 4 (UE4, by Epic Games) and Unity (by Unity Technologies) are arguably two of the most popular game engines for the developments of many well-selling 3D games. But, when it comes to graphics, UE4 is really a next- generation game engine [7, 8]. Another advantage of Unreal Engine (UE) is that it supports a node-based visual scripting method, called Blueprint. C++ is the programming language for UE, but it is possible to create an entire game using only Blueprints. Epic and LM have jointly developed an official plugin for UE4. By taking advantage of this built-in SDK, we are able to extend its capability for developing a novel interactive puppet game.

The goal of this project is to utilize modern technology to carry forward the traditional puppet culture for the young generations. The LM controller is used as a sole input device to detect hand gestures for puppet manipulations. The puppets have been digitalized and represented by 3D models. Figure 1 shows the setup configuration and a snapshot of the game. In order to attract wide range of users, we have adapted the characters from a famous Chinese classical novel, *Journey to the West* [9], namely Sun Wukong (also known as the Monkey King) and Nezha (also known as Third Lotus Prince). Figure 2 illustrates the 3D models of these two Chinese well-known mythology characters. Two characters can interact with or fight to each other controlled by two separated hands, either from a single user or from two different users. The digital puppets react in real time following the location and orientation of users' hands like traditional puppets do.

Figure 2. 3D models of Sun Wukong with his Somersault Cloud and Nezha with his Wind-fire Wheels.

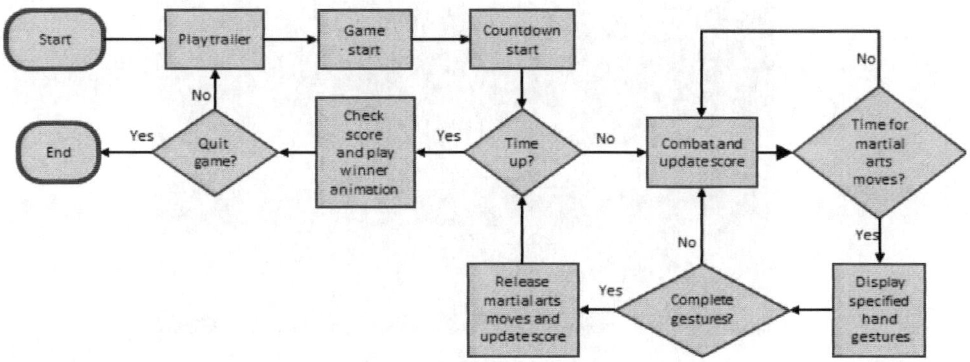

Figure 3. The game flow of the developed interactive puppet game.

2. Methods

This section explains the design of our digital puppet game and elaborates the methods developed using UE4.

2.1. Design of the Interactive Digital Puppet Game

For all forms of puppet shows, audiences are most likely to be attracted to some humorous, rhythmic dialogue and gorgeous battle scenes. In order to deliver excitement and tension during the game play, we have decided to create a combat game. One of the famous battles of "Wukong havocs the Heaven Kingdom" in *Journey to the West* has been engaged, namely the "Sun Wukong fighting against Nezha". The Monkey King's special weapon is called golden cudgel and Nezha in this game is holding a spear. The game is designed that, if one character hit the opponent with his weapon, then it gains points. Within a time-limit, whoever gets the higher point wins the game.

In order to make the game more interesting, some special fighting tricks or martial arts moves for each character can be released by special hand gestures. This idea cannot be realized by traditional puppetry, and thus can be considered as the exclusive extension of traditional puppet play. Other glamorous features only realizable in digital form are all kinds of special visual effects during fighting. Moreover, we have developed particle systems to emulate the well-known accessories belonging to these two characters, they are Wukong's Somersault Cloud and Nezha's Wind-fire Wheels (see Fig. 2).

The game flow is depicted in Fig. 3. As the game starts, a countdown is initiated. While the special martial arts moves haven't been enabled, each player should try to hit the opponent with his weapon. Fold the thumb or pinky to swing the weapon. The special martial arts moves become enabled in a random time manner. When they do, players should compete with each other by correctly completing a set of specified hand gestures to release the special martial arts moves. This would gain more points.

2.2. Introduction of Game Development Component in UE

In UE, the term "level" in UE has similar meaning to a particular status of a virtual world, which can be made up of a collection of game development entities, such as blueprints, static meshes, lightings, particle effects, and volumes etc. Within a defined level, all included entities work together to bring the desired experience to the players.

An "actor" is a generic term for any object that can be placed into a level. There are many different types of actors in UE, for examples, there are player start, camera, skeletal mesh, static mesh, trigger, and particle emitter actors. Actors are a class that support 3D transformations such as translation, rotation, and scale. They can be created and destroyed through gameplay code (blueprints).

A "pawn" in UE is an actor that can be possessed by a Controller, which is set up to easily accept input, and it can do various player-like things. A pawn is not assumed to be humanoid. On the other hand, a "character" extends pawn for a specific human style pawn with specific built-in behaviors.

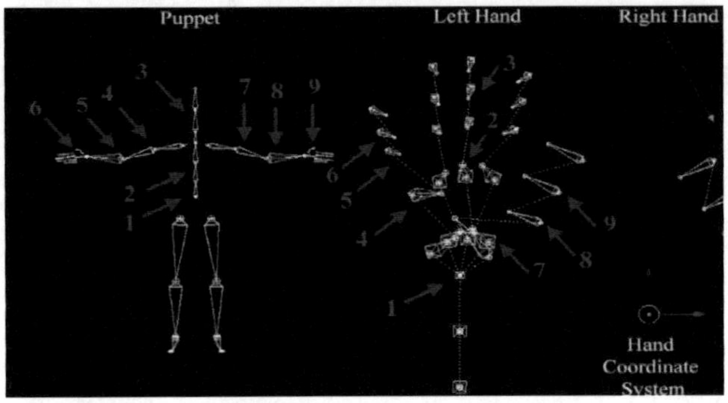

Figure 4. The proposed binding method between a hand and a puppet's skeleton structures.

LM plugin for UE4 by default provides two convenience rigged hand-shape characters, one is a yellow robot hands and the other is white human hands named Echo. We used solely the LeapRiggedEchoHands actor belonged to the Echo character in our project.

2.3. Game Development Method

The developed interactive puppet game involved only one level. A player start actor is included by default placed in the level to designate where the player should start out when the game begins. Two skeletal mesh actors for the characters were newly created in the game for various animations and interactions. Several trigger actors were designed to enable the collision detection and hand gesture determination. Many particle emitter actors were generated in order to create visual effects during game play. This subsection explains some major developments and methods used in this game.

2.3.1. Starting the Game. The game starts when LM device senses two hands (left and right hands) within its sensing range. Two Boolean variables were newly created in the blueprint of LeapRigged-EchoHands actor provided in the LM plugin, one for each hand. While both Boolean variables are true, meaning both hands are recognizable to the LM sensor, the game begins (i.e., the countdown begins).

2.3.2. Replace Echo Hands with Humanoid Characters. By default, Echo hands would be shown, but it is necessary to replace the original hand models by the defined 3D character models. Due to that the rigging (i.e., bone structure) of human hand and humanoid figure are fundamentally different, this is not a trivial task. Making a humanoid figure, in our case the puppet, to follow the user's hand poses and yet ensuring reasonable human movements has not been discussed in the literatures nor in the developer forums.

The internal technology of LM sensor interprets the rotation value of each finger joint in real time. We have proposed to transform these values into puppet's joints in order to manipulate the puppet. However, the rotation values do not represent the absolute rotations/orientations with respect to the world coordinate system. Instead, they represent the relative rotation with respect to the hand coordinate system at the middle of two hands. Moreover, due to that fingers' postures are very different from human's standard poses. Before those rotation values can be assigned to puppet's joints, some recalculations (or calibrations) and additional constraints are required. Figure 4 illustrates the proposed binding between a default hand and a puppet's structures. Only the labeled joints are bound together; moreover, rotation angles were tuned in blueprints to meet the expected puppet movements.

2.3.3. Puppet Animation by Hand Gestures. Two new skeletal actors were created, one for each hand, for the purpose of animating the characters. First of all, they are defined to follow user's hand location and orientation in real time. Moreover, the poses of the characters should also follow the hand gestures in real time as mentioned in the above paragraphs. New blueprints for these actors were developed to enable more variety of hand gestures for different fighting tricks or martial arts moves. We set up trigger actors for all fingers and palm as illustrated by red ellipses in Fig. 5 (left). Different combinations of those triggers define different gestures written in blueprints as shown in the background. Figure 5 (right) depicts the additional gestures specially designed for the game.

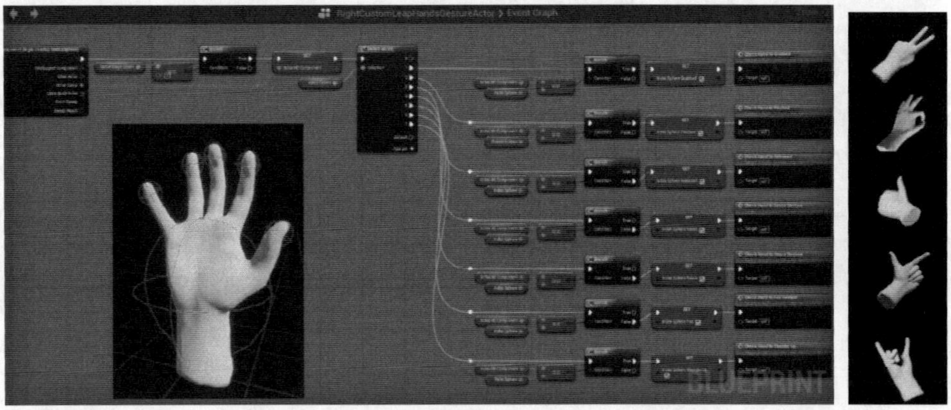

Figure 5. (Left) Six ellipsoid triggers are added to palm and fingers in accompanied with the created blueprints (only partial view is shown) to recognize hand gestures. (Right) A set of new hand gestures were designed for the game.

3. Results

An interactive digital puppet game has been developed from scratch using UE4 (ref. [10]). The 3D models were created mainly in Blender 2.79 in accompany with ZBrush 4R7 for detailed sculpturing. The texturing (UV maps) were enhanced in Photoshop. Normal maps were outputted from Blender. The materials and rendering condition were set up in UE4. Moreover, the animation sequences were generated in Blender and all special effects were created using the particle emitter actors provided in UE4. Figure 6 shows some of the intermediate working details.

The computer is equipped with Windows 7 (64-bit operating system), Intel Xeon E3-1231 V3 (3.4GHz) processor, 16 GB DDR3-1600 memory, and a ASUS GTX660 2GB graphics card. The performance is above 60 fps.

Latest version of the game has been exhibited to public in order to receive more variety of players' feedbacks. Figure 7 shows some photos of the exhibition. The interaction method has been successfully approved by players of a wide range of ages. Some users have suggested to eliminate few difficult hand postures, and adding more visual effects during the combat. All the advices have been taken into account in the current version. A demo of the current version of the game is available at YouTube (https://youtu.be/Wt8yHfj678M).

4. Conclusion

In this paper, a newly developed interactive digital puppet game using recent technology, Leap Motion Controller and Unreal Engine, has been presented. Based on the limited functionalities provided by Leap Motion SDK, extended capabilities were implemented in order to manipulate two humanoid puppet characters played by two players. Additional hand gesture recognitions have been developed to make the puppet interaction more interesting. To the best of our knowledge, the concept and the method of the proposed puppet game is unique and has not been seen or discussed in the literatures nor in the developers' forums. For the future work, we plan to combine it with one of the available virtual reality displaying methods to bring the players an engaging and more realistic gaming experience.

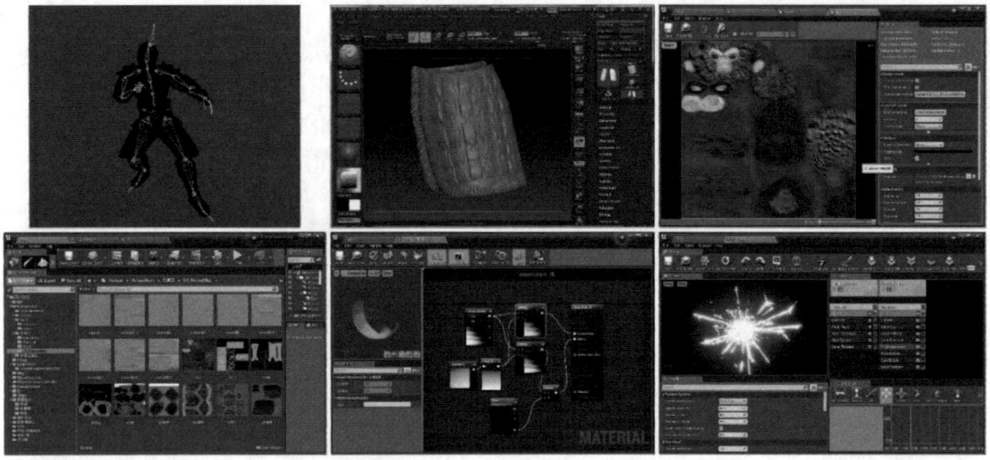

Figure 6. The interactive digital puppet game has been developed from scratch. Here are some of the intermediate working details,

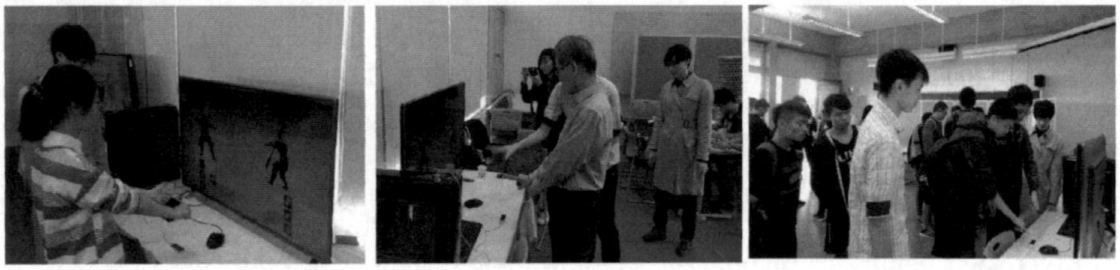

Figure 7. Some photos from the exhibition.

Acknowledgments
Project mainly funded byJinjiang Multimedia Technology company (In Chinese:晶將數位多媒體科技); partially supported by MOST 106-2221-E-197-003.

References

[1] Seixas M, Cardoso J and Dias M T G 2015. The Leap Motion movement for 2D pointing tasks: Characterisation and comparison to other devices. In *Proc. PECCS'15*.

[2] Sutton J 2013. Air painting with Corel Painter Freestyle and the leap motion controller: a revolutionary new way to paint. In *Proc. ACM SIGGRAPH '13*, Studio talks article No. 21.

[3] Vasudevan S K, Vivek C and Srivathsan S 2015. An Intelligent Boxing Application through Augmented Reality for two users – Human Computer Interaction Attempt. *Indian Journal of Science & Technology* **8(34)** pp 1–7

[4] Weichert F, Bachmann D, Rudak B and Fisseler D 2013. Analysis of the Accuracy and Robustness of the Leap Motion Controller. *Sensors* **13(5)**, pp 6380–6393

[5] Guna J, Jakus G, Pogačnik M, Tomažič S and Sodnik J 2014. An Analysis of the Precision and Reliability of the Leap Motion Sensor and Its Suitability for Static and Dynamic Tracking. *Sensors* **14(2)**, pp 3702–3720

[6] Coelho J C and Verbeek F J 2014. Pointing Task Evaluation of Leap Motion Controller in 3D Virtual Environment. In *Proc. Chi Sparks'14*. pp 78–85

[7] Petridis, P, Dunwell, I, De Freitas, S and Panzoli, D 2010. An Engine Selection Methodology for High Fidelity Serious Games. In *Proc. IEEE VS-GAMES'10*, pp 27–34

[8] Mishra, P and Shrawankar, U 2016. Comparison between Famous Game Engines and Eminent Games. *International Journal of Interactive Multimedia & Artificial Intelligence*, **4(1)**, pp 69–77

[9] Wu, C E 2012. *The Journey to the West* (Vol. 1). University of Chicago Press.

[10] Kanode, C and Haddad, H 2009. Software Engineering Challenges in Game Development. In *Proc. IEEE ITNG'09*, pp 260–265

IOP Conf. Series: Journal of Physics: Conf. Series **1004** (2018) 012026 doi:10.1088/1742-6596/1004/1/012026

Ink Wash Painting Style Rendering With Physically-based Ink Dispersion Model

Yifan Wang[1,*], Weiran Li[1], and Qing Zhu[1]

[1]Beijing University of Technology, China

*chaldea_wyf@foxmail.com

Abstract. This paper presents a real-time rendering method based on the GPU programmable pipeline for rendering the 3D scene in ink wash painting style. The method is divided into main three parts: First, render the ink properties of 3D model by calculating its vertex curvature. Then, cached the ink properties to a paper structure and using an ink dispersion model which is defined by referencing the theory of porous media to simulate the dispersion of ink. Finally, convert the ink properties to the pixel color information and render it to the screen. This method has a better performance than previous methods in visual quality.

1. Introduction

Ink wash painting is a traditional East Asian type of brush painting style originating in China that only uses black ink in various concentrations. By controlling the proportion of water and ink and the strength of the pen, artists can use different shades of black colour to painting different scenes on the rice paper. This artistic form can perform distinctive visual effect and unique aesthetic features of orientalism, meanwhile possess an irreplaceable position and incalculable value in Chinese art history.

The ink wash painting style is a demanding art-form requiring great skill, concentration, and years of training. In the 1960s, Shanghai Animation Film Studio produced a series of Chinese ink wash style animated film such as "Baby Tadpoles Look for Their Mother" and "Mu Di", which were highly praised worldwide and received many awards. However, because of the dispersion of ink on the rice paper is difficult to control, it takes a lot of time to make an ink wash style animated film. A 20 minutes ink wash style animated film usually takes 2 years to be completed. At present, with help of technologies such as 3D rendering and digital effects, the working process can be reduced to a few months. Therefore, this paper endeavor to develop a real-time system in order to improve the efficiency for producing ink wash paintings style animations. For real-time ink wash painting style rendering, the game "Okami" by CAPCOM is a very representative piece of work which realizes the real-time rendering of Japanese ink wash painting style through contour rendering and texture mapping techniques. Yuan et al.[1] and Cao[2] also provided Chinese ink painting real-time rendering systems. However, these techniques have a poor performance on simulating the ink diffusion effect, which is an important defect for these simulate system because of it neglects some significant artistic features.

The method proposed in this paper aims to improve the ink diffusion effect caused by the ink dispersion in the rice paper and the colour sense of ink due to the change of ink and water concentration. The paper will focus on the dispersion model in section 3, including paper structure, ink properties and specific algorithms of dispersion simulation. The ink wash painting style rendering method is introduced in section 4. The experiment results and conclusion is presented in section 5 and section 6.

2. Related work

For the ink wash painting simulation, Chu and Tai[3] proposed a fluid model based on Lattice Boltzmann Equation (LBE) to simulate the diffusion of ink on rice paper in real time. Way et al.[4] and Sun et al.[5] also proposed and improved the physically-based approach to simulate ink diffusion. These approaches are more suitable for real-time ink wash drawing system but too inefficient for 3D real-time rendering engines. Dong et al.[6] presented an image processing approach to render the picture to the Chinese ink painting style in real-time. Xie et al.[7] achieved an image-based ink wash painting style rendering system by converting the extracted abstract stroke of the photo with predefined ink wash stroke texture. However, the process of stroke extraction requires some user interaction to assist manually, which makes it difficult to realize in real-time rendering. Yang et al.[8][9] proposed a paper-ink simulating model based on porous media and capillary mechanism for simulating ink dispersion in rice paper, which is fast in convergence and suited to take advantage of the GPU's parallel processing features to improve performance for real-time rendering.

For 3D ink wash painting style Rendering, Chan et al.[10] reconstructed the 3D object to generate a 3D geometric model that is specially used for ink wash painting style rendering, use for combined with illumination, viewpoint and other information to simulate the contour, fly white and other techniques of drawing ink wash painting. Tianchen Xu et al.[11] established a particle form of the stroke model, through the stroke model fetch the information from the silhouette of 3D model and then transform into ink brush strokes. Yuan et al.[1] presented a GPU-based real-time ink wash painting rendering system, which proposes a rendering framework for real-time ink wash painting style rendering, and provides a basis for the application of ink wash painting style rendering. Cao[2] summary the techniques of real-time rendering that can be used to Chinese ink painting rendering, and provided an improved rendering system for Chinese ink painting rendering.

This paper mainly refers to the rendering framework of Yuan et al. [1] and combine with the ink dispersion simulation method proposed by Yang et al.[8][9] to propose a real-time ink wash painting style rendering method based on physically-based ink dispersion model.

3. Physically-based ink dispersion model

According to the paper-ink simulating model proposed by Yang et al.[9], rice paper is abstracted into a two-dimensional finite element group consisting of anisotropic capillary tubes, as same as dividing a piece of rice paper into numerous square grids. Each grid corresponds to an element that contains many capillary tubes called paper cells. As shown in figure 1(a), grid *i* is called a paper cell, and *in*, *ie*, *is*, *iw* is the adjacent paper cells of *i*. The main procedure of ink dispersion simulation includes ink adding on the surface of the paper, the ink penetrating into the paper from the paper surface, and the ink dispersing between adjacent paper cells.

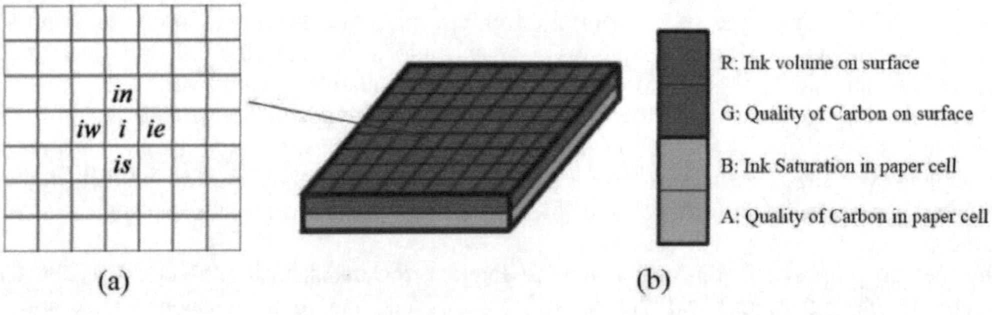

Figure 1. (a) Paper cell i and its surrounding paper cells (b) Paper structure and ink properties

3.1. Paper structure and ink properties

For simulating the dispersion of ink in rice paper, this paper defines a double layer rice paper structure which is used to store the ink properties. The paper structure consists of a surface layer and a paper layer. The surface layer is used for recording the ink properties on the rice paper surface. The paper layer is used for recording ink properties in the rice paper. The ink properties stored in the surface layer include the ink volume on the surface and the quality of carbon in the ink on the surface. The ink properties stored in the paper layer include the ink saturation in the paper cell and the quality of carbon in the ink in the paper cell. The paper structure is represented by a render target texture, with a pixel represents a paper cell, and the ink properties are stored as RGBA value, as shown in figure 1(b). Furthermore, the properties of rice paper are also defined, including the diffuse color of paper and porosity of the rice paper. The porosity is a fraction of the volume of voids over the total volume of the paper cell, between 0 and 1. The diffuse color and porosity of rice paper are represented by two texture map respectively, as shown in figure 2.

(a) (b)

Figure 2. (a) Diffuse color texture of rice paper (b) Porosity texture of rice paper

3.2. Paper penetration

Paper penetration describes the process of ink flowing into the interior of paper from the surface of rice paper. When a drop of ink drops on the surface of the paper cell *i*, it gradually flows into the inside of paper from the surface, the flow rate is influenced by ink density, gravity, the porosity of paper cell ***i***, and the kinetic viscosity of ink:

$$\Delta V_i = \frac{\rho g \Phi(i)^2}{3\mu} \quad (1)$$

Where ρ, g, $\Phi(i)$, μ denote the ink density, the acceleration of gravity, the porosity of paper cell ***i*** and the kinetic viscosity of ink respectively. In order to reduce the calculation error, the kinetic viscosity of the ink is an approximate value influenced by the concentration of ink:

$$\mu = \mu_{water} + \frac{\omega_i}{\omega_{\max}}(\mu_{ink} - \mu_{water}) \quad (2)$$

Where μ_{water} and μ_{ink} denote the kinetic viscosity of pure water and pure ink respectively, which are defined by the real measurement. ω_i is the concentration of ink on the surface of paper cell i. ω_{max} is maximum dilution concentration of ink defined by the user. If the concentration of ink is greater or equal to the ω_{max}, it means that the kinetic viscosity of the ink can be equivalent to water.

3.3. Dispersion in the paper

Dispersion in the paper is the process of transferring the ink in the paper cell to the adjacent paper cell. When a paper cell *i* containing ink and has no ink in its adjacent paper cell *j*, like the boundary area of an ink stroke, the paper cell *i* in the ink will be transferred to the paper cell *j*. With the ink transferring, the boundaries of the ink will expand and form a diffusion effect. the rate of ink transferring from paper cell *i* to paper cell *j* is calculated by the equation as follow:

$$Q_{ij} = \begin{cases} 0 & S_i < S_t \\ -\dfrac{\Phi(i)^2}{\mu}(S_i - S_j) & S_i \geq S_t \end{cases} \tag{3}$$

Where S_i and S_j are the ink saturation in the paper cell i and paper cell j respectively, S_t is a constraint threshold value. When the ink saturation in paper cell i is lower than S_t, the ink in the paper cell i no longer transfers to the adjacent paper cell. As a threshold, the size of S_t determines the ink's ability of diffuse. According to the actual observation, it can be found that different concentrations of ink have different diffusion on rice paper. Depending on the concentration of ink in each paper cell, the threshold will also be different. the S_t is calculated by the equations as follow:

$$S_t = S_{water} + e^{-\omega_{mean}}(S_{ink} - S_{water}) \tag{4}$$

$$\omega_{mean} = \frac{\omega_i S_i + \omega_j S_j}{S_i + S_j} \tag{5}$$

$$\omega_i = \frac{\rho_{ink} V_i}{m_i} - 1 \tag{6}$$

$$V_i = \Phi(i) V_{cell} S_i \tag{7}$$

Where S_{water} and S_{ink} denote the constraint threshold value of pure water and pure ink respectively, which are defined by the user. V_i and V_{cell} are the volume of voids in paper cell i and the total volume of paper cell respectively.

4. Ink wash painting style rendering

This section mainly introduces the ink wash painting style rendering system. The detailed rendering framework is described in section 4.1 and illustrated in figure 3.

4.1. Rendering framework

The framework of the rendering method in this paper extends on the basis of the method of Yuan et.al[1] that adds an image processing shader as an additional step by utilizing programmable GPU vertex and fragment shaders. The rendering method mainly consists of three parts, including interior shading, dispersion simulation and final output.

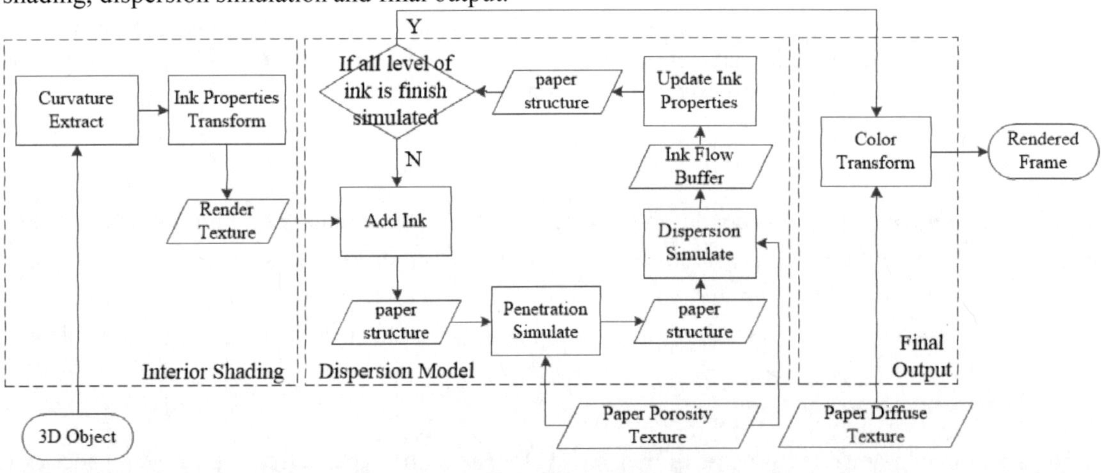

Figure 3. Rendering process

In the part of interior shading, at the first utilizing a vertex and fragment shader to render the curvature of 3D objects and then convert the curvature into the ink properties. The ink properties will be cached into a render target texture which is delivered to next stage.

In the part of dispersion simulation, the procedure of dispersion will be simulated in fragment shader that is used for image processing. There are four passes to process the steps of dispersion simulation as shown in figure 3. First, read the ink properties from the render target texture which is delivered from the last stage, and add them to a render target texture for the paper structure which is defined in section 3.1. Second, input the paper structure to the 'Penetration Simulate' pass which is used to simulate the penetration and described in section 3.2. Then, output the results to the paper structure and input the paper structure to the 'Dispersion Simulate' pass which is used to simulate the dispersion and described in section 3.3. The result of dispersion simulation which is the volume of ink that stream between the paper cell and its adjacent paper cells is cached in a render target texture for an ink flow buffer. Finally, input the buffer and paper structure to the 'Update Ink Properties' pass to update the value of ink properties in the paper structure. In order to present the color sense of ink and the different diffusion effects due to the different ink concentrations, we add ink properties in five times according to the five levels of the quantized curvature. the lightest ink properties are added to the surface layer in first, after a certain number of iterations, the next level of ink properties are added. When all level of ink is finish simulated, the final result of ink properties will be cached into the render target texture for paper structure and delivered to next stage.

In the last part, the ink properties of paper structure are transformed to the pixel color information and rendered to screen.

4.2. Interior shading and color transform

In a general rendering method, the object was usually rendered base on diffuse lighting, such as Lambertian shading. But in our method, the curvature of the model vertices are calculated and mapped to the pixels as color values directly. According to the method proposed by Penner et al.[12], we calculate the vertex curvature value of pixel i by solving the Eqn.(8), the example of calculation result is shown in figure 4.

$$k_i = \frac{\Delta \mathrm{N}}{\Delta P} = \frac{|\, fwidth(normal)\,|}{|\, fwidth(position)\,|} \tag{8}$$

Figure 4. Curvature of the model

Then, according to the vertex curvature, the ink properties adding to the surface layer are calculated by the equations as follow:

$$V_{surface} = \begin{cases} 0 & k_i \le 0.1 \\ V & k_i > 0.1 \end{cases} \tag{9}$$

$$m_{surface} = \rho_{ink} V_{surface} \left(1 + \frac{\omega}{k}\right)^{-1} \tag{10}$$

$$k = \begin{cases} 0.1 & 0.1 < k_i \le 0.3 \\ 0.3 & 0.3 < k_i \le 0.45 \\ 0.45 & 0.45 < k_i \le 0.6 \\ 0.6 & 0.6 < k_i \le 0.8 \\ 0.8 & 0.8 < k_i \end{cases} \tag{11}$$

Where k_i represents the curvature value in pixel i. ρ_{ink} is the density of pure ink, V and ω are the volume of ink and the concentration of ink respectively defined by the user. In order to simulate the gradation of ink color, We quantify the curvature to correspond to the five levels of ink color in the ink painting, the part of $k_i \leq 0.1$ is blank, which means that no ink is added to this part.

When the calculated ink properties are stored in the surface layer, the dispersion model simulates the diffusion of ink by iterating the process of paper penetration and dispersion. Finally, we convert the ink properties of paper layer to screen pixel color value by solving the equation as follow:

$$C(i) = 1 - (1 + (0.01C_i)^2)^{-\frac{1}{2}} \quad (12)$$

5. Experimental results

Our render method is implemented in Unity, and our test machine is equipped with the Graphic Card Nvidia R Geforce GTX1060 and Intel (R) Core (TM2) i5-6300HQ CPUs.

In the experiment, we use vertex and fragment shader to render the model in curvature and simulate ink dispersion in the fragment shader. The render texture is used for the buffer to store and deliver the ink properties during the simulating. The rendering method in this paper is tested with the terrain model generated by terrain editor in Unity.

Figure 5. Mountain rendered in ink wash painting style: the method of Yuan et,al (top); our method(bottom)

Compare the visual quality of our method with the previous method. As shown in the figure 5, the top of the figure is rendered with the method of Yuan et.al[1]. Our result showed on the bottom of the figure comes from this paper. In the simulation of ink diffusion, the effect of ink diffusion generated by our method is more natural, and the color variation of the image is softer. By shading model with the curvature of vertex, our method can show the aesthetic of irregularity in shapes and forms commonly seen in ink wash painting.

Table 1. The performance of rendering method

Resolution of paper structure	FPS
800×600	120 - 140
1024×768	75-90
1280×720	65-80

Table 1 present the rendering speed of our method statistics for various resolutions. Because the simulation requires traversing all the pixels of the paper structure, the performance of our method is obviously influenced by the resolution of paper structures.

6. Conclusion and Future works

This paper proposes a physically-based ink dispersion model is proposed. By defining the paper structure model and ink properties, the theory of the porous media and capillary mechanism are used to abstract the flowing of ink in rice paper. Then, explore the method of ink wash painting style rendering by using dispersion model. Finally, utilizing Unity to build a rendering system and test the terrain model to generate some images in ink wash painting style. Comparing with previous method, this method has a better performance on visual quality.

The future work could be focused on the interior shading of 3D models and improves the performance of dispersion models further more expected to combine the method with fluid simulation to improve temporal coherence and generate more artistic effects.

Acknowledgments

This research is supported by National Key R&D Program (2017YFF0211801), China Postdoctoral Science Foundation funded project (No.2017M620558) and Beijing Postdoctoral Research Foundation (No.2017-ZZ-025).

References

[1] Yuan, M., Yang, X., Xiao, S., & Ren, Z. 2007. GPU-based rendering and animation for Chinese painting cartoon. graphics interface.
[2] Cao, M. 2016. A study of how Chinese ink painting features can be applied to 3D scenes and models in real-time rendering.
[3] Chu, N. S., & Tai, C. 2005. MoXi: real-time ink dispersion in absorbent paper. international conference on computer graphics and interactive techniques, 24(3), 504-511.
[4] Way, D. L., Huang, S. W., & Shih, Z. C. 2003. Physical-based model of ink diffusion in Chinese paintings. Journal of Wscg.
[5] Sun, M., Sun, J., & Yun, B. 2005. Physical Modeling of "Xuan" Paper in the Simulation of Chinese Ink-Wash Drawing. International Conference on Computer Graphics, Imaging and Vision: New Trends (pp.317-322). IEEE.
[6] Dong, L., Lu, S., & Jin, X. 2014. Real-time image-based Chinese ink painting rendering. Multimedia tools and applications, 69(3), 605-620.
[7] Xie, N., Laga, H., Saito, S., & Nakajima, M. 2010. IR2s: interactive real photo to Sumi-e. non-photorealistic animation and rendering.
[8] Yang, Y., Zhu, Q., & Liu, Y. 2015. Simulating water dispersion in Chinese absorbent paper with capillary tubes model. international conference on computer graphics and interactive techniques.
[9] Yang, Y. C. 2016. Research on Temporal coherent oriental ink style 3D rendering. (Doctoral dissertation, Beijing University of Technology).
[10] Chan, C. C., Akleman, E., & Chen, J. 2002. Two methods for creating Chinese painting. pacific conference on computer graphics and applications.
[11] Xu, T., Yang, L., & Wu, E. 2012. Stroke-based real-time ink wash painting style rendering for geometric models. international conference on computer graphics and interactive techniques.
[12] Penner, E., & Borshukov, G. (2011). Pre-Integrated Skin Shading. GPU Pro 2.

IOP Conf. Series: Journal of Physics: Conf. Series **1004** (2018) 012027 doi:10.1088/1742-6596/1004/1/012027

Applying Gradient Descent in Convolutional Neural Networks

Nan Cui[1, *]

[1]Electrical and Computer Engineering, University of Massachusetts Lowell, Lowell, 01854, United States

*834450215@qq.com

Abstract. With the development of the integrated circuit and computer science, people become caring more about solving practical issues via information technologies. Along with that, a new subject called Artificial Intelligent (AI) comes up. One popular research interest of AI is about recognition algorithm. In this paper, one of the most common algorithms, Convolutional Neural Networks (CNNs) will be introduced, for image recognition. Understanding its theory and structure is of great significance for every scholar who is interested in this field. Convolution Neural Network is an artificial neural network which combines the mathematical method of convolution and neural network. The hieratical structure of CNN provides it reliable computer speed and reasonable error rate. The most significant characteristics of CNNs are feature extraction, weight sharing and dimension reduction. Meanwhile, combining with the Back Propagation (BP) mechanism and the Gradient Descent (GD) method, CNNs has the ability to self-study and in-depth learning. Basically, BP provides an opportunity for backward-feedback for enhancing reliability and GD is used for self-training process. This paper mainly discusses the CNN and the related BP and GD algorithms, including the basic structure and function of CNN, details of each layer, the principles and features of BP and GD, and some examples in practice with a summary in the end.

1. Introduction

Human beings have a complex visual recognition system. People distinguish and classify objects independently. Whenever people see an object or scene, it is a picture that is reflected in the brain; then the brain will go through a series of operations and processing to tell people what was seen. The whole process is object recognition. For the computer field, object detection is a very large area of research. This piece of content involves neural networks and machine learning, and deep learning also helps people build more rigorous and efficient identification algorithms. Artificial neural network refers to a computer algorithm that is generated based on animal nerve propagation, which is mainly used to improve the computing speed. From the 1960s, many scientists have begun to study the neural network [1, 2, 3, 4]. Until 1986, with the use of the Back Propagation mechanism (BP), the neural network began to have practical significance at the practical level [5, 6]. Until today, the neural network has become the core algorithm in machine learning and in-depth learning to solve problems such as picture recognition, pattern recognition, speech recognition, natural language processing and video processing [7]. Convolutional Neural Networks (CNNs) is one of the most common algorithms in neural networks. Its biggest characteristics are feature extraction, weight sharing and dimension reduction. The Convolutional Neural Network utilizes the mathematical properties of the convolution, extracts the useful information from the target picture, as well as the propagation characteristics of the neural network, to finally output the result through the classifier. The convolutional neural network combines a variety of algorithms and has the ability of self-machine learning. Nowadays, it has become the most crucial neural network algorithm in depth learning. At the same time, with the development of new technology, Convolutional Neural Network has also been strengthened, where Region with CNN (R-CNN), Fast R-CNN, and Faster R-CNN are the representatives. With the development and progress of CNN, it has been widely used in many fields. To better understand the structure and algorithm of CNN, this paper will focus on CNN itself and its associated BP algorithm and Gradient Descent (GD). The article then further introduces each level of CNNs, mainly including its structure and function. Detailed information about BP and GD is presented as well. The last part is the practical examples and the summary.

2. Convolutional Neural Networks

The Convolutional Neural Networks (CNNs) is a kind of mathematical structure for analysis datasets, images and so on. The primary function of CNNs is extracting features from samples with different requests in fast speed.

2.1. Layers

The Convolutional Neural Network has so many layers which like a long shelf. There are enough quantity computing units or elements in each layer, dealing with datasets at the same time. Computing units or elements in the same layer have the same function working on input data.

Although the CNN is a complex structure, it has three kinds of main layers: convolutional layers, pooling layers, and output layers. Firstly, the features of input data transmitted into the convolution layer will be extracted, and also the size of input data is shrunk. Then, after passing through several convolution layers, the data will be delivered to pooling layers, for enhancing features, and the data would be truncated again. Finally, results of CNNs come out from the output layers. The following will discuss the three layers respectively.

2.1.1 Convolutional Layers. Here is a single computing unit of a convolutional layer, as shown in Figure 1.

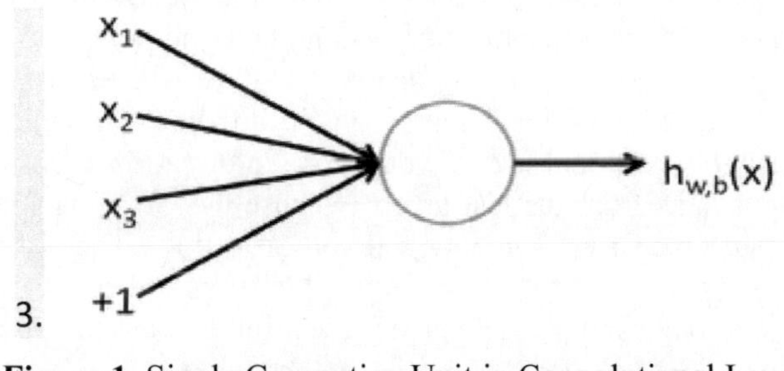

Figure 1. Single Computing Unit in Convolutional Layer

"Xn" refers to the input dataset. The circle is the computing unit which includes a weighted function in it. The weighted function can be regarded as a filter. The equation of the filter is following.

$$h_{W,b}(x) = f(W^T x) = f\left(\textstyle\sum_{i=1}^{3} W_i x_i + b\right) \tag{1}$$

In the equation (1), "W" is weighted functions designed by programmers, for some time it is a "convolutional window", which is a small matrix and elements of the matrix are depended on processes requests. "b" is a fixed element belonged to weighted function, which can be removed sometimes.

Then, if there is a two-layer structure, the first layer can be seen as the input items and the second layer is a convolutional layer, which is shown in Figure 2.

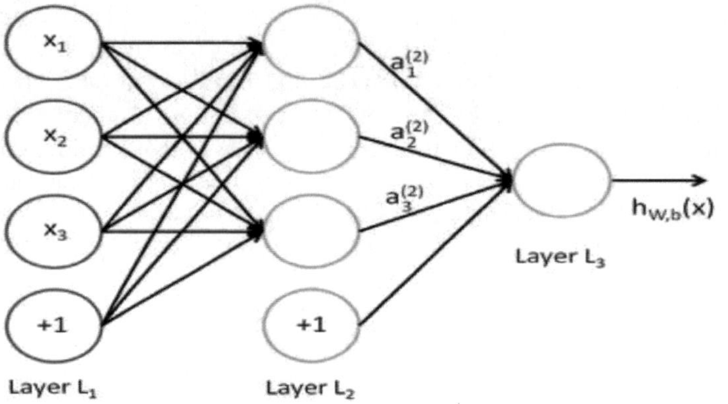

Figure 2. Two-Layer Basic Convolutional Network

In the second layer, each circle has a different weighted function. Therefore, every piece of input data should be calculated with all weighted functions in layer 2, and transmit the results to the layer 3, which is called fully connected convolutional layer. The expressions in details are followings:

$$a_1{}^{(2)} = f\left(W_{11}{}^{(1)}x_1 + W_{12}{}^{(1)}x_2 + W_{13}{}^{(1)}x_3 + b_1{}^{(1)}\right) \quad (2)$$

$$a_2{}^{(2)} = f\left(W_{21}{}^{(1)}x_1 + W_{22}{}^{(1)}x_2 + W_{23}{}^{(1)}x_3 + b_2{}^{(1)}\right) \quad (3)$$

$$a_3{}^{(2)} = f\left(W_{31}{}^{(1)}x_1 + W_{32}{}^{(1)}x_2 + W_{33}{}^{(1)}x_3 + b_3{}^{(1)}\right) \quad (4)$$

$$h_{W,b}(x) = a_1{}^{(3)} = f(W_{11}{}^{(2)}a_1{}^{(2)} + W_{12}{}^{(2)}a_2{}^{(2)} + W_{13}{}^{(2)}a_3{}^{(2)} + b_1{}^{(2)} \quad (5)$$

The fully connected convolutional layer can guarantee all information can be transmitted precisely. However, it also produces huge computational burden. For instance, there are 10000 pieces of data as input units and 100 computed units in a convolutional layer at second row in the network. Then, the calculation will be about 10^6 times, from layer 1 (input layer) to layer 2.

To save operating time and cost, the input items will be passed through a part of convolutional layers units instead of all convolutional units, which is called non-connection convolutional layer. While non-connection convolutional layers would lose information when it is extracting features, the layers miss some input units from the last layer. Because the CNNs is a structure for extracting data features. It can be suffered from some degree of distortion. The dropping information will be affordable in CNNs. People set the first convolutional layer as full connected. After several non-fulled connection convolutional layers, it also needs to place a full connected convolutional layer.

2.1.2Pooling Layers. Before discussing pooling layers, the concept of the classifier will be introduced, which is a standard step in programming, classifying data statistically into various categories. The data cannot be analysed directly, without categorising. To convince classifiers operating, we need to extract features further by using pooling layers and also pooling steps can save the space of data.

Pooling means picking up features from different parts of the results at high layers. It is not similar to convolutional layer's function. For convolutional layers, the weights in filtering equations in units should be fixed again and again by Back Propagation and input data can overlap in different weighted units. For pooling layers, it just wants to pick a sample feature fastly in various groups, such as maximum and mean. And sampling data should not overlap in each pooling groups. The brief working process of pooling layers is shown in Figure 3.

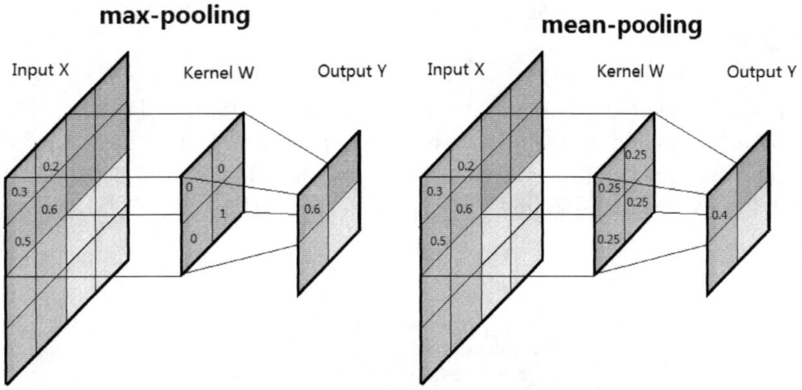

Figure 3. Pooling Works

2.1.3 Output Layers. After filling with so many convolutional layers, the results should be output by classes. Convolutional layers and pooling layers are focusing on extracting features and reducing unnecessary factors. Then, a fully connected layer will work as an output layer to produce suitable quantity and variety results. The categorization cannot be realized only by convolutional layers. There is always a classifier and computing unit for calculating loss function in the output layer. The loss function is for amending weighted factor in above convolutional layers. Once the CNNs finishes forward passway, the loss function also begins to work on backward passway which fixes error and loss during weighted functions.

2.2. Process
As the CNNs are always applied on image processing field, data is displayed as pixel map like a huge matrix.

2.2.1. Convolution. For most weighted functions in convolutional layers, they are designed as smaller pixel maps according to need, which is called Kernel or convolution window. Figure 4 shows how the pixel map works with the kernel in a convolutional way.

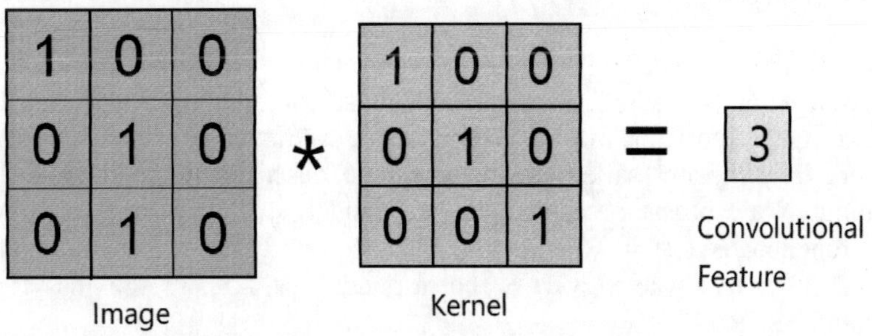

Figure 4. An Example of Convolution

The convolution is the sum of the products between each image pixel and kernel pixel. For example, in Figure 4, each of 9 pixels can be extracted as 1 pixel to stands feature value. The higher value of result means this area's data has a higher relationship with a convolutional window so that kernel can be regarded as a filter. Besides various values in kernel's pixels, the convolution process is also changed by edge and stride. Edge is easy to understand it is just the length of the side of convolution window. It describes the size of kernels. When a kernel finishes a convolution process with a part of input data, it should move to another place for doing the next convolution. The moving distance of kernels depends on the value of stride that the distance of kernels is moving during each convolutional processes.

2.2.2 Pooling. When the picture is too large to pass through the classifier, the pooling layers will be used. The only function of pooling layers is reducing the size of input data or picture. The normal way is grouping the input data and picking the maximum in each group which is called max-pooling. Figure 5 displays the max-pooling.

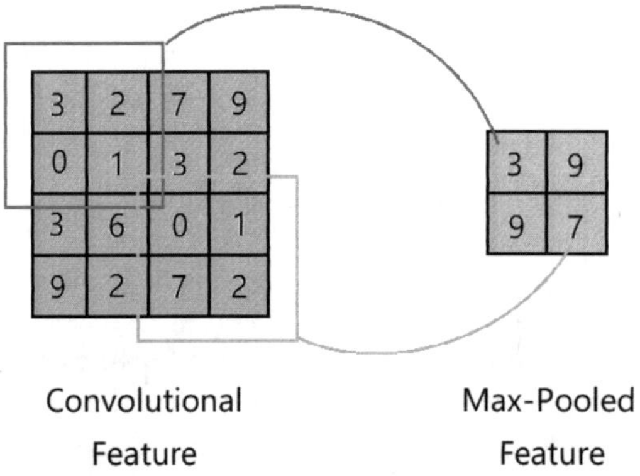

Figure 5. An Example of Max-Pooling

3. Gradient Descent and Backpropagation

There are two passways in the Convolutional Neural Networks: forward passway and backward passway. The first two sections in the paper are talking about the forward passway. In this section, backward passway will be discussed. Because of the initialization of the weighted function at first, in the CNNs process, the weighted function will not be confirmed, whether satisfying the request of structure's precision. The function should be fixed again and again. The Back Propagation (BP) is for transmitting the fixing error to the lower layers from the upper layers. And then the lower layers can maintain the weighted functions by fixed errors. How can we find "the fixed errors"? Here comes the Gradient Descent method.

3.1. Gradient Descent method

Gradient Descent (GD) method is one of the most common optimisation algorithms in machine learning field. As we know, the most of the mathematical models have errors because of the estimating factors. There is a concept called loss function, to describe the total error between samples and function output. We can assume a mathematical model is a linear regression equation and give it a hypothesis function: $h_\theta(x) = \theta_0 + \theta_1 x$, θ_0 and θ_1 are factors of the equation. Its sample is (x_i, y_i) $(i = 1,2, \ldots, n)$, where every x_i corresponds a y_i. Then, the loss function of hypothesis fucntion should be:

$$J(\theta_0, \theta_1) = \sum_{i=1}^{m} (h_\theta(x_i) - y_i)^2 \tag{6}$$

Gradient Descent method aims to focus on optimising mathematical model and loss function, which means amending the factors of original functions and reduce the value of loss functions.

Now, let us make the hypothesis function more complicated than before, such as $h_\theta(x_1, x_2, \ldots . x_n) = \theta_0 + \theta_1 x_1 + \cdots + \theta_n x_n$ $[\theta_i = (i = 1,2, \ldots, n)]$, we can add a feature value $x_0 = 1$, then we simplify it as: $h_\theta(x_1, x_2, \ldots . x_n) = \sum_{i=0}^{n} \theta_i x_i$.

Then, the loss function is:

IOP Conf. Series: Journal of Physics: Conf. Series **1004** (2018) 012027 doi:10.1088/1742-6596/1004/1/012027

$$J(\theta_0, \theta_1, \dots, \theta_n) = \frac{1}{2m}\sum_{i=0}^{m}(h_\theta(x_0, x_1, \dots, x_n) - y_i)^2 \tag{7}$$

Because we want to reduce the value of loss function, we should make microcosmic it at first to analyse changes of functions. The partial derivatives are good chances.Here we should mention some concepts for initialising the Gradient Descent method.

Gradient, which measures the trend of aim function, we regard the partial derivative result of θ_i to loss functions as a gradient, the equation is: $\frac{\partial}{\partial\theta_i}J(\theta_0, \theta_1, \dots, \theta_n)$. Recursion terminal distance ε is the functions' precision usually a difference between mathmetical model output and sample. When the difference is less than or equals to the recursion terminal distance, the weighted functions in the model are satisfied to the request of process. Then, the process can stop. Step size α, a factor contrals how much percent of gradient used to update new θ_i. The updating fucntion expression is: $\theta_{i\,new} = \theta_{i\,old} - \alpha\frac{\partial}{\partial\theta_i}J(\theta_0, \theta_1, \dots \theta_n)$. If the goal function is a convex function, finally it can be optimizied. The step size is also a significant element during programming. When the step size is too small, a process can figure out the most optimal option, however, the convergence speed is very slow. Large step size can improve convergence speed extremely, while it cannot garuantee obtaining the optimal option, because one large step size sometimes already covers a small convergence space. Then it cannot detect a optimal point in this convergence space. Figure 6 shows the comparison among different step size's loss functions convergence situations.

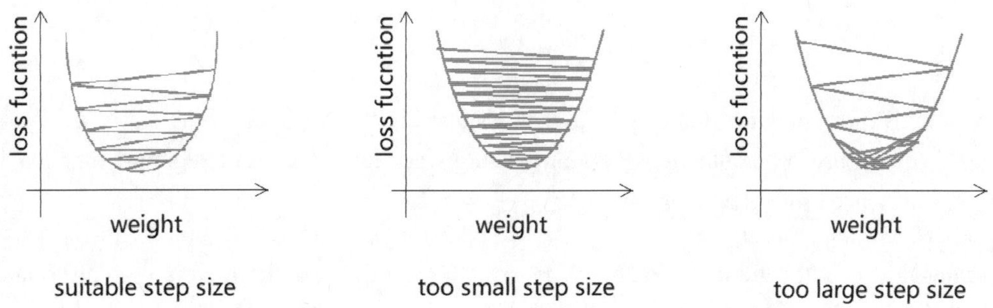

Figure 6. The Comparison of Different Step Size

Therefore, the general gradient descent algorithm is following:

Initialise $\theta_0, \theta_1, \dots . \theta_n$ in aim function, recursion terminal distance ε, step size α.

Calculate the gradient, the partial derivative of the loss function of aim function, $\frac{\partial}{\partial\theta_i}J(\theta_0, \theta_1, \dots . \theta_n)$.

If $J(\theta_0, \theta_1, \dots . \theta_n) \leq \varepsilon$, stop the algorithm, otherwise, continue the process.

Use the gradient multiplying with step size α.

Renew all θ by $\theta_i = \theta_i - \alpha\frac{\partial}{\partial\theta_i}J(\theta_0, \theta_1, \dots . \theta_n)$, and then back to step II.

3.2. Backpropagation

Besides updating weight functions, the updating fixed errors should be passed through the layers in the CNNs. It seems like combining Gradient Descent algorithm, the chain rule and Back Propagation algorithm.

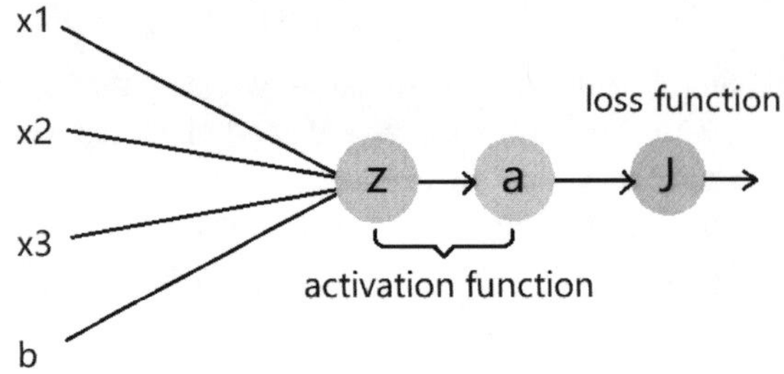

Figure 7. An Example of Backpropagation in Single Calculating Unit

Figure 7 is a diagram of a single neural cell. We assume training sample is(x, y), x is input, which passes through activation function$h_{w,b}(x)$. And the unit gets an output"a", then the result "a" goes through the loss function, the output of loss function is the cost J.

The expression of activation function or hypothesis function is:

$$h_{w,b}(x) = a = sigmoid(z) = sigmoid(\sum_{i=0}^{n}(x_i w_i + b)) \quad (8)$$

The equation of loss function is:

$$\mathrm{J(W, b, x, y)} = \frac{1}{2}\left\|y - h_{w,b}(x)\right\|^2 \quad (9)$$

In the activation function (equation (8)), the sigmoid function $\sigma(z) = \frac{1}{1+e^{-z}}$, is a hypothesis function. We can use other functions as an activation function either. The sigmoid function has a property is convinced for calculating: $a = \sigma(z)$; $\frac{\partial a}{\partial z} = a(1 - a)$.

During the training, we should figure out the gradients of the cost J by partial derivatives. There are two gradients; one is for the factor W, another one is for the factor b. Then we are going to take the partial derivative concerning W and b. However, we cannot calculate the partial derivative from cost J to W and b. The chain rule would help to solve this problem. It can display as $\frac{\partial}{\partial W}J = \frac{\partial J}{\partial z}\frac{\partial z}{\partial W}$. Now, we should figure out the partial derivative from cost J to intermediate variable "a" and z at first:

$$\delta^{(a)} = \frac{\partial}{\partial a}J(W, b, x, y) = -(y - a) \quad (10)$$

$$\delta^{(b)} = \frac{\partial}{\partial z}J(W, b, x, y) = \frac{\partial J}{\partial a}\frac{\partial a}{\partial z} = \delta^{(a)}a(1 - a) \quad (11)$$

Then, according to the chain rule, obtain factor W and b's gradients:

$$\nabla_W J(W, b, x, y) = \frac{\partial}{\partial W}\mathrm{J} = \frac{\partial J}{\partial z}\frac{\partial z}{\partial W} = \delta^{(z)}x^T \quad (12)$$

$$\nabla_b J(W, b, x, y) = \frac{\partial}{\partial b}\mathrm{J} = \frac{\partial J}{\partial z}\frac{\partial z}{\partial b} = \delta^{(z)} \quad (13)$$

During the above processes, we take $\frac{\partial J}{\partial a}$ firstly, then figure out $\frac{\partial J}{\partial z}$, finally we get $\frac{\partial J}{\partial W}$ and $\frac{\partial J}{\partial b}$. Therefore, we can find that this is a procedure which propagates the increment $\partial \mathrm{J}$ of the cost function from the back forward, it called back propagation.

4. Applications

4.1 License Plate Recognition
The most important application field of CNN is in picture recognition, and a very common application scene in picture recognition is vehicle license plate recognition. In the same area, the number and style of vehicle license plate numbers or characters are fixed, so the training system cost is greatly reduced. At the same time, the characteristics of vehicle license plate, such as the unified size of plate shape, make the license plate easily to be located and recognized. However, there are also problems and limitations in license plate recognition. For instance, as for high-speed vehicles, it is difficult to use the ordinary camera for image recognition. Similarly, under some extreme weather conditions, such as fog, rain and snow weather, it is extremely hard to obtain clear vehicle license plate images. These problems increase the difficulty of license plate recognition. Therefore, the vehicle license plate recognition can only be applied in a specific scene, such as the underground garage. With the upgrading of hardware technology, vehicle license plate recognition technology is bound to have more development and application.

4.2 Semantic Recognition
In addition to image recognition, CNN can also be used for semantic recognition. Most of the applications of CNN in semantic recognition exist in meaning detection, such as spam detection or topic classification. By using word2vec to translate semantic words into word vectors, an input matrix is constructed, followed by a convolution layer consisting of several filters, as well as a maximal pool layer and a Softmax classifier. Training mainly exists in the adjustment of word2vec, with the increase of training set, the semantic differentiation becomes more obvious.

5. Conclusion
This paper introduces the Convolutional Neural Network algorithm, which is very popular in the field of machine learning and image recognition. It combines the Back Propagation mechanism and the Gradient Descent method to discuss its basic structure and function. At the same time, the practical application and role are discussed.

With the continuous development of technology, the neural network is bound to become an important tool in people's lives. The structure of the neural network will be more complete and efficient. For the neural network, the biggest challenges in the future may not only exist in the more optimized system, but also in the quality and quantity of the training set. A high-quality training set is an efficient neural network guarantee. Therefore, in the subsequent research of machine learning and in-depth study, the competition of algorithms is likely to become a competition of the quality and scale of data.

References
[1] Rosenblatt, F. (1962). Principles of neurodynamics.

[2] Kabrisky, M. (1964). a Proposed Model for Visual Information Processing in the Human Brain.

[3] Giebel, H. (1971). Feature extraction and recognition of handwritten characters by homogeneous layers. In Zeichenerkennung durch biologische and technische Systeme/Pattern Recognition in Biological and Technical Systems (pp. 162-169). Springer, Berlin, Heidelberg.

[4] Fukushima, K. (1975). Cognition: A self-organizing multilayered neural network. Biological cybernetics, 20(3-4), 121-136.

[5] Rumelhart, D. E., Hinton, G. E., & Williams, R. J. (1985). Learning internal representations by error propagation (No. ICS-8506). California Univ San Diego La Jolla Inst for Cognitive Science.

[6] LeCun, Y., Boser, B., Denker, J. S., Henderson, D., Howard, R. E., Hubbard, W., & Jackel, L. D. (1989). Backpropagation applied to handwritten zip code recognition. Neural computation, 1(4), 541-551.

[7] Koushik, J. 2016. Understanding convolutional neural networks.

IOP Conf. Series: Journal of Physics: Conf. Series **1004** (2018) 012028 doi:10.1088/1742-6596/1004/1/012028

Understanding the Convolutional Neural Networks with Gradient Descent and Backpropagation

XueFei Zhou[1,] *

[1]School of Computer Science and Technology, Soochow University, Suzhou 215006, China

*821573861@qq.com

Abstract. With the development of computer technology, the applications of machine learning are more and more extensive. And machine learning is providing endless opportunities to develop new applications. One of those applications is image recognition by using Convolutional Neural Networks (CNNs). CNN is one of the most common algorithms in image recognition. It is significant to understand its theory and structure for every scholar who is interested in this field. CNN is mainly used in computer identification, especially in voice, text recognition and other aspects of the application. It utilizes hierarchical structure with different layers to accelerate computing speed. In addition, the greatest features of CNNs are the weight sharing and dimension reduction. And all of these consolidate the high effectiveness and efficiency of CNNs with idea computing speed and error rate. With the help of other learning altruisms, CNNs could be used in several scenarios for machine learning, especially for deep learning. Based on the general introduction to the background and the core solution CNN, this paper is going to focus on summarizing how Gradient Descent and Backpropagation work, and how they contribute to the high performances of CNNs. Also, some practical applications will be discussed in the following parts. The last section exhibits the conclusion and some perspectives of future work.

1. Introduction

The human has a complex visual recognition system. We distinguish and classify objects independently. The structure of the neural network is imitated from the biological neural network. Each neuron in the neural network is connected with other neurons. When it is "excited", it is necessary to send the next level of neurons to send the chemical substance to change the potential of these neurons. If the potential of a neuron exceeds a threshold, it is activated; otherwise it will not be activated.

One of the most common research fields in computer science is image recognition. And Convolutional Neural Networks (CNNs) are hierarchical neural networks for recognitions, by using significant process, such as Gradient Descent [1] and Backpropagation [2]. The idea of the earliest neural network originated in the 1943. McCulloch and Pitts (MCP) [3] raised an artificial neural model, which was intended to use a computer to simulate the process of human neuron response. It simplifies neurons into two processes: linearization of the input signal, and non-linear activation (threshold method). The first use of MCP for machine learning was the perceptron algorithm invented by Rosenblatt in 1958 [4]. The algorithm uses the MCP model to classify the input multidimensional data and uses the gradient descent method to automatically learn the update weights from the training samples. In 1962, the method proved to be able to converge; the theoretical and practical effects caused the wave of the first neural network [5]. However, the history of the development of disciplines is not always plain sailing. In 1969, the American mathematician and artificial intelligence pioneer Minsky [6] proved that the sensor is essentially a linear model, which can only deal with linear classification problems; even the simplest XOR (or) problems cannot be correctly classified. This is equivalent to a direct sentence of the death of sensor, neural network research also stuck in nearly 20 years of stagnation. The first time to break the nonlinear curse was undoubtedly the research conducted by Rumelhart, Hinton, et al. [7], who invented a multi-layer sensor (MLP) BP algorithm in 1986, and using Sigmoid for non-linear mapping, an effective solution to the nonlinear classification and Learning problems. This method causes the second wave of neural networks. In 1989, Robert Hecht-Nielsen [8] proved MLP's universal approximation theorem, that is, for a continuous function f in any closed interval, it can be approximated by a BP network with a hidden layer. This discovery greatly encouraged the researchers. Also in 1989, LeCun [2] invented the convolution of the neural network - LeNet, and its use for digital identification. In 2012, in order to respond to others' query about the deep learning, Hinton and his students applied deep learning to ImageNet (the largest database in image recognition), and ultimately achieved considerably impressive results [9].

This paper is focusing on the theory of CNNs, including its structure and learning process as well as the understanding of Gradient Descent and Backpropagation. Then some practical applications with CNNs will be displayed.

2. Convolutional Neural Networks

2.1. Layers

In a typical CNN, the beginning layer is convolution layer, and the last layer is output layer. The layers between them are called hidden layers. Then main purpose of convolution layer is to extract image features, then drive them into the hidden layers for computing, and output the results via output layer. Layers among hidden layers usually, such as pooling layers (sub-sampling layers) are partially connected, while the output layers are fully connected. Partial connection could reduce the computing load by using shared weights, namely kernels.

2.2. Process

For example, the input layer consists of 32×32 sensing nodes, receiving the original image. The calculation then alternates between convolution and sub-sampling, which stated as follows:

The first hidden layer is convoluted, which consists of eight feature maps, each feature map consists of 28×28 neurons, each neuron specifies a 5×5 acceptance domain;

The second hidden layer implements sub-sampling and local averaging. It is also composed of eight feature maps, but each feature map consists of 14×14 neurons. Each neuron has a 2×2 acceptance field, a training coefficient, a training bias, and a sigmoid activation function. The training factor and the bias control the operating point of the neuron.

The third hidden layer is the second convolution, which consists of 20 feature maps, each consisting of 10×10 neurons. Each neuron in the hidden layer may have a synaptic connection to several feature maps of the next hidden layer, which operates in a similar manner to the first convolution layer.

The fourth hidden layer performs the second sub-sampling and local average calculation. It consists of 20 feature maps, but each feature map consists of 5 x 5 neurons, which operate in a similar manner to the first sample.

The fifth hidden layer implements the final stage of convolution, which consists of 120 neurons, each of which specifies a 5×5 accepted domain.

Finally, a full connection layer, get the output vector.

The successive computational layers alternate between convolutional layers and sampling layers, as spatial resolution decreases in each convolutional layer or sampling layer, the number of feature mappings increases compared to the corresponding previous layer.

3. Gradient Descent

In mathematics, gradient is a vector, referring to the direction with largest changing value at one point. The method Gradient Descent used in machine learning is to define the distinction between two values to train the neural networks.

The gradient descent method is based on the following process:

1) First, assign the value of θ, this value can either be random or an all-zero vector.

2) Second, change the value of θ so that J (θ) decreases in the direction in which the gradient decreases.

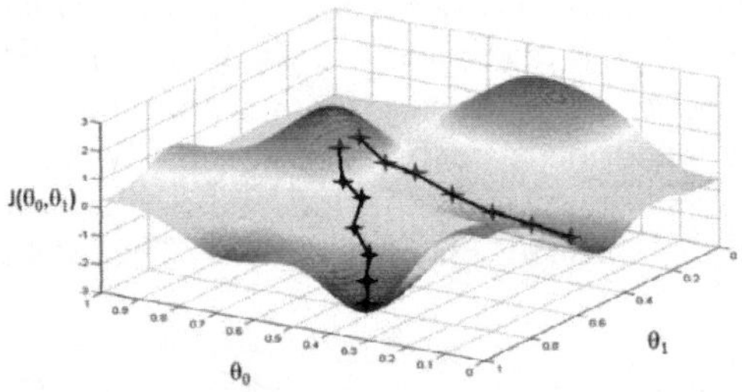

Figure 1. Gradient Descent

Figure 1 is a diagram showing the relationship between the parameter θ and the error function J (θ). The red part is where J (θ) has a relatively high value. We need to be able to make the value of J (θ) as low as possible, where the dark blue part is. Θ0 and θ1 represent two dimensions of the θ vector. The first step of gradient descent is to initiate the value of θ, assuming that the initial value given to the graph is the cross at the graph.

Then we adjust θ in accordance with the direction of the gradient, it will make J (θ) shifts toward a lower direction, as shown in the figure, the algorithm will stop by θ dropping to an extreme point.

Gradient reduction process is stated as follows:

$$\frac{\partial}{\partial\theta}J(\theta)=\frac{\partial}{\partial\theta}\frac{1}{2}\sum_{i=1}^{m}(h_\theta(x)-y)^2=(h_\theta(x)-y)x^{(i)} \tag{1}$$

The following is the process of updating, that is, θi will be reduced according to the direction of the minimum gradient. Θi represents the value before the update, the latter part represents the amount of decrease in the gradient direction, α represents the step size, which is the variation in the direction of the gradient reduction each time.

$$\theta_i=\theta_i-\alpha\frac{\partial}{\partial\theta}J(\theta)=\theta_i-\alpha(h_\theta(x)-y)x^{(i)} \tag{2}$$

A very important thing to be mentioned is that the gradient is directed. For a vector θ, each dimension component θi can find a gradient direction; thus we can find a holistic direction. It is possible to reach a minimum point by following the direction which descends in the quickest way, whether it is local or global.

4. Backpropagation

The Backpropagation error can be seen as the sensitivity of each neuron's base, which defined as follows:

$$\frac{\partial E}{\partial b}=\frac{\partial E}{\partial u}\times\frac{\partial u}{\partial b} \tag{3}$$

$$\text{Since} \frac{\partial u}{\partial b}=1, \frac{\partial E}{\partial b} = \frac{\partial E}{\partial u} = \delta \tag{4}$$

The sensitivity of layer l:

$$\delta^1 = (W^{l+1})^T \times \delta^{l+1} \circ \int{}'(u^l) \tag{5}$$

" o" means that each element is multiplied. The sensitivity of the neurons in the output layer is different:

$$\delta' = \int{}'(u^l) \circ (y^n - t^n) \tag{6}$$

Finally, the weights are updated for each neuron using the delta rule. Specifically, for a given neuron, CNNs get its input, and then use the delta of the neuron to be scaled. In the form of a vector, for the first layer, the derivative of the error for each weight of the layer (combined into a matrix) is the cross-product of the input of the layer (equal to the output of the previous layer) and the sensitivity of the layer (the delta of the neuron is combined into a vector form). And then get the weight update of this layer of neuron by multiplying partial derivative by a negative learning rate:

$$\frac{\partial E}{\partial W^l} = x^{l-1} \times (\delta^l)^T \tag{7}$$

$$\Delta W^l = -\eta \times \frac{\partial E}{\partial W^l} \tag{8}$$

In fact, there is a specific learning rate η_{l_j} for each weightW_{l_j}.

5. Applications

In 2015, SUN Yan-Feng and her team proposed a deep learning algorithm based on Fisher criterion. In the feed forward spread, the method used convolution neural network to extract automatically image features such as structural information. And it used convolution network of sharing weights and pooling, subsampling methods to reduce the weight number, which considerably reduced the model complexity. When the Back Propagation adjusted the weights, the method adopted the constraints based on Fisher criterion. At the same time, it kept the samples in small distance with-class and large distance between-class; so that the weights could be closer to the optimal value for classification. It improved the recognition rate effectively when the sample size was insufficient or when it had few training iterations.

In 2013, an article entitled "Deep Feature Learning for Knee Cartilage Segmentation Using a Triplanar Convolutional Neural Network" by Adhish Prasoon et al. [10] was published. In the paper, CNN was used to segment the articular cartilage of the knee. The traditional CNN is two-dimensional. The direct expansion to the three-dimensional need more parameters, thus the network is more complex and requires longer training time as well as more training data. And the use of two-dimensional data alone does not utilize the three-dimensional features, may lead to decreased accuracy. To this end, Adhish [10] used a compromise: use xy, yz and xz three 2D CNNs and combine them together.

Google's DeepMind team made a major breakthrough in the computer go by using deep convolution neural network. In Google team's paper, they mentioned "We use 19X19 images to pass the board position" to "train" two different depth neural networks, which is "Policy network" and "value network". Their task is to "pick" out those more promising moves, and to abandon those obviously inferior moves, so that the amount of calculation by computer is confined to a rational range.

6. Conclusion

This paper briefly introduces the process of CNNs. The significance of Gradient Descent and Back Propagation makes CNNs to be trained in a more efficient way.

6.1. Strong non-linear mapping capabilities

BP neural network essentially implements a mapping function from input to output, and mathematical theory proves that the three-layer neural network can approximate any nonlinear continuous function

IOP Conf. Series: Journal of Physics: Conf. Series **1004** (2018) 012028 doi:10.1088/1742-6596/1004/1/012028

with arbitrary precision. This makes it particularly suitable for solving the complex problem of internal mechanism.

6.2. High degree of self-learning and adaptive ability

When BP neural network is being trained, it can learn to automatically extract "reasonable rules" between input and output data.

6.3. Fault tolerance

BP neural network's overall training results will not be greatly affected even when its local or part of the neurons get damaged, that is to say even if the system is subject to local damage, it can still work properly.

References

[1] Bottou, L. (2010). Large-scale machine learning with stochastic gradient descent. In Proceedings of COMPSTAT'2010 (pp. 177-186). Physica-Verlag HD.

[2] LeCun, Y., Boser, B., Denker, J. S., Henderson, D., Howard, R. E., Hubbard, W., and Jackel, L. D. (1989). Backpropagation Applied to Handwritten Zip Code Recognition. In Neural Comput., 1(4):541-551.

[3] McCulloch, W. S., & Pitts, W. (1943). A logical calculus of the ideas immanent in nervous activity. The bulletin of mathematical biophysics, 5(4), 115-133.

[4] Rosenblatt, F. (1962). Principles of neurodynamics.

[5] Hubel, D. H., & Wiesel, T. N. (1962). Receptive fields, binocular interaction and functional architecture in the cat's visual cortex. The Journal of physiology, 160(1), 106-154.

[6] Minsky, M., & Papert, S. (1969). Perceptrons.

[7] Rumelhart, D. E., Hinton, G. E., & Williams, R. J. (1988). Learning representations by back-propagating errors. Cognitive modeling, 5(3), 1.

[8] Hecht-Nielsen, R. (1988). Theory of the backpropagation neural network. Neural Networks, 1(Supplement-1), 445-448.

[9] Krizhevsky, A., Sutskever, I., & Hinton, G. E. (2012). Imagenet classification with deep convolutional neural networks. In Advances in neural information processing systems (pp. 1097-1105).

[10] Prasoon, A., Petersen, K., Igel, C., Lauze, F., Dam, E., & Nielsen, M. (2013, September). Deep feature learning for knee cartilage segmentation using a triplanar convolutional neural network. In International conference on medical image computing and computer-assisted intervention (pp. 246-253). Springer, Berlin, Heidelberg.

IOP Conf. Series: Journal of Physics: Conf. Series **1004** (2018) 012029 doi:10.1088/1742-6596/1004/1/012029

Understanding of Object Detection Based on CNN Family and YOLO

Juan Du[1,*]

[1]New Research and Development Center of Hisense, Qingdao 266071, China

*dqxwpl@sina.com

Abstract. As a key use of image processing, object detection has boomed along with the unprecedented advancement of Convolutional Neural Network (CNN) and its variants since 2012. When CNN series develops to Faster Region with CNN (R-CNN), the Mean Average Precision (mAP) has reached 76.4, whereas, the Frame Per Second (FPS) of Faster R-CNN remains 5 to 18 which is far slower than the real-time effect. Thus, the most urgent requirement of object detection improvement is to accelerate the speed. Based on the general introduction to the background and the core solution CNN, this paper exhibits one of the best CNN representatives You Only Look Once (YOLO), which breaks through the CNN family's tradition and innovates a complete new way of solving the object detection with most simple and high efficient way. Its fastest speed has achieved the exciting unparalleled result with FPS 155, and its mAP can also reach up to 78.6, both of which have surpassed the performance of Faster R-CNN greatly. Additionally, compared with the latest most advanced solution, YOLOv2 achieves an excellent tradeoff between speed and accuracy as well as an object detector with strong generalization ability to represent the whole image.

1. Introduction

Humans glance at an image and instantly know what objects are in the image, where they are, and how they interact. If algorithms for image processing could be accurate and fast enough, the computers would be able to drive cars without specialized sensors, and assistive devices would be able to convey real-time scene information to users. Likewise, if these algorithms could complete Deep Learning (DL) tasks with high efficient and excellent performance like human beings do, it would be real Artificial Intelligent (AI). Thus, the core tasks of image processing are a serial of recognition: classification, localization and object detection; and the key challenges are: accuracy, speed, cost and complexity.

To achieve an ideal effect, algorithms of recognition have gone a long way before the significant boom of Convolutional Neural Network (CNN) brought by AlexNet [1] in 2012. From the early concept of the receptive fields [2] and the improved Neural Network (NN) models (Rosenblatt [3] in 1962; Kabrisky [4] in 1966; Giebel [5] in 1971; Fukushima [6] in 1975) to Back Propagation Neural Netowrk (BP_NN) [7, 8] in 1986. The convolutional framework came out in Fukushima's Neocognitron [9] in 1980s, a biologically inspired hierarchical and shift-invariant pattern recognition model with shortage of lacking supervised training algorithm. It first grew into classical model LeNet [10] in 1998. CNNs saw heavy use in 1990s [11] but had been overshadowed by shallow machine learning models (eg. Support Vector Machines (SVM), Boosting, Maximum Entropy Methods). Until 2012, AlexNet via CNN reduced the error rate from 26% to 15.3%, which dramatically fueled the development of DL, rekindled CNNs and GPU.

To approach higher accuracy and speed, CNNs have been developed in two directions after being popular in 2012. One direction is optimization, such as batch normalization, VGGNet [12] (22 layers; Error rate is 7.3%) of Visual Geometry Group (VGG), GoogLeNet [13] (22 layers, error rate is 6.67%) which gained the champion of ImageNet Large Scale Visual Recognition Challenge 2014 (2014 SVRC), and Deep Residual Learning [14] (ResNet, 152 layers, error rate is 3.57%). The other direction is to improve the framework of image processing. CNN's strong ability of feature extraction and classification has good effect on dealing with image with single object, while the problem of window selection was highlighted when the image contains multi-objects. To solve the problem, Regions with CNN (R-CNN) introduced new techniques, such as region proposals, crop/warp, SVM classification and bounding-box regression. R-CNN achieved excellent object detection accuracy with the Mean Average Precision (mAP) of 54% on Visual Object Classes (VOC) 2010 compared to 33% for the Deformable Part Model (DPM) [15, 16] which is based on Histogram of Oriented Gradients (HOG). However, R-CNN's pre-extraction and multi-stage pipeline using cross-entropy loss for each object proposal consume expensive space and time. Then, with the help of warping proposal behind the R-CNN's last convolutional layer, Spatial Pyramid Pooling Network (SPPNet) [17] accelerated its test speed by 10 times and training speed by 3 times than those of R-CNN. And SPPNet shared similar drawbacks with R-CNN as its multi-stage pipelines. Fast R-CNN solved the problems with a single-stage and multi-task loss; added Region of Interest (ROI) to enhance the whole training; and used Singular Value Decomposition (SVD) algorithm to speed up. Fast R-CNN advanced with 10 times' test speed and 3 times' training speed than SPPNet, gained mAP value of 70. Even though Fast R-CNN seemed to have achieved nearly real-time rates using very deep networks [12]. Its proposals are the computational bottleneck in detection systems as well. Therefore, Faster R-CNN effectively introduced Region Proposal Networks (RPN) to share convolutional layers with object detection networks and computed proposals with deep net. It gained a nearly cost-free computation of the detection network. Faster R-CNN with ResNet [18] gained mAP value of 76.4 and Frame Per Second (FPS) of 5. On the accuracy of image recognition, Faster R-CNN has indeed achieved fairly good effect, while dealing speed of Faster R-CNN still has space to quicken in the state-of-the-art detection systems.

In this paper, a new direction of solving the bottleneck of proposals You Only Look Once (YOLO) will be discussed after the background descriptions of CNNs (especially the key incarnation Faster R-CNN). Two versions of YOLO will be introduced: YOLO V1 and V2; YOLO's layers, algorithms and characteristics will all be listed. By comparing YOLO with Faster R-CNN on accuracy, speed, cost and complexity, their advantages and shortages would be exposed. At last, the conclusion of YOLO and Faster R-CNN will be summarized with the outlook to the CNN future.

2. The Introduction of CNN

CNN [19] started with LeNet in 1998 and prospered in 2012 because of AlexNet, with the error rate 15.3%. Then, ZF-net [20], GoogLeNet and VGGNet brought lower and lower error rate. By the end of 2015, ResNet has gained an error rate 3.6% which was even less than that of the human eyes (5.1%). And this is the first time that the recognition capability of deep learning models has surpassed that of the human beings.

2.1. Layers of CNN

The framework of the network layers was kept from NN to CNN, while the function and formation of the network have been changed. The layers of CNN are shown as following.

2.1.1. Input Layer The main task of input layer is to initialize the input image data to make all the dimensions of the input data zero-centered. Then, it normalizes the scale of all the input data within [0,1] to accelerate the converging speed, and whitens the data to decrease the redundancy. It utilizes Principal Components Analysis (PCA) to degrade and decorrelate the dimensions of the data to make them focused on just several key factors [1].

2.1.2. Convolutional (CONV). Layer Convolutional layer is the core of CNNs. It uses CONV kernel as a filter to slide on the original image. The values of each pixel of the local correlated data within the filter will be multiplied and added as the convolutional results. This type of rule is called convolution which enables the features of the image extracted with designed CONV kernel. Besides, the method of filtering different parts of an image with same CONV kernel is called shared weight, which enables that the neutral cells with same feature can be recognized and classified into same object type. The parameters include: kernel size, depth, stride, zero-padding, and filter quantity. The calculation algorithm of the output size is:

$$\mathbf{H_{out}} = \mathbf{1} + \frac{\mathbf{H_{in}+(2*pad)-K_{height}}}{\mathbf{S}} \quad ; \quad \mathbf{W_{out}} = \mathbf{1} + \frac{\mathbf{W_{in}+(2*pad)-K_{width}}}{\mathbf{S}} \tag{1}$$

2.1.3. Active Layer. To make the result of the CONV layer nonlinear, active Layer is set to solve the vanishing gradient problem caused by the underfitting. There are many functions for it: Sigmoid, Tanh, the rectified Linear Unit (ReLU), Leaky ReLU, the Exponential Linear Unit (ELU), and Maxout. Recently, Leaky ReLU is the most popular as it converges faster than Sigmoid and Tanh, calculates in a simpler and more efficient way, and contains no dead area.

2.1.4. Pooling Layer. Pooling layer is used to decrease the dimension of the results from CONV layer, located between two CONV layers. There are three types of pooling: general pooling, overlapping pooling and Spatial Pyramid Pooling (SPP). The general pooling's width is often as the same as stride. It's typical ways are max pooling and average pooling. The overlapping pooling normally has longer width than the stride. SPP can transform the convolutional features of images with any size, into the same number of dimensions. This advantage of SPP enables that CNN could not only deal with multi types of images but also avoid the information loss caused by cropping and warping, which makes it the core of SPPNet.

2.1.5. Fully Connected Layer. Fully connected layers are often the last layer at the end of CNNs. They transmit the data to the output, as well as simplify and speed up the data calculation.

2.1.6. Other Layers. Some of CNNs models still have dropout layers and regression layers. The former layer is to solve the overfitting. To avoid the weight is too subjective, it typically updates the weight of the neural cell knot with a certain probability decided by the stochastic policy. Regression layer is to classify the features. There are many methods: Logistic Regression (LR), Gaussian Processes for Regression (GPR), Bayesian Linear Regression (BLR). The regression gives the probabilities of all the possible object types, which is the base of final judgement.

2.2. Advantages and Shortages of CNN

CNN offers an effective mathematical way to express and extract the features of the input data. The weight sharing mechanism can recognize the data with same feature, which enables CNNs dealing with high dimensions' data and laying foundation for the final excellent classification. Its pooling layer reduces the dimensions easily. These all make CNNs very competitive and flourishing on intelligent recognition in many areas of data processing. However, to get better results, the main barrier is the scope of the parameters and datasets, as well as the computing capability of the hardware. These high physical requirements and CNN's obscure physical meaning are all current challenges.

3. CNN Family

The development process of CNN is to build more stable and more efficient system for object detection. As along with the definition of proposal, Region with CNN, Fast Region with CNN (Fast R-CNN), and Faster Region with CNN (Faster R-CNN) have been proposed. This section is going to introduce these algorithms in detail.

3.1. R-CNN

R-CNN [21] is an important milestone of CNN for object detection. It is the first neural network propounding the region proposal to realize the object detection based on the excellent ability of CNN's feature extraction and classification. The region proposal gets the potential objects by sliding the proposals with different width and height, such as selective search. Before CNN, R-CNN proposes crop/warp to normalize the candidate images with fixed size to standard. After CNN, it adds SVM classification and bounding-box regression to get the exact object detection result. But region proposal makes R-CNN consume massive data, time, computation and energy; crop/warp brings information loss. Both of them leave space for further improvement to the R-CNN.

3.2. Fast R-CNN

Fast R-CNN [22] is proposed based on SPPNet. SPPNet deletes the crop/warp of R-CNN, replaces the last pooling layer before FC layer with SPP, and keeps the output image m*n parts no matter what resolution of the input image is. These features accelerate the test speed by 24 to 102 times. Fast R-CNN profferes Region of Interest (ROI) pooling and proposal reflection, which solve the whole training problem of SPP. Besides, it uses multi-task loss layer: SoftmaxLoss replaces SVM, and SmoothL1Loss replaces Bouding-box. These new methods raise the precision of the algorithm by combining the classification and regression. Additionally, it speeds up the fully connected layers with SVD. Therefore, the training and test speed of Fast R-CNN are respectively 3 times and 10 times faster than that of SPP. VOC07 dataset shows the mAP of Fast R-CNN is 70.

3.3. Faster R-CNN

Based on Fast R-CNN, Faster R-CNN [23] solves the regional proposal problem by adding RPN, which is the Key contribution of Faster R-CNN. It gets the regional proposal not on the original image but the final feature image which will be input into the ROI pooling. As the resolution of the feature image is lower than that of the original image, the computation of Faster R-CNN is sure much less than that of all the former models of CNN.

The core feature of Region Proposal Network (RPN) is to get the proposal by sliding it. Each sliding proposal will generate 9 candidate anchors with different scales, widths and heights. Faster R-CNN's course of extracting the anchors' features is similar with that of the Fast R-CNN, while its object classification is only to recognize that the features are foreground or background. And its proposal regression is only to find out a more precise location of the target object. For each location of the proposal, RPN uses two full-connected layers (object classification and proposal regression) to judge and abandon the anchors. It never does explicit regional proposal. The main rules of selecting the anchors are: (1) dismissing the anchors on the boundary; (2) the anchors whose overlapping area with the sample are greater than 0.7 would be classified as foreground, and those whose overlapping areas are smaller than 0.3 would be classified to be background. In this way, RPN chooses about 300 anchors for each sliding proposal.

Faster R-CNN uses the alternating training mode to train the shared features. It uses the recent network to initiate the weight of the RPN, extracts the right proposals from the training dataset and trains the Faster R-CNN model with the proposals repeatedly until the result converges well.

Faster R-CNN has already offered a result with perfect precision of recognition. Its only left space for further improvement is speed, which is the main reason for the later algorithms doing.

4. YOLO

A new approach to object detection is called You Only Look Once (YOLO) which means that an image can be predicted what the objects are and where they are at one glance. As the first method entirely throwing out pipeline, it frames object detection as a regression problem to spatially separated bounding boxes and associated class probabilities, which are predicted with a single neural network from full images in one evaluation. Besides, YOLO selects GoogLeNet but not VGG-16 as base network, as their rates of the forwarding transport computation versus precision are $\frac{852}{88}$ VS $\frac{3069}{90} \approx \frac{1}{3.52}$, the former is far faster than VGG. So, YOLO is fast by design and indeed real-time while keeping good accuracy.

4.1. YOLOv1

Base YOLO [24] is also called YOLO Version 1 (YOLOv1). It models detection as a regression problem. A single convolutional network simultaneously predicts multiple bounding boxes and class probabilities for those boxes.

YOLOv1 divides the input image into an $S \times S$ grid. If the center of an object falls into a grid cell, that grid cell is responsible for detecting that object. Each grid cell predicts B bounding boxes, confidence scores for those boxes and C class probabilities of the grid. These predictions are encoded as an $S \times S \times (B \times 5 + C)$ tensor. In the test process, YOLOv1 multiplies the conditional class probabilities and the individual box confidence predictions which give us class-specific confidence scores for each box:

$$\mathbf{Pr(Class_i|Object) \times Pr(Object) \times IOU_{pred}^{truth} = Pr\ (Class_i) \times IOU_{pred}^{truth}} \quad \mathbf{(2)}$$

(IOU stands for Intersection Over Union). The scores encode both the probability of that class appearing in the box and how well the predicted box fits the object. Each bounding box consists of 5 predictions: x, y, w, h and confidence. The(x, y) coordinates represent the center of the box relative to the bounds of the grid cell. The width w and the height h are predicted relative to the whole image. And that is why YOLOv1 uses $B \times 5$ for calculating tensor. For evaluating YOLO on Pascal VOC, $S = 7, B = 2$ are normally used. Pascal VOC has 20 labelled classes, so $C = 20$. YOLOv1's final prediction is a $7 \times 7 \times (5 \times 2 + 20) = 7 \times 7 \times 30$ tensor. It uses only 98 bounding boxes per image versus 2000 from Selective Search.

YOLOv1's network has 24 convolutional layers followed by 2 fully connected layers. Instead of the inception modules used by GoogLeNet, YOLOv1 simply uses 1×1 reduction layer followed by 3×3 convolutional layers, similar to Lin et al [25]. On Pascal VOC2007, YOLOv1 processes images at 45 frames per second (FPS) which is twice to nine times faster than that of Faster R-CNN. Especially, Fast YOLO, a fast version of YOLO designed to push the boundaries of fast object detection, has reaches an astounding 155 FPS.

4.2. YOLOv2

YOLO Version 2 (YOLOv2) [18] is a variously improved model to YOLO which keeps the advantage on speed and try to rise the mAP value from YOLOv1's 63.4. Using a novel, multi-scale training method the same YOLOv2 model can run at varying sizes, offering an easy tradeoff between speed and accuracy.

YOLOv2 adds a list of significant solutions to increase mAP. Batch Normalization preprocesses the input data. High Resolution Classifier from YOLOv1's 224×224 to 448×448 raises the mAP by 4%. YOLOv2 also uses a new network with the thought "network in network", predicts with global average pooling, and compresses the features by putting 1×1 convolutional core among between the 3×3 convolutional cores. Its neural network has fewer convolutional layers (19 instead of YOLOv1's 24) and fewer filters plus 5 max pooling layers. Besides, YOLOv2 adopts convolutional with anchor boxes and increases each grid's resolution of from YOLOv1's 7×7 to 13×13. It has only 1 bounding box for each grid. Additionally, after researching the IOU function curve by counting the Ground Truth statistics on VOC and COCO with the K-means algorithm, YOLOv2 finds a solution to the best quantity of anchor boxes "Dimension Clusters" and final decides the number of anchor boxes to be 5, whose Average IOU (61.0) is tantamount with Faster R-CNN's 60.9 with 9 anchor boxes. Meanwhile, direct location prediction is also used to mitigate the instability brought by the anchor boxes, and it finally raises the mAP by 5% together with the good effect of dimension clusters. Finally, YOLOv2 adds a pass-through layer to get the extracted features from the former 26×26 layer and combine them with the original final output features, so that the ability of detecting the small object would be enhanced. In this mean, YOLOv2 raises the mAP by 1%.

4.3. Advantages and Shortages of YOLO

YOLO breaks through the max-speed limit of CNN and realizes excellent balance of speed and accuracy. It is extremely fast and real time. YOLOv2 has achieved 76.8 mAP at 67 FPS and 78.6 mAP at 40 FPS, outperforming state-of-the-art methods like Faster R-CNN with ResNet while still running precisely. Secondly, YOLO reasons globally and encodes contextual information about the image. Thus, it's less likely to predict false positives on background. Finally, YOLO learns very general representations of objects, outstripping other detection methods including DPM and R-CNN by a wide margin, when generalizing from natural images to other domains like artwork.

However, YOLO has limitations. It imposes strong spatial constraints on bounding box predictions such as recognizing small objects in group. It still struggles to generalize to objects in new or unusual aspect ratios or configurations. It's also not perfect that YOLO's loss function treats errors the same in small bounding boxes versus large boxes.

5. The Compare of Faster R-CNN and YOLOv2

YOLO suggests a completely new way to process image, varying too much from not only Faster R-CNN but also R-CNN and all its variants. Here only the key differences between YOLO and Faster R-CNN will be discussed as following.

5.1. The Framework

Though both Faster R-CNN and YOLO use CNN as core and their key purposes are all to find a better way of dividing proposals based on CNN, their frameworks differ from each other greatly. Faster R-CNN keeps the traditional general frame of R-CNN: using CNN dealing with the whole input image at first and dividing proposals later, maintaining regional proposal and ROL pooling. But its key contribution is adopting Regional Proposal Network to specially quicken the computation of processing proposals. YOLO divides the whole image at the very beginning and later uses CNN to process. Besides, YOLO gives up sliding windows and regional proposal completely and divides the input image into $S \times S$ grids. Meanwhile, it innovates confidence IOU mechanism to each grid and class probabilities for making decision. Though YOLOv2 reuses anchor boxes, YOLOv1 did give up it at the inception of YOLO; and YOLOv2 uses 5 anchor boxes while Faster R-CNN has 9 anchor boxes. In addition, YOLOv2 creates many new techniques to improve precision, such as dimension clusters, these are all what Faster R-CNN doesn't support.

5.2. Outside Performance

Faster R-CNN focuses on speeding up the R-CNN framework by sharing computation and using neural networks to propose regions instead of Selective Search. While they offer speed and accuracy improvements over R-CNN, both still fall short of real-time performance. Instead of trying to optimize individual components of a large detection pipeline, YOLO throws out the pipeline entirely and is fast by design. So YOLOv2 can achieve high precision and keep real-time for pictures with high resolution. While for pictures with low resolution, it shows outstanding high speed with excellent mAP value. Additionally, none of the detection systems has surpassed Fast YOLO's unparalleled performance with 155 FPS and mAP value 52.7 until now.

Table 1. The Performance Compare of the Detection Systems.

		Detection Frameworks	**Train**	**mAP ↑**	**FPS**	**PS (mAP×FPS)**	**PS Order**
Less Than Real-Time Detectors	1	Fastest DPM [26]	2007	30.4	15	456	11
	2	R-CNN Minus R [27]	2007	53.5	6	321	13
	3	Faster R-CNN ZF[20]	2007+2012	**62.1**	**18**	1118	9
	4	YOLO VGG-16[24]	2007+2012	**66.4**	**21**	1394	8
	5	Fast R-CNN[22]	2007+2012	70.0	0.5	35	14

	6	Faster R-CNN VGG-16[20]	2007+2012	73.2	7	512	10
	7	Faster R-CNN ResNet[20]	2007+2012	76.4	5	382	12
Real-Time Detectors	1	Fast YOLO [24]	2007+2012	52.7	155	8169	1
	2	YOLO(YOLOv1)[24]	2007+2012	63.4	45	2853	7
	3	YOLOv2 288×288[24]	2007+2012	69.0	91	6279	2
	4	YOLOv2 352×352[24]	2007+2012	73.7	81	5970	3
	5	YOLOv2 416×416[24]	2007+2012	**76.8**	**67**	5146	4
	6	YOLOv2 480×480[24]	2007+2012	**77.8**	**59**	4590	5
	7	YOLOv2 544×544[24]	2007+2012	**78.6**	**40**	3144	6

Table 1 shows theadvancement of YOLO. The index PS is actually created to show the general valueof each detection system based on regarding the precision and speed have same weight.PS is only for reference; its definition can be changed with different weight scale formAP and FPS. The weight scale shows the relative importance of mAP and FPS to target.

6. Conclusion

This paper briefly discusses about the current algorithms in object detection, including the CNN family and YOLO. Compared with CNNs, YOLO has more advanced application in practice. YOLO is a unified object detection model. It's simple to construct and can be trained directly on full images. Unlike classifier-based approaches, YOLO is trained on a loss function that directly corresponds to detection performance and the entire model is trained jointly. Fast YOLO is the fastest general-purpose object detector. And YOLOv2 provides state-of-the-art the best tradeoff between real-time speed and excellent accuracy for object detection than other detection systems across a variety of detection datasets. Furthermore, YOLO's better generalizing representation of object than other models making it ideal for applications that rely on fast, robust object detection. These preeminent and precious advantages make it worthy of being strongly recommended and popularized. Except the structure of each algorithm, the most urgent up-coming challenge for machine learning is the scope of the dataset. The availability of suitable training data could be the vital part in the learning process, to achieve an idea results.

Reference

[1] Krizhevsky, A., Sutskever, I., & Hinton, G. E. (2012). Imagenet classification with deep convolutional neural networks. In Advances in neural information processing systems (pp. 1097-1105).
[2] Hubel, D. H., & Wiesel, T. N. (1962). Receptive fields, binocular interaction and functional architecture in the cat's visual cortex. The Journal of physiology, 160(1), 106-154.
[3] Rosenblatt, F. (1962). Principles of neurodynamics.
[4] Kabrisky, M. (1964). a Proposed Model for Visual Information Processing in the Human Brain.
[5] Giebel, H. (1971). Feature extraction and recognition of handwritten characters by homogeneous layers. In Zeichenerkennung durch biologische und technische Systeme/Pattern Recognition in Biological and Technical Systems(pp. 162-169). Springer, Berlin, Heidelberg.
[6] Fukushima, K. (1975). Cognitron: A self-organizing multilayered neural network. Biological cybernetics, 20(3-4), 121-136.
[7] Rumelhart, D. E.,Hinton,G. E.,&Williams,R.J.(1985). Learning internal representations by error propagation (No.ICS-8506). California Univ San Diego La Jolla Inst for Cognitive Science.
[8] LeCun, Y., Boser, B., Denker, J. S., Henderson, D., Howard, R. E., Hubbard, W., & Jackel, L. D. (1989). Backpropagation applied to handwritten zip code recognition. Neural computation, 1(4), 541-551.

[9] Fukushima, K., & Miyake, S. (1982). Neocognitron: A self-organizing neural network model for a mechanism of visual pattern recognition. In Competition and cooperation in neural nets(pp. 267-285). Springer, Berlin, Heidelberg.
[10] LeCun, Y., Boser, B., Denker, J. S., Henderson, D., Howard, R. E., Hubbard, W., & Jackel, L. D. (1989). Backpropagation applied to handwritten zip code recognition. Neural computation, 1(4), 541-551.
[11] LeCun, Y., Bottou, L., Bengio, Y., & Haffner, P. (1998). Gradient-based learning applied to document recognition. Proceedings of the IEEE, 86(11), 2278-2324.
[12] Simonyan, K., & Zisserman, A. (2014). Very deep convolutional networks for large-scale image recognition. arXiv preprint arXiv:1409.1556.
[13] Szegedy, C., Liu, W., Jia, Y., Sermanet, P., Reed, S., Anguelov, D., ... & Rabinovich, A. (2015). Going deeper with convolutions. In Proceedings of the IEEE conference on computer vision and pattern recognition (pp. 1-9).
[14] He,K.,Zhang,X.,Ren,S.,&Sun,J.(2016). Deep residual learning for image recognition. In Proceedings of the IEEE conference on computer vision & pattern recognition (pp. 770-778).
[15] Felzenszwalb, P. F., Girshick, R. B., McAllester, D., & Ramanan, D. (2010). Object detection with discriminatively trained part-based models. IEEE transactions on pattern analysis and machine intelligence, 32(9), 1627-1645.
[16] Girshick, R. B., Felzenszwalb, P. F., & McAllester, D. (2012). Discriminatively trained deformable part models, release 5.
[17] Hosang, J., Benenson, R., Dollár, P., & Schiele, B. (2016). What makes for effective detection proposals?. IEEE transactions on pattern analysis and machine intelligence, 38(4), 814-830.
[18] Redmon, J., & Farhadi, A. (2016). YOLO9000: better, faster, stronger. arXiv preprint arXiv:1612.08242.
[19] Sharif Razavian, A., Azizpour, H., Sullivan, J., & Carlsson, S. (2014). CNN features off-the-shelf: an astounding baseline for recognition. In Proceedings of the IEEE conference on computer vision and pattern recognition workshops (pp. 806-813).
[20] Ren, S., He, K., Girshick, R., & Sun, J. (2015). Faster R-CNN: Towards real-time object detection with region proposal networks. In Advances in neural information processing systems (pp. 91-99).
[21] Girshick, R., Donahue, J., Darrell, T., & Malik, J. (2014). Rich feature hierarchies for accurate object detection and semantic segmentation. In Proceedings of the IEEE conference on computer vision and pattern recognition (pp. 580-587).
[22] Girshick, R. (2015). Fast r-cnn. In Proceedings of the IEEE international conference on computer vision (pp. 1440-1448).
[23] Ren, S., He, K., Girshick, R., & Sun, J. (2015). Faster R-CNN: Towards real-time object detection with region proposal networks. In Advances in neural information processing systems (pp. 91-99).
[24] Redmon, J., Divvala, S., Girshick, R., & Farhadi, A. (2016). You only look once: Unified, real-time object detection. In Proceedings of the IEEE Conference on Computer Vision and Pattern Recognition (pp. 779-788).
[25] Lin, M., Chen, Q., & Yan, S. (2013). Network in network. arXiv preprint arXiv:1312.4400.
[26] Yan, J., Lei, Z., Wen, L., & Li, S. Z. (2014). The fastest deformable part model for object detection. In Proceedings of the IEEE Conference on Computer Vision and Pattern Recognition (pp. 2497-2504).
[27] Lenc, K., & Vedaldi, A. (2015). R-cnn minus r. arXiv preprint arXiv:1506.06981.

IOP Conf. Series: Journal of Physics: Conf. Series **1004** (2018) 012030 doi:10.1088/1742-6596/1004/1/012030

Rerating the Movie Scores in Douban through Word Embedding

Mingyu Cui[1,*]

[1]School of Computer Science and Engineering, SouthEast University, Nanjing 211189, China

*mingyuc36@163.com

Abstract. The movie scores in the social networking service website such as IMDb, Totten Tomatoes and Douban are important references to evaluate the movies. Always, it will influence the box office directly. However, the public rating has strong bias depended on the types of movies, release time, and ages and background of the audiences. To fix the bias and give a movie a fair judgement is an important problem. In the paper, we focus on the movie scores on Douban, which is one of the most famous Chinese movie network community. We decompose the movie scores into two parts. One is the basis scores based on the basic properties of movies. The other is the extra scores which represent the excess value of the movies. We use the word-embedding technique to reduce the movies in a small dense subspace. Then, in the reduced subspace, we use the k-means method to offer the similar movies a basis scores.

1. Introduction

The social networking service websites such as IMDb, Totten Tomatoes and Douban are an important approach for common people to express the point of view of the movies. The final movie scores are the average of all personal ratings. It is difficult to cheat through bribing a small group of movie experts or critics. Thus, it is always deemed to be a fair way to evaluate the values of movies. However, most of the people know little professional standard to judge a movie. Since they provide the scores just followed their hearts, the public rating has strong bias depended on the types of movies, release time, and ages and background of the audiences. It is usually unfair to compare a commercial film to a literary film directly. Sometimes, it is also difficult to compare an action movie to a documentary. On the other hand, people are unlikely not see a small minority movie which they are not interested in. It means that if one person has watched a small minority movie, then with high probability, they would give it a high score. On the contrary, most people have the experience to see popular movies even do not know the movie title clearly.

Thus, it is important to provide a fair evaluation. Professional accreditation, such as Oscar is one approach. But usually, it is unrelated to the public rating. In the paper, we provide an alternative way to give a fair evaluation totally dependent on the public ratings themselves. Briefly speaking, we decompose the movie scores into two parts. One is the basis scores based on the basic properties of movies. The other is the extra scores which represent the excess value of the movies.

We use the word embedding technique to reduce the movies in a small dense subspace. The movies with similar basis score are located in the close locations in the subspace. Thus, we can use the (scaled) average scores among its neighbors as its basis scores. Then, according to the real scores of the movie,

IOP Conf. Series: Journal of Physics: Conf. Series **1004** (2018) 012030 doi:10.1088/1742-6596/1004/1/012030

we can obtain is extra scores. After many attempts, we conclude a formula to calculate the real score of a movie. Formally speaking,

$$\text{Real Score} = \alpha * \text{Basis Score} + (1-\alpha) * \text{Extra Score}...... \quad ..(1)$$

Word embedding is the collective techniques in natural language processing (NLP) where words or phrases from the vocabulary are mapped to vectors of real numbers. Representation of words as embedding vectors has a long history. Bengio et al. [1] provided a feedforward neural network with a linear projection layer and a non-linear hidden layer to learn jointly the word vectors representation. Roweis and Saul [2] published in Science how to use "locally linear embedding" (LLE) to discover representations of high dimensional data structure. In 2013, Tomas Mikolov et al.[3] [4] introduced the Continuous Bag-of-Words model (CBOW) and the Skip-Gram model which achieved large improvements in accuracy at much lower computational cost compared to the previous methods. Based on the methods, they created word2vec, a toolkit which is particularly computationally-efficient predictive model for learning word embedding.

Although considerable research has been devoted to the sentiment analysis of the linguistic contexts, rather less attention has been paid to use word embeddings to analyze the potential laws in movies. In the paper, we focus on the movie scores on Douban, which is one of the most famous Chinese movie network community. We randomly choose 1000 movies from Douban. Collect the scores, basic informations such as directors, editors, actors and types, and film reviews of these movies. Based on the data, we use word2vec to construct the embedding vectors. Then, according to the labels such as actors and movie types, we embedding the movies in a small subspace. Finally, we applied K-means to cluster movies in the space and calculate the basis scores for each one.

2. Methodology

2.1. Embedding the Movies in Small Spaces.

As we all know, how to represent the precise features of the data is a key problem in machine learning. In audio, we can use the matrix which contains the audio spectrums to solve the problem and we can use some distance and rate to judge whether there exist similarities between two audios. While in Natural Language Processing (NLP), all we can use are words which only consist of characters. Therefore, how to express a word and judge the similarities between two words in program are important.

One-hot representation is one of the easiest way to present words which are to use a long vector to stands for a word and the length of the vector is the number of all words. In that vector, we use 1 to represent the word and the rest of elements in the vector are 0. In fact, this method can solve many tasks but the dimension of the vector will increase if the number of total words increases. The representation is extreme sparse which will waste many spaces. Besides, any two vectors are independent which means that we can't obtain semantic information from any two vectors.

Therefore, reduce the one-hot vectors in a small, dense subspace is a key problem. We call the new representation distributed representation and call the process word embedding. In order to avoid the disadvantages brought by one-hot representation, the word embedding changes data types from integer to float and pushes the big dimension vectors into a smaller dimension vectors. There are three main ways to achieve distributed representation which are respectively based on matrix [5], clustering [6] and neutral network [7].

The neutral network is the most efficient way for word embedding. The basic network is illustrated in Figure 1. See [3] [4] for more details.

Word2vec, a computationally-efficient toolkit for learning word embeddings, involves statistical language model, CBOW model and Skip-gram model. The statistical language model is to use probability to judge whether a sentence is reasonable which means that calculating the probability that a term appears in the premise of some given words. And the CBOW model and Skip-gram model both contains three layers which are input layer, projection layer and output layer. CBOW model is to predict a possible word in the premise of some context of this word. And Skip-gram model is a reverse process of CBOW.

When we are training this model, we can obtain the word vectors which means that word vectors are the by-product of training language neutral network model.

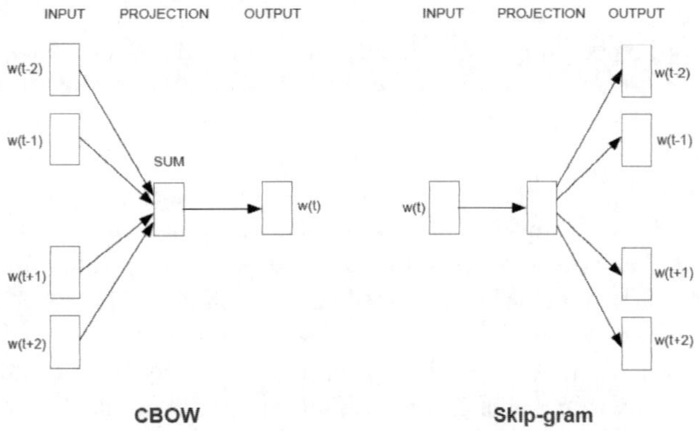

Figure 1: The CBOW architecture predicts the current word based on the context, and the Skip-gram predicts surrounding words given the current word.

2.2. Compute the Basis Scores

The distributed representation can retain the semantic information of a word. we assume that the information of movie can be executed by distributed representation and the information of movie is saved to some extent and by this way, we can revise the point of movie according to the labels it has. That is to say if a director shot a movie and received a high evaluation, by clustering all the embeddings of director, we can obtain some groups that only contain the excellent director who shot a good movie once and some groups that only contain the director who had a bad performance once. Then, we can obtain a point of movie according to the director label. Therefore, we want to find out whether a good movie still has a good performance in our model and revise the point of movie using the word embeddings.

Following the idea, we maintain several hash-tables to map the movies to the word vector spaces. In each space, we group the movies in several clusters. For each movie in any cluster, we use the average scores of the movies in the cluster as one components of the basis score of the movie. Denote the number of spaces as m. Then, following the process, each movie will have m components of basis scores. We use the average scores of the components as the final basis scores of the movies. Then, we discuss the details of the hash tables and the cluster algorithm.

Table 1. Examples of one information of movies in the movie set.

Name	Director	Editor	Actor	Type	Point
Farewell My Concubine	Kaige Chen	Pik Wah Lee	Leslie Cheung	Drama	9.50

In the paper, we construct four hash mappings, i.e., m = 4. Concretely speaking, they are (movie - the word vector of director), (movie - the word vector of editor), (movie - the word vector of actor), (movie - the word vector of type). See Table1 for an example. Note that, here the word vector is the word-embedding vector of the corresponding word obtained through word2vec.

The cluster algorithm used in the paper is K-means algorithms. The K-means algorithm is to regard the average data of every subgroup as the representative point and obtains some groups which they are intra-class compactness and inter-class separability. Formally spreaking, the algorithm clusters the data set into different groups by iterating and minimize the within-cluster sum of squared criterion function as follows.

IOP Conf. Series: Journal of Physics: Conf. Series **1004** (2018) 012030 doi:10.1088/1742-6596/1004/1/012030

$$argmin_S \sum_{i=1}^{k} \sum_{x \in S_i} \|x - \mu_i\|^2 \ldots \qquad \ldots..(2)$$

Where μ_i is the mean of points in S_i.

2.3. Experiments

The problem was solved in two main steps. At first, we used word2vec to train the information of movies and obtained word embeddings of each label. The contexts of movie contain the name, director, editor, type and 5 short comments of each movie. What's more, we applied a word segmentation tool to do segmentation. After word embedding, we obtained 28000 word vectors approximately which are partly shown in Figure 2.

After that, we applied the K-means algorithm to cluster the word embeddings to ten groups according to the label of word embeddings. As to the director label, the point of one group was calculated by calculating the average point of the total movies in this group. And the rest of the groups were also calculated by this method. After the whole process, the four labels of each movie have a new point. Then, we applied Alpha Coefficient to calculate the comprehensive point of each movie and after many attempts we set Alpha=0.7 to calculate real score. There is an example about the executed information of a movie shown in Table 2. Finally, all we obtained that the information and the real scores of movies are shown in Table 3.

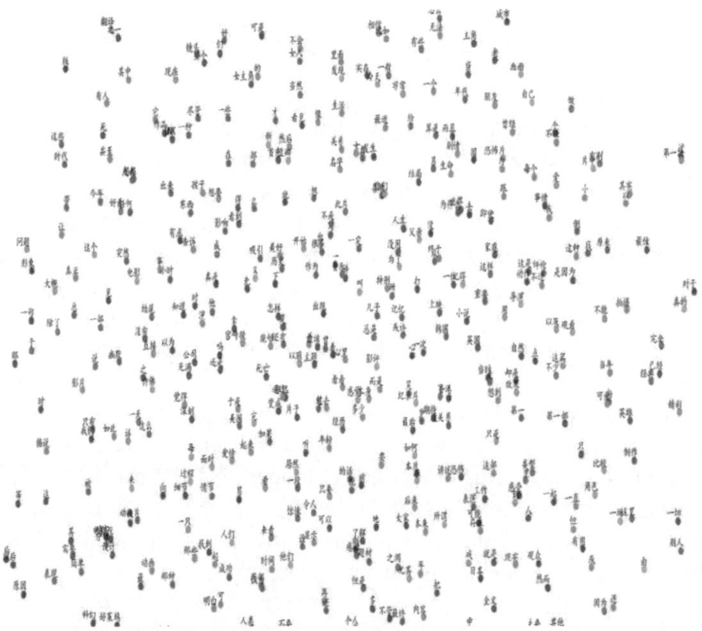

Figure 2. Example of embedding words.

Table 2. Example of the executed information of a movie.

Name	Director Of Point	Editor Of Point	Actor Of Point	Type Of Point	Point(Basis score)	Point(Extra score)	Real Score
Farewell My Concubine	7.38	7.30	7.56	7.57	7.45	9.50	8.88

Table 3. Example of executed movies which contains the new point of four labels and final point.

Name	Director of	Editor of	Actor of	Type of	Point(	Point(Extra	Real

	Point	Point	Point	Point	Basis score)	score)	Score
Life is beautiful	7.58	7.45	7.32	7.57	7.48	9.5	8.89
Farewell My Concubine	7.38	7.3	7.56	7.57	7.45	9.5	8.88
Laputa:Castle in the sky	7.42	7.39	7.46	8.41	7.67	9.1	8.60
The Matrix	7.42	7.39	7.59	6.71	7.27	8.8	8.34
X-men	7.58	7.45	7.32	6.71	7.26	7.6	7.49
The void	5.80	5.80	5.80	6.83	6.05	5.8	5.87

3. Conclusion

In the paper,it is now possible to state that in our experiment a movie that possesses a high point in DOUBAN after word embedding still possesses a high point and a movie that does not have a good performance in DOUBAN still has a terrible performance in our study. In our process of revising the point of movie, our experiment results are consistent with the movie point form DOUBAN. Therefore we can assure that the word embeddings about the information of movies still contains semantic information of movies, which proves that analyzing movies by distributed representation method is feasible.

Our study shows that the mechanism of DOUBAN is convincing and word embeddings can be applied to correct the semantic information. In the future, it would be interesting to analyze which director is a good director or he always shoots good movies. Similarly, we can analyze which actor is a good actor or he always presents to us good movies.

References

[1] Bengio, Yoshua, et al. "A neural probabilistic language model." Journal of machine learning research 3.Feb (2003): 1137-1155.

[2] Roweis, Sam T., and Lawrence K. Saul. "Nonlinear dimensionality reduction by locally linear embedding." science 290.5500 (2000): 2323-2326.

[3] Mikolov, Tomas, et al. "Efficient estimation of word representations in vector space." arXiv preprint arXiv: 1301.3781 (2013).

[4] Mikolov, Tomas, et al. "Distributed representations of words and phrases and their compositionality." Advances in neural information processing systems. 2013.

[5] Pennington J, Socher R, Manning C. Glove: Global vectors for word representation[C]//Proceedings of the 2014 conference on empirical methods in natural language processing (EMNLP). 2014: 1532-1543.

[6] Turian J, Ratinov L, Bengio Y. Word representations: a simple and general method for semi-supervised learning[C]//Proceedings of the 48th annual meeting of the association for computational linguistics. Association for Computational Linguistics, 2010: 384-394.

[7] Brown P F, Desouza P V, Mercer R L, et al. Class-based n-gram models of natural language [J]. Computational linguistics, 1992, 18(4): 467-479.

IOP Conf. Series: Journal of Physics: Conf. Series **1004** (2018) 012031 doi:10.1088/1742-6596/1004/1/012031

A Distributed Cache Update Deployment Strategy in CDN

Xinhua E[1,*] and Binjie Zhu[2]

[1]Faculty of Information Technology, Beijing University of Technology, Beijing, 100124, China
[2]China Mobile Group Beijing Company Limited, Beijing, 100007, China

*Elson1900@163.com

Abstract. The CDN management system distributes content objects to the edge of the internet to achieve the user's near access. Cache strategy is an important problem in network content distribution. A cache strategy was designed in which the content effective diffusion in the cache group, so more content was storage in the cache, and it improved the group hit rate.

1. Introduction

Nowadays services such as Video-on-Demand and live TV programming available over Internet. Users want to have access to high quality video at any times. Content Delivery Network (CDN) is a solution that has been used to delivery traditional Web page content and Video content [1-7]. The CDN management system distributes content objects to the edge of the internet to achieve the user's near access. Cache strategy is an important problem in network content distribution. A cache strategy was designed in which the content effective diffusion in the cache group, so more content was storage in the cache, and it improved the group hit rate.

2. Cache update deployment strategy

The cache server group can only store partial content objects. However, if contents requested by Client are not available, a distributed cache group can pull those content objects on a globe scale. The cache server that stores content objects at the first time is called as an intra-group primary source server. The system will spread this content object to other servers in the group via certain methods according to dynamic changes requested by clients. Then, they also become intra-group servers of this content object after storage. The rules and methods of how servers in the distributed cache group to pull content objects from intra-group primary source servers or origin servers determine their deployment in the cache group, which is called as a cache update deployment strategy.

Based on previous analysis, most of current researches on cache strategies aim at cache replacement algorithms or cache deployment methods with certain architectures. This paper proposed a cache deployment strategy that is suitable to cache server groups from the perspective of improving degree of collaboration. This method allows various cache server nodes to make cache update decisions instead of adopting centralized control modes.

The cache server group is composed by some close servers. If contents are not available in the group, delays will happen when pulling them from outside the server group. Therefore, requests are better to be completed within the group, which shall improve degree of collaboration of cache servers and optimize deployment globally.

EPP-CPS takes expiration time (ET), potential acceleration effect (PAE) and pull gate (PG) as parameters of cache update deployment so as to be more suitable to the optimized goal in the cache collaboration group.

This article firstly introduced concepts of expiration time, potential acceleration effect and pull gate, then explains cache update methods. The expiration time is divided into two concepts including data object's expiration time and cache expiration time.

$$\mathrm{Det(D,C)} = \mathrm{T_R} - \mathrm{T_i} \tag{1}$$

A data object's expiration time is defined as the difference between $\mathrm{T_R}$ (time that a data object is being removed from the cache) and $\mathrm{T_i}$ (time of the last hit). The cache expiration time refers to average numbers of expiration time of data objects that are being removed from the cache over a period of time. The data objects that are being removed from cache C within T is graded as a set of S, then the cache expiration time Cet is shown as the formula below. Dnum is the number of data objects in the set S.

$$\mathrm{Cet(T,C)} = \frac{\sum_{\mathrm{D}\in\mathrm{S}} \mathrm{Det(D,C)}}{\mathrm{Dnum}} \tag{2}$$

The cache expiration time is one of parameters to measure cache competition. If the cache expiration time is large, then the degree of competition is relatively small; vise versa. If one content is being stored in a cache server with relatively small competition, then it will have a longer life cycle in the cache group.

The CDN environment is characterized by particularity. Thus, some special factors in CDN needs to be considered. First of all, the acceleration effect (AE) is defined as a kind of evaluation on increase of access speed (decrease of delays) after a content object is being stored in the cache. If a content object is being cached from S (content source node) to D (destination node). The delay from S to D is defined as $\mathrm{Delay_{SD}}$. If the access time of content visitors that are very close to D is $\mathrm{l_d}$, then the acceleration effect of delivering contents to D is shown as formula below.

$$\mathrm{AE_{SD}} = \mathrm{Delay_{SD}} \times \mathrm{l_d} \tag{3}$$

$$\mathrm{AE_{SD}} = \mathrm{Delay_{SD}} \times \mathrm{l_d} \tag{4}$$

The potential acceleration effect (PAE) between two points is defined on the basis of acceleration effects. There are two points including C and A in the network. C stores contents requested by clients, while A is an access point of Client nodes that are close to server A. If Client requests content object $\mathrm{C_r}$ from server A, but it is not in that server. In this case, server A can not satisfy Client's request, but server C can do that. The delay between C and D is taken as $\mathrm{Delay_{ca}}$. If the forwarding time of content request from A to C is $\mathrm{Rd_{ac}}$, then the potential acceleration effect is defined as formula below.

$$\mathrm{PAE_{ca}} = \mathrm{Delay_{ca}} \times \mathrm{Rd_{ac}} \tag{5}$$

The potential acceleration effect is also used to evaluate content cache values. However, different from the acceleration effect, the potential acceleration effect is used to evaluate decrease of acceleration effect due to the fact that the content object is not available in the local cache.

Every server will record content objects that are being accessed lately. Those records are a kind of structural organization in FIFO queue. When a request is delivered, please check FIFO queue firstly. If there are same contents in the recent records, then it is regard as up to the pull gate. The cache update will only be started once up to the pull gate.

The server does not need to pull missed contents to the local space in the collaborative cache server group, because it will increase the pull cost. Only pulling missed contents when necessary by determining certain rules via intragroup methods has two advantages: first of all, decrease the pull cost; secondly, utilize "group space" in the distributed cache group so as to store more contents in the limited space to improve the group hit rate effectively.

Cache update deployment strategy is shown in Figure 1.

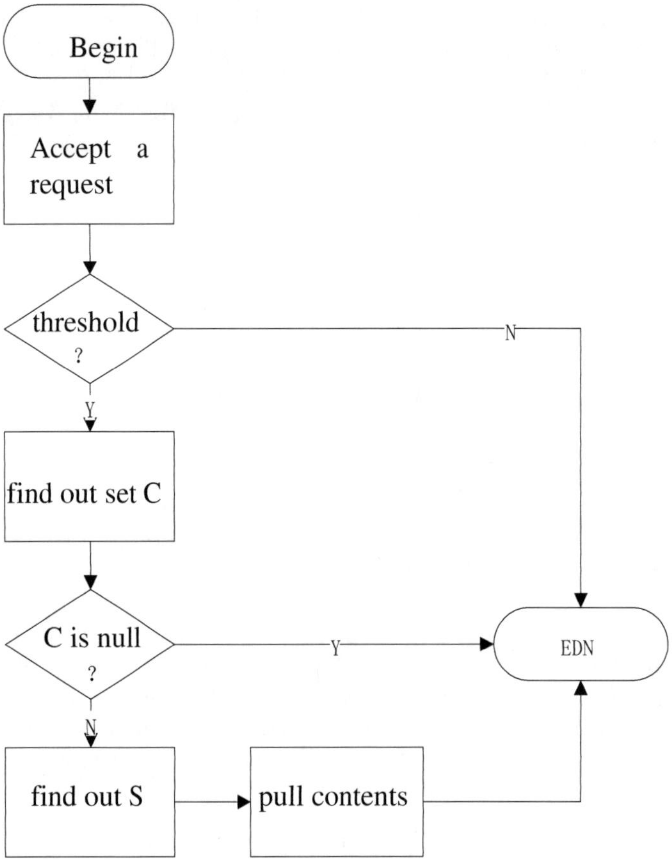

Figure 1. Cache update deployment strategy

Step 1: Accept requests outside cache servers; if other servers do not have the requested content, then route the request to the located server via local routing system; the located server will accept the request accepted from other servers;

Step 2: Judge whether it is up to the pull gate; if yes, then execute the next step, or else "end";

Step 3: Find out the set C with ET numbers larger than the local cache server;

Step 4: Find out server S with the biggest PAE value from set C;

Step 5: Pull contents from server S, end;

The threshold judgement module, PAE module, ET module, update deployment decision module, local content update module and cache replacement algorithm module are involved in cache update deployment.

The PAE module calculates the potential acceleration effect by accessing historical data. There are n servers in the network, including $S_1, \ldots S_i, \ldots, S_n$. The server S_i records the times Rdnum aiming at the request to C_l forwarded from other servers S_j. The potential acceleration effect (PAE) from C_l to various nodes is obtained based on parameters of delay between the request time and nodes. The pseudo-code of PAE module function is shown in figure 2.

```
PAE_arry(Content)// PAE
  {
        For every server i
      {
        Delay=SelectDelay(server i);
        Rdnum=SelectRdnum(content);
        PAE_arry[i]= Delay*Rdnum;
            }
    Retrun PAE_arry;
  }
```

Figure 2. Pseudo-code of PAE module function

The ET module is used to calculate ET numbers and exchange for ET information with other nodes. The ET module calculates ET numbers of local cache servers via formula. It not only needs to calculate the local ET numbers, but stores ET numbers of other servers. The ET numbers shall also be forwarded to the target server at the time when a server forwards content requests to another one. In this way, ET information of other server will also be obtained.

The pull gate module records information of content objects that are being accessed lately. When a new request is delivered to the server, the content information will be recorded at the tail of the queue. Meanwhile, the record at the head of the queue will be deleted. This access record queue is used to judge whether it is up to the update deployment threshold: when a content request is delivered, it shall be judged whether this is being stored in the recent access records. If yes, then start the cache update decision process.

The update deployment decision module selects alternative nodes of content update deployment through parameters in PAE and ET modules. Then the content object is being updated and deployed in nodes. The local content update module will execute its update process after accepting relevant information.

3. Analysis and verification

In order to verify EPP-CPS, a simulation framework of an edge cache server group is established in Opnet network software. There are 9 servers in every cache group. Every server has their own user request nodes that will access the cache group according to Poisson's distribution, which includes 1,000 document types. The group hit rate between EA method and EPP-CPS is further compared. T/S refers to the ratio of storage space in every node to single file space. Five testing scenes are established with their T/S values of 10, 30, 50, 70 and 100. The experimental results are shown in the figure 3.

It can be seen from figure 3 that EPP-CPS update strategy has higher group hit rate than EA strategy. The average GHR of EPP-CPS is 43.8, while that of EA is 30.4, which is improved by 31%. EPP-CPS optimizes the update strategy based on EA so as to spread contents in the cache group effectively. Therefore, it can store more contents in a limited storage space so as to improve the group hit rate.

IOP Conf. Series: Journal of Physics: Conf. Series **1004** (2018) 012031 doi:10.1088/1742-6596/1004/1/012031

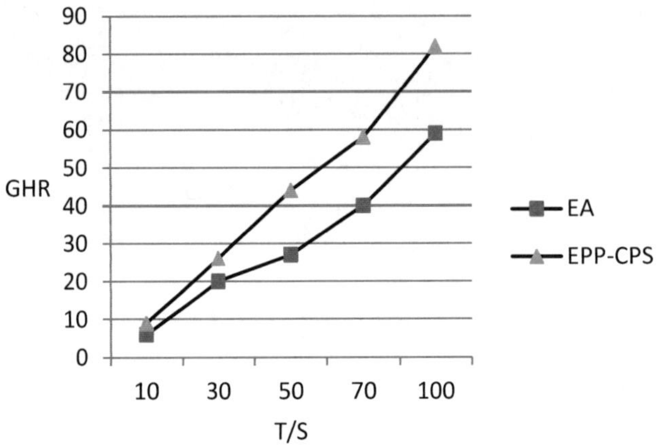

Figure 3. Comparison of EPP-CPS and EA

References

[1] Sanaa Sharafeddine, Karim Jahed, Omar Farhat, et al. Failure recovery in wireless content distribution networks with device-to-device cooperation[J]. Computer Networks, 2017, 128(9): 108-122.

[2] Rami Halloush, Hang Liu, Lijun Dong, et al. Hop-by-hop Content Distribution with Network Coding in Multihop Wireless Networks[J]. Digital Communications and Networks, 2017, 3(1): 47-54.

[3] Xiaoying Zheng, Ye Xia. Optimizing network objectives in collaborative content distribution [J]. Computer Networks, 2015, 91(14): 244-261.

[4] Uttam Mandal, Pulak Chowdhury, et al. Energy-efficient networking for content distribution over telecom network infrastructure[J]. Optical Switching and Networking, 2013, 10(4): 393-405.

[5] Lazaros Gkatzikis, Vasilis Sourlas, Carlo Fischione, et al. Low complexity content replication through clustering in Content-Delivery Networks[J]. Computer Networks, 2017,121(5): 137-151.

[6] Bernd Klasen. Efficient content distribution in social-aware hybrid networks [J]. Journal of Computational Science, Pages 209-218, 2013, 4(4): 209-218.

[7] Tom Bostoen, Sape Mullender, Yolande Berbers. LofoSwitch: An online policy for concerted server and disk power control in content distribution networks[J]. Ad Hoc Networks, 2015, 25(B).

IOP Conf. Series: Journal of Physics: Conf. Series **1004** (2018) 012032 doi:10.1088/1742-6596/1004/1/012032

Research on cardiovascular disease prediction based on distance metric learning

Zhuang Ni[1, 2, a], Kui Liu[1, b] and Guixia Kang[1,2, c]

[1]Key Laboratory of Universal Wireless Communications, Ministry of Education, Beijing University of Posts and telecommunications. Beijing, China.
[2]Wuxi BUPT Sensory Technology and Industry Institute CO.LTD

[a]nizhuang_bupt@163.com; [b]liukui_006@126.com; [c]gxkang@bupt.edu.cn

Abstract. Distance metric learning algorithm has been widely applied to medical diagnosis and exhibited its strengths in classification problems. The k-nearest neighbour (KNN) is an efficient method which treats each feature equally. The large margin nearest neighbour classification (LMNN) improves the accuracy of KNN by learning a global distance metric, which did not consider the locality of data distributions. In this paper, we propose a new distance metric algorithm adopting cosine metric and LMNN named COS-SUBLMNN which takes more care about local feature of data to overcome the shortage of LMNN and improve the classification accuracy. The proposed methodology is verified on CVDs patient vector derived from real-world medical data. The Experimental results show that our method provides higher accuracy than KNN and LMNN did, which demonstrates the effectiveness of the Risk predictive model of CVDs based on COS-SUBLMNN.

1. Introduction

From the World Health Organization (WHO) statistics [1], cardiovascular diseases (CVDs) which are the leading cause of death globally, known as a threat of public health across the world, are estimated to cause about 17 million deaths every year. Moreover, an estimated 23 million new probable cases will die of cardiovascular diseases over the next decades. Obviously, the high mortality and death rate in cardiovascular diseases patients is a major public health concern. Cardiovascular diseases are multifactorial which are really hard to make an accurate diagnosis. Efforts should be taken for considering the risk factors of CVDs and finding a global strategy for prediction and prevention of CVDs remains a tough challenge. Making accurate predictions for the risk of cardiovascular diseases is really meaningful for human.

With the emergence and popularity of intelligent technology equipment like medical wearable devices, medical health data adopted by electronic health records (EHRs) grow rapidly and provide various sources of information for medical research. Based on EHRs, many risk prediction models for CVDs have been developed. The Framingham risk score [2], Reynolds risk score [3] and ASSIGN Scottish algorithm [4] are developed to predict the risk of CVDs by using the risk factors of CVDs such as BMI, smoking history, hypertension status diabetes mellitus and some blood biomarkers[5-6]. Some other variables are added to make the prediction more precise [3, 7]. However, these prediction models did not take personalized medicine decision into consideration, making the predictive task not accurate. For these reason, considering patient similarity of pairs of patients based on EHRs will increase the accuracy of prediction model. Fei Wang et al. [8] proposed the SimProX system which uses low-rank mahalanobis distance with physician's feedbacks to measure patient dissimilarity and the accuracy of this model is nearly 92% better than the performance of other traditional machine learning algorithm. Gottlieb et al [9] combine types of similarity measure to predict the likely discharge diagnosis for a new cases using information from medical history. Nevertheless, they never care about the patient similarity in local scope, which can capture the discriminative feature better.

KNN is a basic and simple distance metric algorithm, which calculates the distance between vectors to measure the similarity and make prediction, but KNN treats each variable of feature vector equally. However, different features have different weight and make different contributions to label classification. Weinberger et al. [10] proposed the large margin nearest neighbour classification (LMNN) to overcome the weakness of KNN and improves the performance. The aim of LMNN is to minimize the distance between samples belonging to the same class and make vectors with opposite label a large margin by a global distance metric. Considering the direction not just the distance, some studies [12-13] discussed distance matric algorithm based on cosine similarity. All algorithm mentioned above, they did not consider locality of data distributions, which is a significant part to capture nuances of local heterogeneity among patients.

In this paper, we propose the COS-SUBLMNN algorithm adopting cosine metric and large margin nearest neighbour to capture differences between local samples to improve the classification accuracy. We will build an automated predictive system based on COS-SUBLMNN that infers the CVDs diagnosis of patients using patient EHRs derived from routine medical data. Medical data we use includes routine physical examination data, medical questionnaire survey results and a return visit followed 1 month after the physical examination for all the candidates [11]. Every features in medical data list are considered as potential risk factors for CVDs. The diagnostic feedback and personalized suggestion will be sent to all patients and the feedback of patients for the model's results will be used to optimize algorithm. The system of predicting CVDs is shown in Figure.1.

Other parts of the paper are organized as follows: Section 2 describe the details of the study subject, the construction of patient profiling and data preprocessing. Part 2 proposes the COS-SUBLMNN algorithm. In part 3, we have experiments of risk predictive models produced by COS-SUBLMNN, KNN and LMNN, respectively, and make comparisons on the performance of three risk predictive model. We draw an overall summary and conclusion of our entire work in section 4.

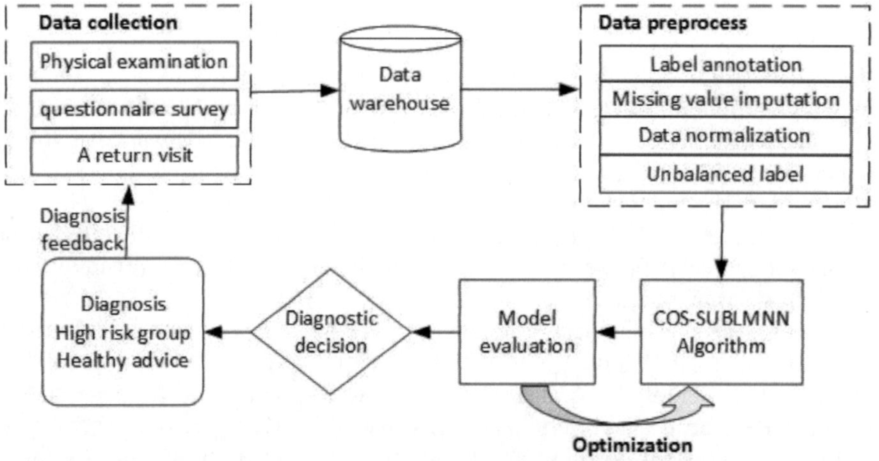

Figure 1. The whole framework of risk predictive model based on COS-SUBLMNN.

2. Materials and methods

In this part, we present the details of our COS-SUBLMNN algorithm and the system of predicting CVDs. Firstly, we show the way we construct the representative vector of every candidates based on EHRs.

2.1. Study Subjects

All medical data for our task are extracted from CiMing Health Checkup Center. Here we present the routine physical examination data briefly in Table 1 which is introduced in detail in our last paper [10]. In addition, some other predictor variables displayed are from medical questionnaire survey, such as family medical history, medical history, education level, occupation, dietary law and so on. We use data warehouse, hive to save all medical data. The data table in hive is comprised of some columns: candidate ID, inspection item and its value and all tables in hive are combined by the primary key, candidate ID. The label annotation is still a tough work for us. Some criteria used to define the cases of CVDs is introduced below: (i) The candidate was diagnosed with CVDs, such as cerebral stroke, hypertension, coronary disease, hyperglycaemia, hyperlipoidemia, cerebral infarction and so on in 30-day interval return visit. (ii) The diagnosis of CVDs appeared on the corresponding list of ICD-9 codes. So the label (class variable) in our model is health status based on the criteria of label annotation. If a candidate met the criteria above, the label is 1, otherwise it is 0.

Table 1. Typical variables in physical examination data.

routine physical examination data	
Demographic information	Age
	gender
Vitals	Body Mass Index (BMI)
	Systolic Blood Pressure (SBP)
	Diastolic Blood Pressure (DBP)
	…
Blood routine examination	White Blood Cell (WBC), , …
	Red Blood Cell (RBC)
	Hemoglobin (HGB)
	…
Urine routine examination	Hydrogen (PH)
	Specific Gravity (SG)
Biochemical examination	Creatinine (Cr),
	Uric Acid (UA)
	Fasting Blood-Glucose (FBG)
	…
Other risk predictors	alpha fetoprotein (AFP) and
	carcino-embryonic antigen (CEA)

2.2. Data preprocessing

Incomplete data is a common problem for most tasks in the area of data mining. Missing values is a typical characteristic for real-world medical data. The approach and specific procedure for handling the missing values is the same as [10]. We use different normalization methods for variables according to the feature type. the discrete variables such as age and gender are normalized by one-hot encoding and all the numerical variables like blood pressure are normalized by Z-Score and. The formula of Z-Score is:

$$Z = (X-\mu) / \sigma \qquad (1)$$

Where X is the raw value of variables. μ and σ is the mean and standard deviation, respectively.

As mentioned above, a representative vector for every candidate is constructed, which is a high dimensional vector combined by risk factors from different data sources and label, serving as the input to distance metric algorithm. The purpose and advantage of vector construction is to capture sufficient difference among candidates and distinguish the group of patients used by distance metric algorithm.

Next, our purpose is to design a distance metric algorithm to make fully use of patient feature vector and get the candidate feedback of whether the result of predictive model is right or not.

2.3. COS-SUBLMNN distance metric algorithm
In this section, we present a local distance metric learning method adopting cosine metric and LMNN called COS-SUBLMNN, which first use LMNN to get a global distance metric, mapping samples to new feature space, then get K local cluster by k-means and use LMNN in the specific target local cluster which target vector exists in to make prediction.

2.3.1. Constructing COS-LMNN. The feature vector of each candidate is denoted by an n-dimensional feature vector x, and z is the corresponding label of x. So X presented below is the dataset we use.

$$X = \{(x_i, z_i)\}(i = 1,2, \dots, m, x_i \in \mathbb{R}^n, z_i \in \{1,2, \dots, c\}).$$

In this part, we introduce COS-LMNN which learns a global matrix using cosine metric:

$$COS(x_i, x_j) = \|A(x_i, x_j)\|^2 = \frac{{x_i}^T M(x_j)}{\sqrt{{x_i}^T M(x_i)}\sqrt{{x_j}^T M(x_j)}} \tag{2}$$

Where $M \in R^{nxn} = A^T A$ is a positive Semi-Definite matrix, and A is the matrix that COS-LMNN needs to learn. We define the homogeneous neighbourhood of x_i, $S(x_i)$, which has the same label with x_i and heterogeneous neighbourhood of x_i, $I(x_i)$, which has the different label with x_i. The size of $S(x_i)$ is K_s which should be set by users. $COS(x_i, x_j)$ satisfy the property of cosine, so cosine relationship between x_k, the sample of $I(x_i)$, and input sample x_i is presented below:

$$COS(x_i, x_j) \le 1 - COS(x_i, x_k) \tag{3}$$

COS-LMNN aims at learning a global matrix to lessen the gap between samples with same label and magnify the distance between examples with opposite label. So the objective function is described as:

$$\text{Obj}(A) = \sum_i \sum_{j \in S(x_i)} COS_A(x_i\,,\, x_j) + \mu \sum_{i, j \in S(x_i)} \sum_{k \in I(x_i)} [COS_A(x_i\,, x_j) + COS_A(x_i\,, x_k) - 1]_+ \tag{4}$$

As we can see, the first term $\varepsilon_{pull}(A)$ penalizes big gap between every input and its homogeneous neighbourhood.

$$\varepsilon_{pull}(A) = \sum_i \sum_{j \in S(x_i)} COS_A(x_i\,,\, x_j) \tag{5}$$

The second term $\varepsilon_{push}(A)$ magnify the distance by penalizing the small distance between examples with opposite label. The indicator variable $y_{ik}=0$ if $z_i = z_k$, otherwise $y_{ik}=1$.

$$\varepsilon_{push}(A) = \sum_{i, j \in S(x_i)} \sum_{k \in I(x_i)} y_{ik} [COS_A(x_i\,, x_j) + COS_A(x_i\,, x_k) - 1]_+ \tag{6}$$

Where $[z]_+ = \max(0,z)$ is the hinge loss. The hinge loss indicates that if the input x_i keep a safe distance with its heterogeneous neighbourhood, then its hinge loss has no influence on the objective function.

So the optimization function combining two penalty term and the slack variable ξ_{ijk} is shown:

$$\max \sum_{i, j \in T(x_i)} COS(x_i, x_j) + \mu \sum_{\substack{i, j \in S(x_i), \\ k \in I(x_i)}} \xi_{ijk} \tag{7}$$

$$\text{s.t. } COS(x_i, x_j) + COS(x_i, x_k) \le 1 - \xi_{ijk}$$

$$\xi_{ijk} \ge 0,$$

$$M = A^T A \geq 0$$

2.3.2. COS-SUBLMNN algorithm. COS-SUBLMNN algorithm has three step: Firstly, a global matrix A is learned by COS-LMNN, and then K local clusters are determined by k-means and the target cluster which obtains input sample is determined. Finally, a local matrix A_s is learned in the target cluster by COS-LMNN. We introduce COS-SUBLMNN in detail below.

(1) The calculation of global distance

Considering input sample x_i dataset X and global matrix A, $M = A^T A$, we calculate the distances between xi and each sample in dataset X:

$$D(x_i, x_j) = COS(x_i, x_j) = \|A(x_i, x_j)\|^2 = \frac{x_i{}^T M(x_j)}{\sqrt{x_i{}^T M(x_i)}\sqrt{x_j{}^T M(x_j)}} \quad (8)$$

D_{ij} denotes the distance between x_i and x_j and $D_s = \{D_{i1}, D_{i2}, \dots, D_{im}\}$ is set of distances

(2) The determination of target cluster

K_a local cluster is created by k-means. Based on $D_S = \{D_{i1}, D_{i2}, \dots, D_{im}\}$, choose the cluster which is nearest to input sample xi from K_a local cluster as the target cluster.

(3) The classification in target cluster

COS-LMNN is used to get a local matrix As in the target cluster $L_s = \{x_{s1}, x_{s2}, \dots, x_{sn}\}$ and then the distances between xi and each sample in the target cluster Ks are calculated by cosine distance.

$$COS(x_i, x_{si}) = \|A_s(x_i, x_{si})\|^2 = \frac{x_i{}^T M(x_{si})}{\sqrt{x_i{}^T M(x_i)}\sqrt{x_{si}{}^T M(x_{si})}} \quad (9)$$

Finally, the input sample xi is classified by KNN（K_b is set by users）

3. Experiments and discussions

In this part, the details of the experiments of risk predictive model of CVDs on medical data are presented to demonstrate the effectiveness of COS-SUBLMNN algorithm. To get a fair and persuasive conclusion, we compare the performances of COS-SUBLMNN algorithm with these different classifier models (KNN and LMNN) with 5-fold cross validation which is a good solution to the overfitting problem. Some metrics like accuracy, sensitivity and specificity are proposed to assess the performance of risk predictive model of CVDs. All experiments of risk predictive model of CVDs are run on the platform of MATLAB R2014b. The parameters of COS-SUBLMNN algorithm is K_a, K_s and K_b introduced in detail in chapter 2. All parameters are selected by grid search.

(1)The results of experiments show that the performances and accuracy of COS-SUBLMNN are better when K_a=8 and K_s=7.So firstly we compare the performances of COS-SUBLMNN algorithm with KNN and LMNN when K_b varies and K_a=8 and K_s= 7.

Table 2. The results of 3 classifiers when K_b varies and K_a=8 and K_s= 7.

K_b	KNN			LMNN			COS-SUBLMNN		
	Sen	Spe	Acc	Sen	Spe	Acc	Sen	Spe	Acc
2	0.641	0.828	0.767	0.711	0.862	0.811	0.720	0.849	0.807
4	0.640	0.831	0.769	0.716	0.869	0.819	0.733	0.848	0.813
6	0.667	0.821	0.775	0.721	0.872	0.822	0.719	0.865	0.817
8	0.671	0.822	0.781	0.725	0.863	0.818	0.726	0.875	0.825
10	0.672	0.840	0.787	0.724	0.878	0.823	0.724	0.877	0.830
12	0.679	0.838	0.783	0.723	0.863	0.819	0.735	0.881	0.833
14	0.671	0.848	**0.799**	0.731	0.881	**0.827**	0.746	0.878	**0.835**

16	0.675	0.843	0.797	0.736	0.868	0.825	0.741	0.876	0.831
18	0.670	0.855	0.793	0.744	0.862	0.823	0.734	0.871	0.827
20	0.672	0.828	0.781	0.733	0.857	0.817	0.719	0.861	0.815
22	0.640	0.833	0.772	0.726	0.854	0.812	0.721	0.863	0.820

From figure 2 and table 2, when K_b=14, the accuracy of COS-SUBLMNN is up to 0.835 higher than KNN and LMNN, demonstrating that COS-SUBLMNN has advantages of identifying all group of patients. When K_b=14, the accuracy of KNN and LMNN reaches the top. As K_b is bigger, the accuracy of COS-SUBLMNN is smaller than LMNN's and COS-SUBLMNN and LMNN performs better than KNN. The specificity of COS-SUBLMNN and LMNN is up to 0.88 and performs well, indicating that has a strong ability of identifying negative group (normal people).

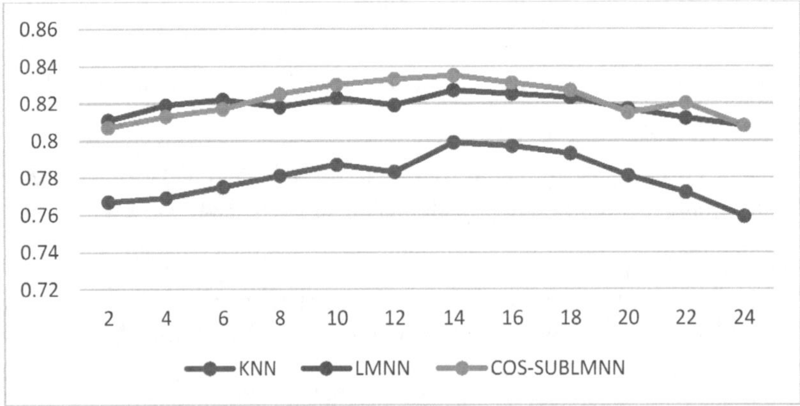

Figure 2. The accuracy of 3 classifiers when K_b varies and K_a=8 and K_s= 7.

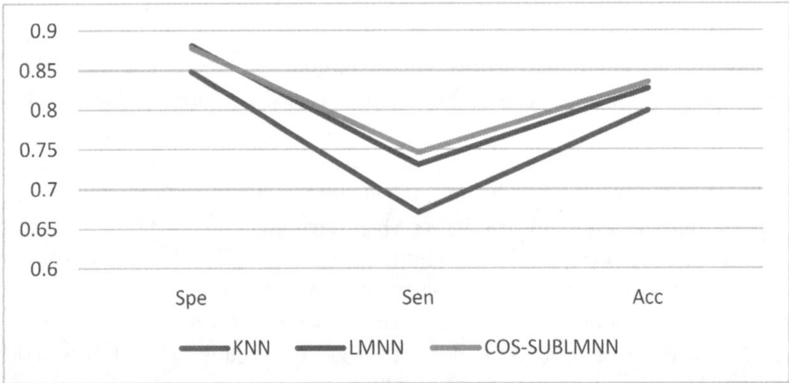

Figure 3. The performance of 3 classifiers in terms of specificity, sensitivity and accuracy.

As figure 3 is shown, the performances of COS-SUBLMNN is the best, which is the purpose and significance of our work and is the reason why we build the risk predictive model using COS-SUBLMNN.

(2)K_a=8 and K_b=14 set, we observe the change of the performance of KNN, LMNN and COS-SUBLMNN with the change of K_S and get the K_S value which makes the accuracy of COS-SUBLMNN better.

Table 3. The results of 3 classifiers when K_S varies and K_a=8 and K_b= 14.

K_S	KNN			LMNN			COS-SUBLMNN		
	Sen	Spe	Acc	Sen	Spe	Acc	Sen	Spe	Acc
2	0.671	0.848	**0.799**	0.745	0.841	0.809	0.752	0.859	0.822

4	0.671	0.848	**0.799**	0.743	0.839	0.803	0.717	0.871	0.826
6	0.671	0.848	**0.799**	0.737	0.853	0.813	0.762	0.859	0.825
7	0.671	0.848	**0.799**	0.731	0.881	**0.827**	0.746	0.878	**0.835**
8	0.671	0.848	**0.799**	0.734	0.857	0.818	0.762	0.867	0.830
9	0.671	0.848	**0.799**	0.723	0.852	0.812	0.717	0.871	0.826
10	0.671	0.848	**0.799**	0.723	0.851	0.807	0.718	0.873	0.820
11	0.671	0.848	**0.799**	0.724	0.835	0.797	0.738	0.856	0.816

According to the introduction of COS-SUBLMNN, when K_s is small, noise decrease the performance of algorithm, otherwise, time complexity is too high. The results show that when K_S =7, the performances of COS-SUBLMNN and LMNN reach the top. Obviously, KNN is not a function of K_S, so the accuracy of KNN stays still no matter how K_S changes.

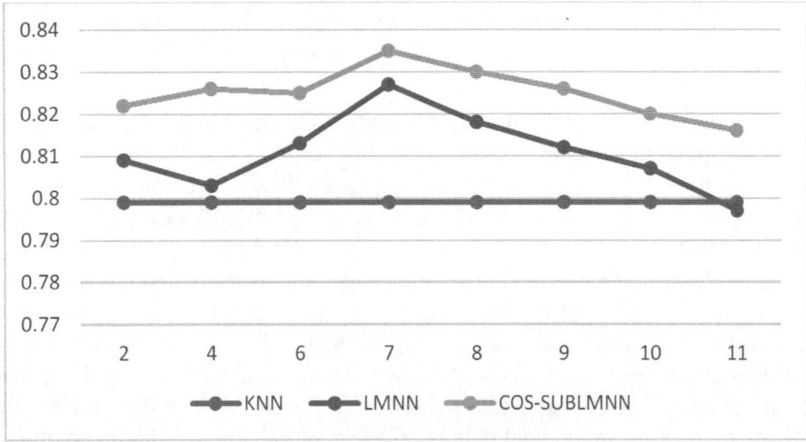

Figure 4. The accuracy of 3 classifiers when K_S varies and K_a=8 and K_b= 14.

(3)K_S=7 and K_b=14 set, we observe the change of the performance of KNN, LMNN and COS-SUBLMNN corresponding to the change of K_a and get the K_a value when COS-SUBLMNN provides the best results.

Table 4. The results of 3 classifiers when K_a varies and K_S=7 and K_b= 14.

K_a	KNN			LMNN			COS-SUBLMNN		
	Sen	Spe	Acc	Sen	Spe	Acc	Sen	Spe	Acc
2	0.682	0.838	**0.787**	0.731	0.881	**0.827**	0.733	0.855	0.817
4	0.682	0.838	**0.787**	0.731	0.881	**0.827**	0.717	0.871	0.826
6	0.682	0.838	**0.787**	0.731	0.881	**0.827**	0.734	0.872	0.829
7	0.682	0.838	**0.787**	0.731	0.881	**0.827**	0.724	0.877	0.830
8	0.682	0.838	**0.787**	0.731	0.881	**0.827**	0.746	0.878	**0.835**
9	0.682	0.838	**0.787**	0.731	0.881	**0.827**	0.734	0.885	0.833
10	0.682	0.838	**0.787**	0.731	0.881	**0.827**	0.721	0.872	0.822
11	0.682	0.838	**0.787**	0.731	0.881	**0.827**	0.718	0.873	0.820

By observing the consequences in table 4 and introduction to COS-SUBLMNN algorithm, higher K_a value reduces the performance of COS-SUBLMNN algorithm with higher time complexity, otherwise, noise dominate the consequence, causing the accuracy of COS-SUBLMNN lower than LMNN. When K_a =8, COS-SUBLMNN provides the best accuracy. As we can see from figure 5, the performance of KNN and LMNN stays still when K_a varies.

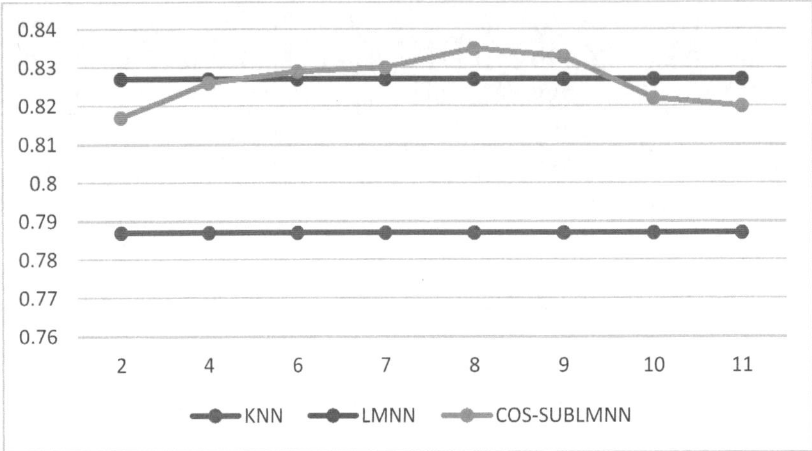

Figure 5. The accuracy of 3 classifiers when K_a varies and K_S=7 and K_b= 14.

4. Conclusions

In our research, we propose a distance metric learning algorithm adopting cosine metric and LMNN named COS-SUBLMNN for risk predictive model of CVDs based on the patient profiling derived from EHR medical data. The proposed algorithm capture nuances of differences on local features among patient vector by considering the locality of medical data. The practical benefits of the optimized algorithm have been assessed by experiments. The model built by COS-SUBLMNN achieved much higher accuracy than KNN and LMNN. Predicting the risk of CVDs accurately plays an important role in daily life for each patient. As our results above shown, the application of COS-SUBLMNN based on real-world medical data for CVDs are valuable for both patients and physicians.

Acknowledgments

This work was supported by National Natural Science Foundation of China (61471064) and National Science and Technology Major Project of China (No.2017ZX03001022-005).

References

[1] Otsu N. Discriminant and Least Square Threshold Selection [J]. Proc 4IJCPR, 1978, 596. Mendis S, Puska P, Norrving B. Global atlas on cardiovascular disease prevention and control [M] World Health Organization, 2011.

[2] Wilson P W F, Levy D, Belanger A M, et al. Prediction of Coronary Heart Disease Using Risk Factor Categories[J]. Circulation, 1998, 97(18):1837-47.

[3] Ridker P M, Buring J E, Rifai N, et al. Development and validation of improved algorithms for the assessment of global cardiovascular risk in women: the Reynolds Risk Score.[J]. Jama, 2007, 297(6):611.

[4] Woodward M, Brindle P, Tunstallpedoe H. Adding social deprivation and family history to cardiovascular risk assessment-the ASSIGN score from the Scottish Heart Health Extended Cohort (SHHEC) [J]. Heart, 2007, 93(2):172.

[5] Baik I, Cho N H, Kim S H, et al. Dietary information improves cardiovascular disease risk prediction models [J]. European Journal of Clinical Nutrition, 2013, 67(1):25.

[6] Björnson E, Borén J, Mardinoglu A. Personalized Cardiovascular Disease Prediction and Treatment-A Review of Existing Strategies and Novel Systems Medicine Tools [J]. Frontiers in Physiology, 2016, 7:2.

[7] Melander O, Newtoncheh C, Almgren P, et al. Novel and conventional biomarkers for prediction of incident cardiovascular events in the community.[J]. Jama the Journal of the American Medical Association, 2009, 302(1):49-57.

[8] F Wang, J Hu, J Sun. Medical prognosis based on patient similarity and expert feedback. International Conference on Pattern Recognition, 2012:1799-1802.

[9] Gottlieb A, Stein G Y, Ruppin E, et al. A method for inferring medical diagnoses from patient similarities [J]. Bmc Medicine, 2013, 11(1):194.

[10] Distance metric learning for large margin nearest neighbor classification [J]. Journal of Machine Learning Research, 2009:207-244.

[11] Kang G, Ni Z. Research on early risk predictive model and discriminative feature selection of cancer based on real-world routine physical examination data[C]// IEEE International Conference on Bioinformatics and Biomedicine. IEEE, 2017:1512-1519.

[12] H.V.Nguyen, L. Bai. Cosine similarity metric learning for face verification [C]. Asian Conference on Computer Vision. Springer, 2010, 709-720.

[13] Kumar M P, Torr P H S, Zisserman A. An Invariant Large Margin Nearest Neighbour Classifier[C]// IEEE, International Conference on Computer Vision. IEEE, 2007:1-8.

IOP Conf. Series: Journal of Physics: Conf. Series **1004** (2018) 012033 doi:10.1088/1742-6596/1004/1/012033

Study on the Secant Segmentation Algorithm of Rubber Tree

Shute Li [1], Jie Zhang [1], Jian Zhang [1], Liang Sun [2] and Yongna Liu [1,*]

[1]Institute of Tropical Agriculture and Forestry, Hainan University, Danzhou, Hainan 571737, China

[2]Rubber Research Institute, Chinese Academy of Tropical Agricultural Sciences, Danzhou, Hainan 571737, China

*yongnaliu@qq.com

Abstract. Natural rubber is one of the most important materials in the national defense and industry, and the tapping panel dryness (TPD) of the rubber tree is one of the most serious diseases that affect the production of rubber. Although considerable progress has been made in the more than 100 years of research on the TPD, there are still many areas to be improved. At present, the method of artificial observation is widely used to identify TPD, but the diversity of rubber tree secant symptoms leads to the inaccurate judgement of the level of TPD. In this paper, image processing technology is used to separate the secant and latex, so that we can get rid of the interference factors, get the exact secant and latex binary image. By calculating the area ratio of the corresponding binary images, the grade of TPD can be classified accurately. and can also provide an objective basis for the accurate identification of the tapping panel dryness (TPD) level.

1. Introduction

Natural rubber has become one of the indispensable materials in the development of national defense and industry because of its good characteristics. However, the tapping panel dryness (TPD) has a certain harm to the planting of rubber tree, which seriously affects the production of natural rubber. Therefore, it is of great significance and value to study the tapping panel dryness of the rubber tree. Grading of rubber bark cut secant is a prerequisite for studying the mechanism of the tapping panel dryness and the prevention and control technology of the tapping panel dryness. Traditional manual classification methods are subjective and have adverse effects on research results. In order to ensure the objectivity of the classification of the tapping panel dryness symptoms and improve the accuracy of classification, we use image processing technology to study the segmentation of rubber tree secant and latex image.

2. Related research

Skin symptom rating is the base of the study on TPD, and currently the classification of TPD is mainly by human eye observation and judging according to a certain standard. Grade 0 (normal tree), grade 1 (TPD length less than 2cm), grade 2 (TPD length 2cm to 1/4 the length of secant, grade 3 (TPD length is the length of the 1/4 secant to 2/4), grade 4 (TPD length is the length of the 2/4 secant to 3/4), grade 5 (TPD length is the length of the 3/4 or more) [1]. With the development of artificial intelligence, pattern recognition and machine learning, image processing technology is widely used in the field of agriculture and has achieved good results. The extraction and recognition of leaf of sunflower and jujube under natural environment using computer vision technology, the research results have certain reference significance for field leaf extraction and leaf lesion segmentation and classification [2]. Kyung-Mo Koo enhances image recognition performance by using the idea of image normalization [3]. Article [4] studied for maize disease diagnosis, and puts forward some methods on leaf spot extraction and lesion type identification. These methods are good for identifying specific samples. Literature [5] uses an automated active shape matching weeds identification (AASM) technique to identify crops and weeds according to their shapes. The above research has a good effect in its corresponding field, but it is not suitable for the study of the TPD of the rubber tree. In this paper, we divide the rubber tree secant and latex by digital image processing technology, so as to reduce the subjectivity and instability of the TPD level by directly observing the secant of human eye.

3. Research framework and image acquisition

3.1 Research framework

The research of image segmentation based on image processing technology can be divided into the following parts: image acquisition, image preprocessing and image segmentation. Image acquisition is the premise and foundation of the follow-up work, collecting equipment selection, acquisition environment and other factors directly affect the segmentation effect; image preprocessing is to as much as possible to eliminate the interference factors, this part mainly used enhancement and denoising technique to eliminate interference factors; image segmentation is to extract the target from the background, the extraction effect directly effect of identification and classification of the late. The framework of image segmentation in this paper is as follow shown in figure 1:

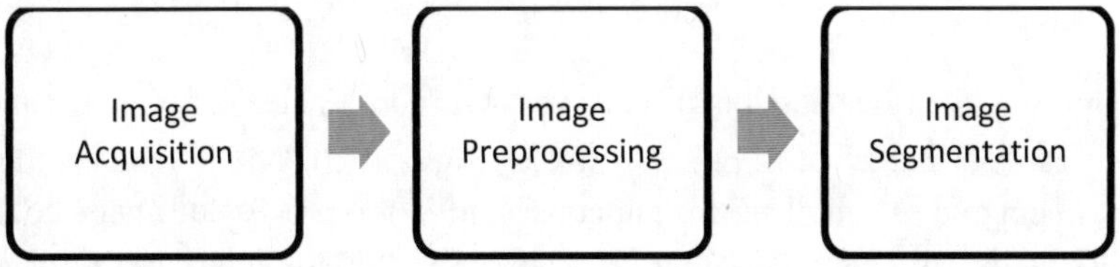

Figure 1. The framework of image segmentation.

3.2 image acquisition

Image acquisition is the basis of the study of the TPD of rubber tree based on image processing technology. The quality of image acquisition directly affects the effect of late treatment. The construction of collection environment is one of the most important links, which mainly involves the selection of light and acquisition equipment.

Illumination is one of the key that affect the quality of image acquisition. In order to avoid the influence of light on image acquisition, we choose clear weather in the morning to collect data between 4:00-7:00. This time is in the lowest temperature and humidity maximum of a day and rubber trees after a night of rest, the body of water is full, water transpiration leaves are lowest, this period is the best time for tapping.

Because the location of the image acquisition is in the field, the acquisition device has to satisfy the characteristics of the field operation, and requires the equipment to be portable, flexible, abundant in power and high in image accuracy, while considering the moderate price of the equipment. Through comprehensive comparison this paper selects the Nikon D90 camera as the acquisition equipment, in order to ensure the position of the camera lens and the secant accords with the acquisition requirements, in the process of collection with three fixed camera tripod, camera and relationship to adjust the position of secant mainly considers three aspects of factors, namely height, distance and angle. The height adjustment is that the lens is parallel to the secant, the distance between the lens and the secant is 40cm, and the lens is on the cut face when the image is imaging. The image resolution is 4288 * 2848 pixels by using the automatic white balance mode.

4. Preprocessing and Segmentation

IOP Conf. Series: Journal of Physics: Conf. Series **1004** (2018) 012033 doi:10.1088/1742-6596/1004/1/012033

4.1 preprocessing

The digital image is obtained directly from optical equipment such as high precision digital camera or camera, which saves the process of image digital conversion. Although various measures have been taken to improve the quality of photography, due to the imaging characteristics of optical devices, there still exist some disturbing factors, which influence the segmentation of secant. Using some preprocessing algorithms in digital image processing, we can greatly improve the symptom expression forms in the image, and create favorable conditions for the accurate extraction of secant features. Before image segmentation, it is necessary to normalize the contrast, angle and scale of the image. The tapping method has strict restrictions, so the general case is operated by trained professionals. Therefore, the length, width, depth and inclination angle of every secant are basically the same. We can use the geometrical characteristics of secant to normalize the angles and scales.

In the process of collection, sometimes the latex will overflow the Secant and attach to the trunk, which will cause serious interference to image segmentation. We must exclude the factors that interfere with it. As shown in Figure 2, according to the 2D coordinates of point A, point O and point B, $A(x_a, y_a)$, $O(x_o, y_o)$, $B(x_b, y_b)$,the position and size of parallelogram can be calculated, and then the parallelogram AOBC can be tailored.

Rotation transformation, the parallelogram AOBC is rotated, so that the OB side and the level of the same level. The angle between the OB and the horizontal line DB is recorded as theta, and using formula 1, the AOBC is rotated to the angle theta in the direction of the DB, thus the image with the same direction as the OB side and the horizontal direction is obtained.

$$\theta = \arctan \frac{|y_o - y_b|}{|x_o - x_b|} \tag{1}$$

Where, x_o and y_o are the horizontal and vertical coordinates of O, x_b and y_b are the horizontal and vertical coordinates of B, $|y_o - y_b|$ is the length of OD and $|x_o - x_b|$ is the length of OB.

Although the camera and secant distances are consistent in the process of image collection for each shot, but inevitably there will be some errors, in order to exclude the interference distance difference brings to the image recognition, the secant image scale normalized, uniform adjustment of the image size of 100*400 pixels.

After processing these three steps, the geometric invariance of secant images is obtained to a certain extent, and the interference factors of the background are basically eliminated. The image after preprocessing is shown in Figure 3.

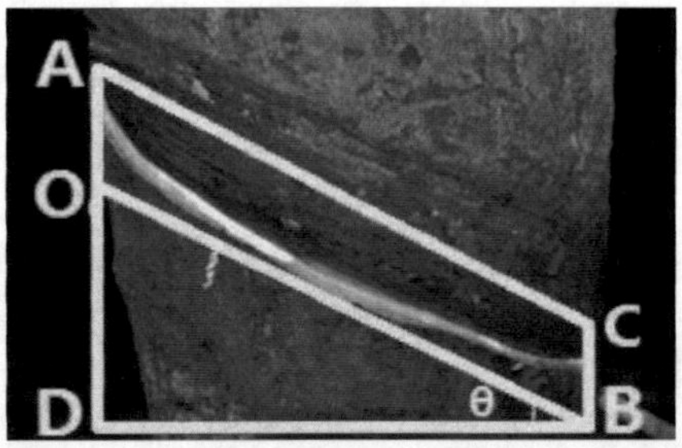

Figure 2. Preconditioning schematic diagram.

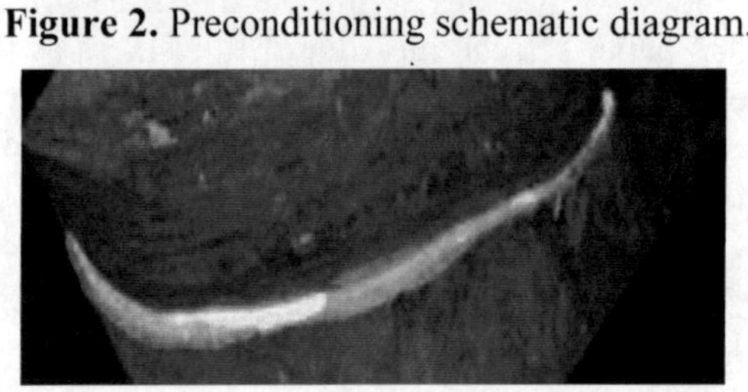

Figure 3. Preprocessed image.

4.2 Secant and latex image segmentation

Image segmentation means that the image is divided into several regions according to different characteristics, and the interest target is extracted from it. The application field of image segmentation is very wide, such as identification of plant diseases and insect pests, forest fire identification, Coastline Extraction, traffic image analysis and so on. Due to the complexity and diversity of segmentation objects, so far, there is no general segmentation method. In many segmentation methods, the largest class variance method has good segmentation effect and strong adaptability. In this paper, the method of OTSU is used to divide the secant of rubber tree and the white latex on the cut secant.

The OTSU method belongs to the threshold segmentation method, and its theoretical basis is to use the least square method to deduce the gray histogram. The criterion of selecting threshold by the maximum inter class variance method is to determine the threshold according to the background and target of the image according to the maximum of the inter class variance. The best threshold selection method is: set the image to contain N pixels, the gray level is $[1,m]$, and the n_i pixels are used to represent the gray level i, then the probability of each gray value is $p_i = n_i / N, (i = 1,2,3,...,m)$. The image is divided into two regions $C_0 = \{1 \sim T\}$ and $C_1 = \{T+1 \sim m\}$ by the threshold value $T \in [1,m]$. The mean of C_0 and C_1 is as follows:

$$u_0 = \sum_{i=1}^{T} ip_i / w_0, u_1 = \sum_{i=T+1}^{m} ip_i / w_1 \tag{2}$$

Among,

$$w_0 = \sum_{i=1}^{T} p_i, w_1 = 1 - w_0 \tag{3}$$

And the OTSU value is,

$$\sigma^2(T) = \omega_0(u_0 - u_T)^2 + \omega_1(u_1 - u_T)^2 = \omega_0\omega_1(u_1 - u_0)^2 \tag{4}$$

T is valued in $[1,m]$. When the formula 4 reaches the maximum, the T is the best threshold. The largest class variance method is used to measure the distribution of grayscale with variance. The larger the variance is, the greater the difference between the two regions is. The lower the probability of mismatch between the target and the background is. This method has the advantages of simple model, stable effect and strong adaptive ability.

5. Experiment and analysis

In order to verify the effectiveness of the method, the experiment was carried out on the secant and the latex image of the rubber tree on the Matlab platform. Because the gray value of secant and white latex is closer to the trunk, the secant and the rubber part are cut out first, and then the rubber part is segmented from the secant line. In the first round, we first normalize the pre-processed images. The purpose of gray normalization is to increase the contrast of images, so that the computer can automatically analyze and process according to the gray value. After the completion of gray normalization, the secant as part of a class, the trunk as part of another class, using variance method based on the pixel value will be separated from line and bark, the secant segmentation denoising, which become a cut or approximate a continuous closed curve. In the second round, the secant area divided by the first round is used as the processing area, and the latex part is taken as a class, and the secant is divided into the other part except the rest of the latex. First, the pixel gray value of the region is normalized, and then the latex part is separated from the secant by the maximum inter class variance method.

For using the OTSU method, should retrieval the gray value T of the image between $[1,m]$ first, and calculate the value of $T \in [1,m]$ when $\sigma^2(T)$ gets its maximum value. Then set T as the segmentation threshold of the image, the target and the background will be segmented. The image of

IOP Conf. Series: Journal of Physics: Conf. Series **1004** (2018) 012033 doi:10.1088/1742-6596/1004/1/012033

different levels of TPD is divided, the size of the image is pixel of 100×400, and the gray level is [0,255]. Figure 4 gives the segmentation results of the normal and 1~5 level TPD images. From the image segmentation effect, the secant and the latex segmentation effect is ideal.

(a1) Normal original diagram

(a2) Normal secant segmentation effect

(a3) Normal latex segmentation effect

(b1) level 1 original

(b2) level 1 secant segmentation effect

(b3) level 1 latex segmentation effect

(c1) level 2 original

(c2) level 2 secant segmentation effect

(c3) level 2 latex segmentation effect

(d1) level 3 original

(d2) level 3 secant segmentation effect

(d3) level 3 latex segmentation effect

(e1) level 4 original

(e2) level 4 secant segmentation effect

(e3) level 4 latex segmentation effect

(f1) level 5 original

(f2) level 5 secant segmentation effect

(f3) level 5 latex segmentation effect

Figure 4. The secant and the latex segmentation effect.

6. Summary and discussion

Aiming at the existing problems of rubber tree's OTSU level, this paper adopts image processing to separate the secant and latex. In order to eliminate the interference factors in secant segmentation process to ensure the segmentation effect, we first precut the secant image, including clipping, angle and scale processing. On this basis, the segmentation of the Secant and the background, the latex and the background are divided by the maximum variance method, and the effectiveness of the method is verified by the experiment. However, due to the complex illumination conditions in the practical application environment and the white marks on the trunk, it will cause some negative effects on segmentation. Therefore, secant and latex segmentation algorithms still need further improvement and optimization.

Acknowledgment
This work was funded by the Natural Science Foundation of Hainan Province of China (20166212),the Natural Science Foundation of Hainan Provincial Department of Education (Hnky2016-5) and the Research on Education and Teaching Reform of Hainan University (hdjy1730).

References
[1] Kouadio Dian, Abdourahamane Sangare, Jacques Kore Diopoh, Evidence for specific variation of protein pattern during tapping panel dryness condition development in Hevea brasiliensis, Plant Science, vol.105, pp. 207-216, 1995.
[2] Wang J, He J, Han Y, et al, An adaptive thresholding algorithm of field leaf image, Computer and Electronics in Agriculture, vol.96, pp. 23-29, 2013.
[3] Kyung-Mo Koo, Eui-Young Cha, Image recognition performance enhancements using image normalization, Decision Sciences, vol. 7, pp. 1-11, 2017.
[4] J.M. Guerrero, M. Guijarro, M. Montalvo, J. Romeo, L. Emmi, A. Ribeiro, G. Pajares, Automatic expert system based on images for accuracy crop row detection in maize fields, Expert Systems With Applications, vol. 7, pp. 1-11, 2017.
[5] Kishore C. Swain. Michael Nørremark. Rasmus N. Jørgensen. Henrik S. Midtiby. Ole Green, Weed identification using an automated active shape matching (AASM) technique, Biosystems Engineering, vol. 110(4), pp. 22-103, 2011.

IOP Conf. Series: Journal of Physics: Conf. Series **1004** (2018) 012034 doi:10.1088/1742-6596/1004/1/012034

The Research on Denoising of SAR Image Based on Improved K-SVD Algorithm

Linglong Tan [1, a], Changkai Li[1,b] and Yueqin Wang [1,c]

[1]Electronic communication engineering college, Anhui Xinhua University, Hefei, 230088, China;

[a]tanlinglong@axhu.edu.cn; [b]545715909@qq.com;[c]wangyueqin@axhu.edu.cn

Abstract. SAR images often receive noise interference in the process of acquisition and transmission, which can greatly reduce the quality of images and cause great difficulties for image processing. The existing complete DCT dictionary algorithm is fast in processing speed, but its denoising effect is poor. In this paper, the problem of poor denoising, proposed K-SVD (K-means and singular value decomposition) algorithm is applied to the image noise suppression. Firstly, the sparse dictionary structure is introduced in detail. The dictionary has a compact representation and can effectively train the image signal. Then, the sparse dictionary is trained by K-SVD algorithm according to the sparse representation of the dictionary. The algorithm has more advantages in high dimensional data processing. Experimental results show that the proposed algorithm can remove the speckle noise more effectively than the complete DCT dictionary and retain the edge details better.

1. Introduction

SAR is a coherent system capable of producing over-resolution remote sensing images. The SAR images are processed by processing the amplitude and phase of the received continuous signals. The SAR has all-weather, all-weather, multi-band and multi-polarization modes of operation, Variable side-view, strong penetration and high-resolution features make SAR images widely used in industry, civil and military applications. SAR images are often polluted by some noises and seriously affect the quality of the images, which makes the SAR image processing very difficult. Therefore, in the image processing, the suppression of noise is particularly important. At present, the commonly used filtering technology mainly includes spatial domain filtering, transform domain filtering and K-SVD noise suppression based on signal redundancy sparse representation [1,2,3]. The processing effect of spatial domain filtering technology depends largely on the selection of the filtering window, and it is difficult for the image to retain the image details after the restoration, which has the problem of edge blurring. Frequency-domain filtering technology first transform the image, Then transform coefficients are filtered in the frequency domain, The last inverse transform to the time domain, The technique is superior to the spatial domain processing technique at the edge of the image, but the Gibbs effect often appears in the uniform area and the edge in the processed image [4,5]. The research of literature [6,7] shows that the K-SVD algorithm based on dictionary learning algorithm can effectively preserve the speckle noise while preserving the edge and detail of the image, and the algorithm has a good adaptive In this paper, we use this algorithm to denoise noisy SAR images.

2. K-SDV algorithm

The K-SDV algorithm is a process of approximately sparse representation of a signal X under a set of bases, the algorithm is implemented by alternating the two steps of the sparse representation of the current dictionary and the updating of the dictionary using the sparse system. The improved K-SDV algorithm also uses the dictionary D as the sparse signal X by the training signal Y. Fixed X unchanged, and then look for a better training dictionary D [8,9].

Algorithm process is as follows:

Let the training sample set $Y = [y_1, y_2, y_3 \dots, y_N]$, the original signal X, the sparsity K, the initialization dictionary $D = D_0$, the number of iterations k = 1;

Step 1: Update the dictionary. Solve equation (1) by greedy algorithm and look for support set w_k.

$$w_k = \{i|1 \leq i \leq K, x_T^k(i) \neq 0\}$$

$$\min_{D.X}\{\| Y - DX \|_F^2\} \quad \text{s.t. } \forall i, \| x_{i\|} \leq K \tag{1}$$

Step 2: Calculate the error E_k.

$$E_k = Y - \sum_{j \neq k} d_j X_T^j \tag{2}$$

Step 3: Remove E_k zero input to get E_R^k.

$$E_R^k = E_R^k \Omega_k \tag{3}$$

Step 4: Perform SVD decomposition on E_R^k to obtain:

$$E_R^k = U\Delta V^T \tag{4}$$

The dictionary is updated, where $d_k$in the first row of U is $\widetilde{d_k}$ optimized, while the sparse vector X_R^k is updated and the product of the first column of V and $\Delta(1,1)$ is selected.

Step 5: After obtaining the dictionary D, k = k + 1. At this time, the update of one column in D is completed, and the columns in D are updated one after another according to the method from Step 1 to Step 4 to obtain the updated dictionary D.

K-SVD algorithm flow shown in Figure 1.

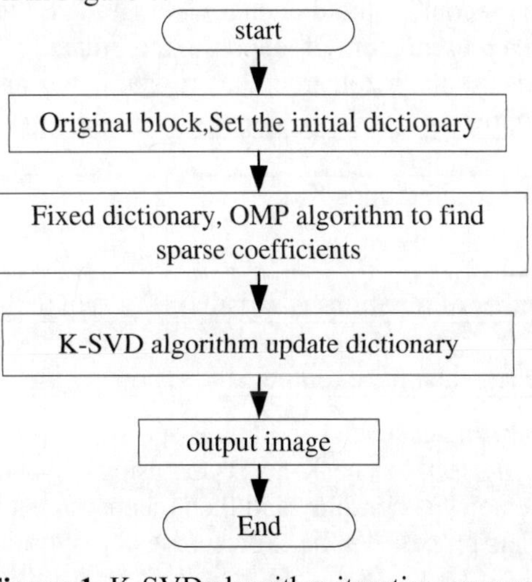

Figure 1. K-SVD algorithm iterative process

3. SAR image speckle model

The K-SVD algorithm is suitable for zero-mean additive speckle noise and has no significant effect on the multiplicative speckle noise existing in SAR images, and smoothes phenomenon. In order to overcome this problem, the SAR image can be logarithmically transformed, the multiplicative noise transformed into additive, then the K-SVD algorithm is used to denoise the changed image, However, the mean of image noise after logarithmic change is non-zero, the result is that the denoised SAR image can not maintain the original radiation characteristics well. this paper, the speckle model of SAR image is used to make up for the shortcomings of the logarithmic transform in converting the multiplicative model into the additive model [10,11,12].

SAR image multiplicative speckle can be expressed as:

$$y = x \cdot n \tag{5}$$

Where n is a coherent noise with mean 1 and variance σ_n^2, y denotes the observed signal, x is the original signal.

From (5) a signal-dependent additive noise model of (6) can be obtained.

$$y = x \cdot n = x + x(n-1) = x + \tilde{n} \tag{6}$$

Where ñ is the phase-add noise of signal x.

Due to the signal and noise are two mutually independent stochastic processes, all of the signal obtained for the mean and variance of x:

$$\mu_x = E(x) = E(y)/E(n) = E(y) = \mu_y \tag{7}$$

$$\sigma_x^2 = \frac{(\sigma_y^2 - \sigma_n^2\sigma_y^2)}{\sigma_n^2 + 1} \tag{8}$$

The average of the signal-dependent noise ñ is:

$$E(\tilde{n}) = E(x)E(n-1) = 0 \tag{9}$$

At this moment, ñ can be regarded as the additive noise of the signal x, the purpose of restraining the speckle is to remove ñ from the observed signal.

4. Improved algorithm for image denoising method

If the original image is contaminated by noise, it can be expressed as Y = X + n, Y is the observed signal, X is the no-noise signal and n is the noise. The goal is to image denoising noise-free signal recovery signal Y from the observation X. The processing of the algorithm is as follows: Firstly, X is initialized, X = Y and D are DCT dictionaries, and the approximate solution of each small image is obtained by OMP algorithm. Secondly, the dictionary is upgraded, the optimal D is found by K-SVD algorithm, and the sparse representation of approximate image is obtained. Finally, Orthogonal Matching Pursuit algorithm is used to reconstruct the original image and remove noise [13].

Algorithm steps are as follows:
Input noisy image;
Loop: initialize the number of iterations K, k = 1; noise image block; training dictionary; judge whether the number of iterations is satisfied; Not satisfied: k = k + 1;
Iteration over; OMP algorithm reconstructs the image; outputs the denoising image.

After partitioning the noisy image, each subgraph is sparsely represented in the dictionary D with a given dictionary a_i, $a = \{a_1, a_2, \ldots a_k\}$. The K-SVD denoising algorithm firstly performs training on the dictionary containing the noisy subgraph y and then reconstructs the non-noisy image X according to the obtained dictionary. This process can be expressed as the optimization process of equation (10).

$$\widetilde{X} = \arg\min_{X.D.A}\{\lambda \parallel Y - X \parallel_2^2 + \textstyle\sum_p \parallel a_p \parallel_0 + \sum_p \parallel Da_p - R_pX \parallel_2^2\} \tag{10}$$

In the formula, $\widetilde{X}$ is a denoised image, λ is a Lagrange factor, and R_p is a matrix of subgraph p.

Iteratively perform sparse coding and dictionary update of K-SVD algorithm. Sparse coding using matching pursuit algorithm to solve equation (11).

$$\forall p: \min \parallel a_p \parallel_0$$

$$\text{s.t.} \parallel \textstyle\sum_p \parallel Da_p - R_pX \parallel_2^2 \parallel \leq (c\sigma^2) \tag{11}$$

Among them, a_p is a sparse representation of subgraphs.

The dictionary update phase is to define a set of representation of any row in the matrix, using equation (2) to calculate the error E_k and applying singular value decomposition equation (3) to update the sparse coefficient.

After obtaining the training dictionary by K-SVD algorithm, each subgraph is sparsely represented. The denoised image $\widetilde{X}$ can be obtained by equation (12).

$$\widetilde{X} = \arg\min_{X}\{\lambda \parallel Y - X \parallel_2^2 + \sum_p \parallel Da_p - R_p X \parallel_2^2\} \tag{12}$$

Calculated:

$$\widetilde{X} = \left(\lambda I + \sum R_p^T R_p\right)^{-1}(\lambda I + \sum_p R_p^T D\tilde{a}_p) \tag{13}$$

Where, I is the identity matrix.

5. Quality evaluation index

In this paper, PSNR (Peak Signal to Noise Ratio) [14], an objective standard for evaluating image quality. PSNR is defined as the mean square error between the original image and the processed image relative to a logarithmic value of $(2^n - 1)^2$, Unit dB, expressed as:

$$PSNR = 10 \times \log\frac{(2^n-1)^2}{MSE} \tag{14}$$

MSE is the mean square error between the original image and the processed image.

$$MSE = \frac{1}{M\times N}\sum_{x=0}^{M-1}\sum_{y=0}^{N=1}[f - f']^2 \tag{15}$$

Among them, $M \times N$ is the image size, f represents the original image, f' represents the degraded image.

Visible, MSE larger value indicates a serious image distortion, PSNR larger the value, the more less distortion.

6. Simulation experiments and results analysis

Two 512 × 512 synthetic aperture radar images were selected as the test images, and the image was smeared with standard noise σ. The K-SVD algorithm was used to verify the denoising effect of the algorithm. The original image is divided into small pieces of 8 × 8, and the base dictionary used by the K-SVD algorithm is a DCT dictionary with a fixed initial value. Under different standard deviations of noise, the noisy images are respectively denoised by the complete DCT dictionary and K-SVD method. The number of K-SVD training iterations is 20, Figure 2 (a), Figure 3 (a) for the training dictionary, The denoising effect is shown in Figure 2 (d), Figure 2 (e), Figure 3 (d) and Figure 3 (e). The PSNR values under two different algorithms are shown in Table 1, and the denoising time is shown in Table 2.

Table 1. PSNR values under two denoising methods with different noise standard deviations

σ/dB			10	20	50	100
PSNR	Original image 1	Noisy picture	39.61	35.91	31.05	27.66
		DCT dictionary	40.01	36.33	31.33	28.08
		K-SVD	40.22	36.52	31.61	28.20
	Original image 2	Noisy picture	31.96	29.17	26.55	20.18
		DCT dictionary	33.42	30.56	27.28	22.56
		K-SVD	33.78	30.85	27.65	22.66

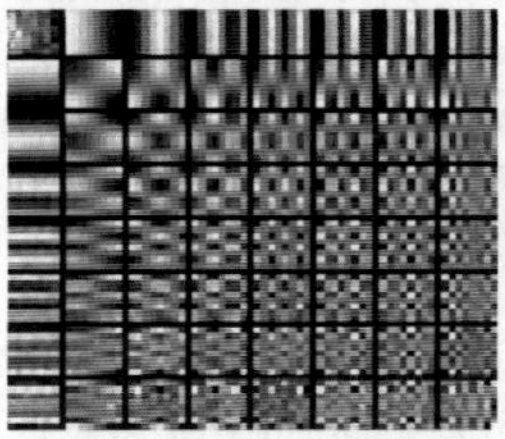

(a) Training matrix

(b) Original image 1

IOP Conf. Series: Journal of Physics: Conf. Series **1004** (2018) 012034 doi:10.1088/1742-6596/1004/1/012034

(c) noisy images (d) DCT (e) K-SVD

Figure 2. SAR image denoising effect 1

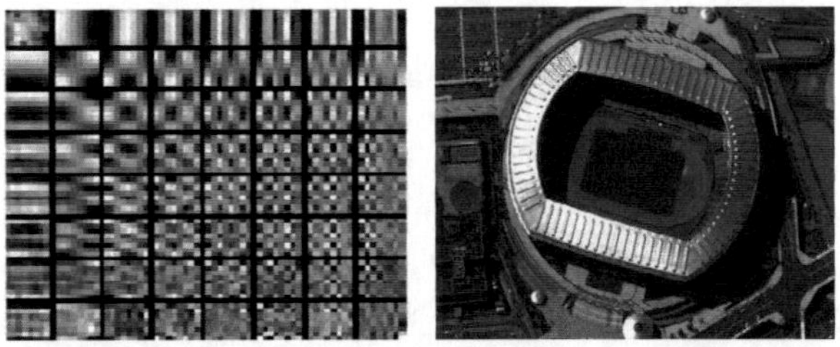

(a) Training matrix (b) Original image 2

(c) noisy images (d) DCT (e) K-SVD

Figure 3. SAR image denoising effect 2

Table 2. Time - consuming Comparison of Two Denoising Algorithms with Different Noise Standard Deviations

σ/dB			10	20	50	100
Running time/s	Original image 1	DCT dictionary	50.23	40.69	14.36	9.21
		K-SVD	312.21	152.32	49.88	26.31
	Original image 2	DCT dictionary	46.34	38.66	10.97	8.22
		K-SVD	296.98	124.58	37.47	20.51

Seen in figure 2 and figure 3, superimposed on the original image after the standard deviation $\sigma = 10$ to obtain speckle noise noisy images, respectively dictionary DCT method and K-SVD algorithm to suppress noise. As can be seen from the figure, compared with the DCT dictionary denoising effect, the K-SVD algorithm can recover the image details well except the speckle noise, and improves the PSNR value of the denoising image. The experimental data in Table 1 also verify this conclusion. The experimental data in Table 2 shows that the denoising time of DCT dictionary is the shortest, and the denoising time of K-SVD algorithm is longer than that of DCT dictionary denoising, but in the meantime, the denoising performance is improved.

7. Conclusion

Compressed sensing theory can meet the needs of a particular transform domain sparse conditions, the signal can be expressed with a small amount of observation data, the theory can be combined with other areas to solve some new problems. In this paper, to solve the problem of the inconvenience caused by the noise in the SAR image to the subsequent processing of the image, the K-SVD algorithm is applied to the image denoising. The reconstructed image after de-manning has a higher peak signal-to-noise ratio, and the edges and details of the image are better preserved, which further improves the visual effect of the image.

Acknowledgments
This work was supported by the Scientific Research Project of Anhui Xinhua College No. 2017zr003 and the domestic outstanding young backbone talents in universities of visiting program No. gxgnfx2018070.

References

[1] Yiming Lei,Ximei Zhao,Guodong Wang,Kexin Yu,Weidong Guo. A novel approach for cirrhosis recognition via improved LBP algorithm and dictionary learning[J]. Biomedical Signal Processing and Control,2017,38.
[2] Matteo Testa,Enrico Magli. Compressive Bayesian K-SVD[J]. Signal Processing: Image Communication,2017.
[3] Helder Sousa,Ying Wang. Sparse representation approach to data compression for strain-based traffic load monitoring: A comparative study[J]. Measurement,2017.
[4] Sidheswar Routray,Arun Kumar Ray,Chandrabhanu Mishra,G. Palai. Efficient hybrid image denoising scheme based on SVM classification[J]. Optik - International Journal for Light and Electron Optics,2018,157.
[5] E. Phaisangittisagul,S. Thainimit,W. K. Chen. Predictive high-level feature representation based on dictionary learning[J]. Expert Systems With Applications,2017,69.
[6] Ziyu Wang,Jianxiong Liu,Jing-Hao Xue. Joint sparse model-based discriminative K-SVD for hyperspectral image classification[J]. Signal Processing,2017,133.
[7] Qiang Zhang,Yi Liu,Rick S. Blum,Jungong Han,Dacheng Tao. Sparse representation based multi-sensor image fusion for multi-focus and multi-modality images: A review[J]. Information Fusion,2018,40.
[8] Xiaohui Yang,Fang Liu,Li Tian,Haifei Li,Xiaoying Jiang. Pseudo-full-space representation based classification for robust face recognition[J]. Signal Processing: Image Communication,2018,60.
[9] Kun Shang,Zheng-Hai Huang,Wanquan Liu,Zhi-Ming Li. A single gallery-based face recognition using extended joint sparse representation[J]. Applied Mathematics and Computation,2018,320.
[10] Seong - Hyeon Lee,In - O Choi,Min - Seok Kang,Kyung - Tae Kim. Efficient sparse representation algorithm for accurate DOA estimation of multiple targets with single measurement vector[J]. Microwave and Optical Technology Letters,2018,60(1).
[11] Yan Zhou,Heming Zhao,Li Shang,Tao Liu. Immune K-SVD algorithm for dictionary learning in speech denoising[J]. Neurocomputing,2014,137.
[12] Tan Chao. Parallel Optimization of K-SVD Algorithm for Image Denoising Based on Spark[A]. IEEE Beijing Section、IET Beijing Local Network、Beijing Jiaotong University.2016 IEEE 13th International Conference on Signal Processing Proceedings（ICSP2016）[C].IEEE Beijing Section、IET Beijing Local Network、Beijing Jiaotong University:,2016:6.
[13] Ribhu. Dictionary Design for Sparse Signal Representations Using K-SVD with Sparse Bayesian Learning[A]. IEEE Beijing Section.Proceedings of 2012 IEEE 11th International Conference on Signal Processing(ICSP 2012)[C].IEEE Beijing Section:,2012:5.
[14] Helder Sousa,Ying Wang. Sparse representation approach to data compression for strain-based traffic load monitoring: A comparative study[J]. Measurement,2017.

IOP Conf. Series: Journal of Physics: Conf. Series **1004** (2018) 012035 doi:10.1088/1742-6596/1004/1/012035

Image Segmentation Method Using Fuzzy C Mean Clustering Based on Multi-Objective Optimization

CHEN Jinlin[1], YANG Chunzhi[1], XU Guangkui[1] and Li Ning[1]

[1]School of Finance, Huainan normal University ,China,country code:457

*E-mail: hustcjl2012@sina.cn

Abstract: Image segmentation is not only one of the hottest topics in digital image processing, but also an important part of computer vision applications. As one kind of image segmentation algorithms, fuzzy C-means clustering is an effective and concise segmentation algorithm. However, the drawback of FCM is that it is sensitive to image noise. To solve the problem, this paper designs a novel fuzzy C-mean clustering algorithm based on multi-objective optimization. We add a parameter λ to the fuzzy distance measurement formula to improve the multi-objective optimization. The parameter λ can adjust the weights of the pixel local information. In the algorithm, the local correlation of neighboring pixels is added to the improved multi-objective mathematical model to optimize the clustering cent. Two different experimental results show that the novel fuzzy C-means approach has an efficient performance and computational time while segmenting images by different type of noises.

1. INTRODUCTION

As we all know, image segmentation is a complex problem in image processing, and it is also an important research topic in computer vision and pattern recognition applications. There are many methods of image segmentation. In the field of clustering, In the field of clustering, The standard fuzzy C- mean was deduced by Dunn in 1973 from the C mean algorithm, and then improved by Bezdek(1981). It has been proved to be an efficient framework for data clustering. This is an unsupervised clustering algorithm which has been successfully applied in the field of medical diagnosis, image analysis, target recognition. [1] However, the FCM algorithm is affected by the constraint condition of membership degree. In order to solve this problem, some researchers have carried on the noise pretreatment to the noisy image, and then use the FCM algorithm to reduce the noise influence.[2] This method will increase the computation time, at the same time, it will also produce unnecessary smoothing.[3] Some researchers have proposed that in the aspect of image processing, using the local information of pixels to modify the objective function of the standard FCM algorithm can reduce the impact of noise on clustering results. [4-6] The algorithm will also directly affect the choice of spatial information and the effect of segmentation time. This algorithm uses a more complicated method to update the cluster center.

Chen [7] et al. (2014) proposed a change form of FCM, called KSSCM, in which the local information was added to estimate the membership function and the original data was mapped to a high imensional Hilbert space. The use of local information can enhance the effect of segmentation, but also increase the computational complexity of the algorithm, which is mainly manifested in the operation time of iteration and the number of iterations. In 2011, Saha et al developed a new multi-objective optimization technique based on fuzzy clustering. [8] This technique uses multi-objective mathematical model to optimize the clustering centre in K dimensional fuzzy space.

We design a new clustering method named fuzzy clustering with multi-objective (MYFCM) in this paper. The sensitivity of the algorithm to noise can be reduced by replacing the single target of the FCM algorithm with the local multi-objective. This approach can reduce the sensitivity of the algorithm to noise, while also taking into account the computing speed advantage.

The main content of the paper is made up of the following four sections. Section 2 introduces the standard fuzzy C-means clustering method. We have improved the multi-objective optimization in Section 3. Next to the fourth section, we design an algorithm of image segmentation processing. Finally, two different experimental results and concluding remarks will be included.

2. FCM ALGORITHM

FCM is an efficient classification method which allows a section of data to belong to two or more clustering centre with a membership value. FCM tries to minimize the following objective function:

$$J_{FCM}(U,V,X)=\sum_{i=1}^{N}\sum_{j=1}^{c}u_{i,j}^{m}d^{2}(x_i,v_j),1<m<\infty \tag{1}$$

Where m (m>1) called the blur exponent. The vector $X=\{x_1,x_2,...,x_N\}$ is consists of N pixels, the vector $V=\{v_1,v_2,...,v_c\}$ is the centre of the cluster, $d^2(x_i,v_j)$ is defined as the Euclidean Distance of the object x_i from the centre of jth cluster, The standard FCM algorithm assigns pixels to each cluster using fuzzy memberships, $u_{i\,j}$ is the degree of memberships.

The degree of membership satisfies the following three conditions:

$$0\le u_{i,j}\le 1 \quad \sum_{j=1}^{c}u_{i,j}=1, \quad \sum_{i=1}^{N}u_{i,j}\le N, i=1,2,\cdots,\mathrm{N} \quad j=1,2,\cdots,\mathrm{c} \tag{2}$$

In the iterative process, the clustering centres and membership degrees are always updating, meanwhile, the value of objective function is closer to the minimum value. The membership matrix updates and the cluster centres by the following formula:

$$u_{i,j}=\frac{1}{\sum_{k=1}^{c}\left(\frac{d_{i,j}}{d_{i,k}}\right)^{\frac{2}{m-1}}} \qquad v_j=\frac{\sum_{i=1}^{N}u_{ij}^{m}x_i}{\sum_{i=1}^{N}u_{ij}^{m}} \tag{3}$$

Related research shows that the membership of the image cannot reflect the nature of the subordinate center in the noisy environment [9, 10] In order to overcome the shortcoming of FCM algorithm, we can use the local information of the data instead of the data, so as to reduce the negative impact of noise on the membership calculation. The calculation of membership matrix and clustering center is optimized by using multiple objects in the local region of the pixel.

3. MULTI-OBJECTIVE OPTIMIZATION

The FCM can be used in the classification of multidimensional data, the point and cluster center distance algorithm in FCM should be improved to adapt to the feasibility of this method. In order to reduce the sensitivity of the FCM image segmentation algorithm to the image noise, with a pixel as the center, to generate 6 dimensional pixel set, then using multi-objective optimization FCM algorithm to cluster the pixel set, achieve the purpose of image segmentation.

The text image X and the cluster centre V can be denoted as follows:

$$X=\{x_1,x_2,...,x_N\},\ V=\{v_1,v_2,...,v_c\} \tag{4}$$

Consider the 6-dimensional data x_i by the following formula: $x_i = (x_{i,1}^1, x_{i,1}^2, x_{i,1}^3, x_{i,2}^1, x_{i,2}^2, x_{i,2}^3)$

The j-th centre of cluster v_i denoted as follows: $v_i = (v_{j,1}^1, v_{j,1}^2, v_{j,1}^3, v_{j,2}^1, v_{j,2}^2, v_{j,2}^3)$

Sadi Nezhad presents a novel fuzzy clustering method based on multi-objective mathematical programming and an efficient fuzzy distance measurement.[11] Taking into account the 6-dimensional data set, improved fuzzy distance measurement is formulated as follows.

$$\begin{cases} d^1_{i,j,p} = \lambda^1_p Max(0, v^1_{j,p} - x^3_{i,p}) \\ d^2_{i,j,p} = \lambda^2_p \,|\, v^2_{j,p} - x^2_{i,p} \,| \\ d^3_{i,j,p} = \lambda^3_p Max(v^3_{j,p} - x^1_{i,p}, x^3_{i,p} - v^1_{j,p}) \end{cases} \tag{5}$$

The parameters λ^i_p is a weight coefficient of the distance measurement, $\lambda^i_p \geq 0$, $i = 1,2,3$, $p = 1,2$. If the parameters λ_i=1, that is, Sadi Nezhad gives a multi-objective fuzzy distance measurement.[11]

$$d_{i,j} = (d^1_{i,j,1}, d^2_{i,j,1}, d^3_{i,j,1}, d^1_{i,j,2}, d^2_{i,j,2}, d^3_{i,j,2}) \tag{6}$$

Then the membership values and the centre cluster are calculated as follows:

$$u_{i,j} = \frac{1}{\sum_{l=1}^{c} \left(\frac{\sum_{p=1}^{2}\sum_{k=1}^{3} d^k_{i,l,p}}{\sum_{p=1}^{2}\sum_{k=1}^{3} d^k_{i,j,p}} \right)^{\frac{1}{m-1}}} \quad k = 1,2,\cdots,c \quad p = 1,2 \tag{7}$$

$$v^k_{j,p} = \frac{\sum_{i=1}^{N} u^m_{i,j} x^k_{i,p}}{\sum_{i=1}^{N} u^m_{i,j}} \quad k = 1,2,\cdots,c \quad p = 1,2 \tag{8}$$

4. FUZZY C MEAN CLUSTERING BASED ON MULTIOBJECTIVE OPTIMIZATION

Let I= $\{i_{i,j} \quad i = 1,2,\cdots,m \quad j = 1,2,\cdots,n\}$ denotes $m \times n$ matrix of the original image. The vector x_i is generated by the neighborhood of the pixels in the text image I. The vector x_i can be simply expressed as:

	$i_{p-1,q}$	
$i_{p,q-1}$	$i_{p,q}$	$i_{p,q+1}$
	$i_{p+1,q}$	

$$x_i = (i_{p,q-1}, i_{p,q}, i_{p,q+1}, i_{p-1,q}, i_{p,q}, i_{p+1,q}) \tag{9}$$

The text image X is composed of the above vector. Meanwhile, the parameters λ^i_p in (6) are defined as follows:

$$\lambda^1_1 = \frac{i_{p,q+1}}{i_{p,q-1} + i_{p,q} + i_{p,q+1} + i_{p-1,q} + i_{p+1,q}} \quad \lambda^1_2 = \frac{i_{p+1,q}}{i_{p,q-1} + i_{p,q} + i_{p,q+1} + i_{p-1,q} + i_{p+1,q}} \quad \lambda^2_1 = \frac{i_{p,q}}{i_{p,q-1} + i_{p,q} + i_{p,q+1} + i_{p-1,q} + i_{p+1,q}}$$

$$\lambda^2_2 = \frac{i_{p,q}}{i_{p,q-1} + i_{p,q} + i_{p,q+1} + i_{p-1,q} + i_{p+1,q}} \quad \lambda^3_1 = \frac{i_{p,q-1} + i_{p,q+1}}{2(i_{p,q-1} + i_{p,q} + i_{p,q+1} + i_{p-1,q} + i_{p+1,q})} \quad \lambda^3_2 = \frac{i_{p-1,q} + i_{p+1,q}}{2(i_{p,q-1} + i_{p,q} + i_{p,q+1} + i_{p-1,q} + i_{p+1,q})} \tag{10}$$

The parameter is used to adjust the weights of the neighborhood of the pixels in fuzzy distance measurement, and the effect of noise on the clustering segmentation method is reduced. There is a brief description of the clustering segmentation method. The text image X is composed of the above vector. Meanwhile, the parameters λ^i_p in (5) are defined as follows:

Step 1: Initialize the 6-dimensional data X set of the original image using (9), set the value of ε;
Step2: Initialize the membership function $U = \{u_{i,j}\}$;
Step3: Calculate the centers vectors using (8);
Step4: Calculate the fuzzy distance using (5) and (6), $d_{i,j} = (d^1_{i,j,1}, d^2_{i,j,1}, d^3_{i,j,1}, d^1_{i,j,2}, d^2_{i,j,2}, d^3_{i,j,2})$;
Step5: Update the membership function $U' = \{u'_{i,j}\}$ using (5).
Step6: If $\|U' - U\| < \varepsilon$ then Stop; otherwise $U = U'$, and return to step 3;

IOP Conf. Series: Journal of Physics: Conf. Series **1004** (2018) 012035 doi:10.1088/1742-6596/1004/1/012035

Step7: Using the 6-dimensional data X and the membership function U to reconstruct a new image.

5. CONCLUSION

This experiment is based on the matlab7.0 programming environment. In order to prove the superiority of our algorithm in dealing with noisy image segmentation problems, experiments are carried out on three kinds of images: Gauss noise, salt and pepper noise. The results show that the MYFCN algorithm has a better effect on the image segmentation and the computation speed compared with other clustering algorithms.

5.1. Text Contrast Test

We first compare the proposed algorithm with other algorithms proposed so far, including standard FCM algorithm, ENFCM algorithm [12] and KSSCM algorithm [7]. The experimental results can be used to compare the segmentation effect, the number of iterations, the computation time and the computational complexity of various algorithms. In all the examples used in this section, the parameters used in algorithm are set to m=2, c=2 and $\varepsilon = 10^{-5}$.

In the process of image segmentation, we employ the rate of pixel misclassification as the measurement tools for the performance of the segmentation method. In this paper, we employ misclassification error(ME) measure. ME can be simply expressed as: [13]

$$ME = 1 - \frac{|B_0 - B_T| + |F_0 - F_T|}{B_0 + F_0} \tag{11}$$

Where F_0 and F_T have been defined as the background and background area pixels of the test image, respectively, B_0 and B_T denote the foreground and foreground area pixels in the test image.

Measure the time complexity (TC) of the algorithm with the time needed for an iteration operation, that is, the total computation time (TCT) divided by the number of iterations(NI).From the following figures and table data show that the algorithm in this paper is better performance in segmentation of the image than three other algorithms.

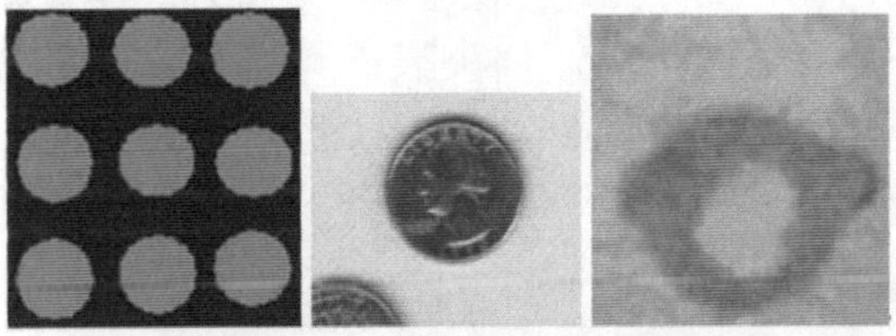

Figure 1. Original image A,B,C

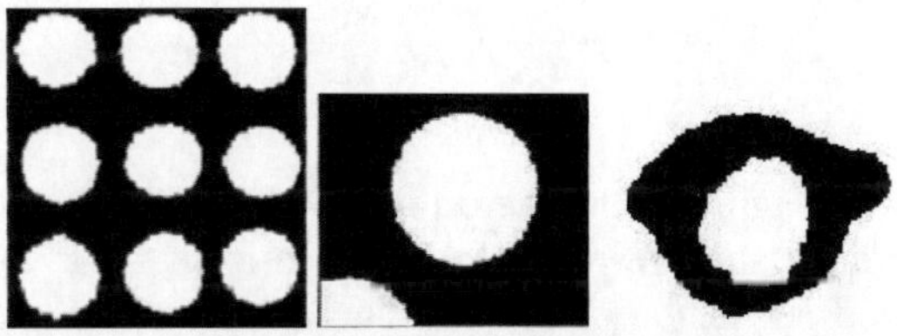

Figure2. Segmentation result of image A,B,C

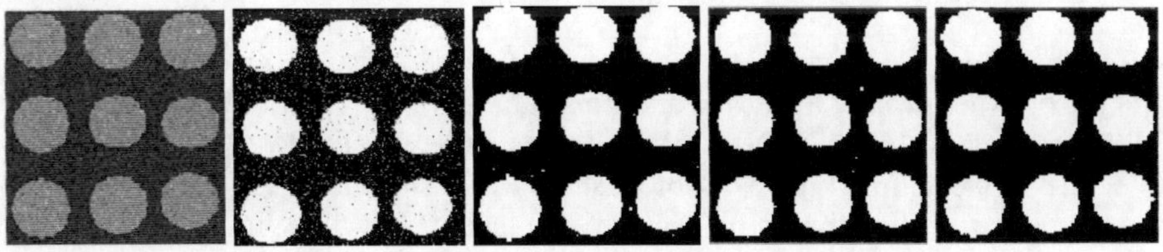

Figure 3. Segmentation results of the image A: results by the Gauss noise (4%) image, FCM, ENFCM, KSSCM and MYFCM respectively

Table 1. The performance comparison of four different algorithms in Figure A

Algorithm	NI	TCT（s）	ME	TC
FCM	21	0.945	0.043	0.045
ENFCM	14	1.552	0.0051	0.111
KSSCM	18	2.346	0.0025	0.130
KSSCM	5	0.891	8.7097e-004	0.178

IOP Conf. Series: Journal of Physics: Conf. Series **1004** (2018) 012035 doi:10.1088/1742-6596/1004/1/012035

Figure4. Segmentation results of the image B: results by the Gauss noise (6%)image, FCM, ENFCM, KSSCM and MYFCM respectively

Table 2. The performance comparison of four different algorithms in Figure B

Algorithm	NI	TCT（s）	ME	TC
FCM	20	0.937	0.054	0.047
ENFCM	13	1.220	0.0021	0.094
KSSCM	14	1.806	0.0017	0.129
KSSCM	5	0.793	9.0033e-004	0.159

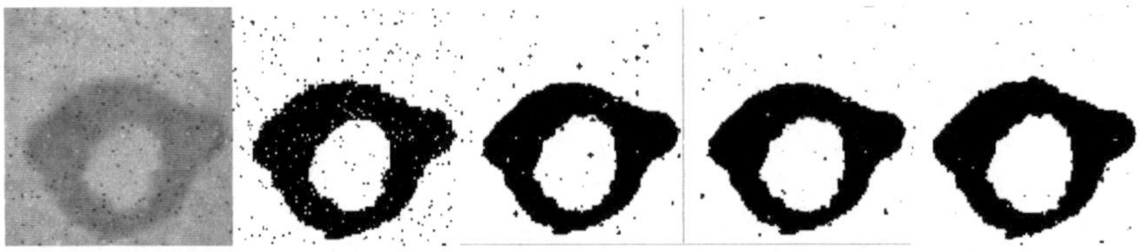

Figure5. Segmentation results of the image C: results by the Salt-pepper noise image, FCM, ENFCM, KSSCM and MYFCM respectively

Table 3. The performance comparison of four different algorithms in Figure C

Algorithm	NI	TCT（s）	ME	TC
FCM	17	0.578	0.102	0.034
ENFCM	11	0.978	0.067	0.094
KSSCM	15	1.232	0.008	0.082
KSSCM	6	0.512	0.003	0.085

In Figure 1, there are three original images A, B, C, and in Figure 2 are the corresponding ideal segmentation result. It can be seen from the Figure 3-5 that the algorithm in this paper has relatively good segmentation results. The data provided in Table 1-3 shows that our algorithm has very good results in terms of the time complexity (TC),the total computation time(TCT), misclassification error (ME) and the number of iterations(NI).

5.2. Text Noise Immunity

The second kind of experiment is to test the anti-noise ability of the algorithm. Add ten different levels (0.01,0.03,0.05,0.07,0.09,0.13,0.17,0.21,0.25) of Gauss noise to figure A, and then use different methods to do the segmentation. Figure 6 show that the ME distribution map of different methods. It is obvious that the proposed algorithm in the Gauss noise level below 13%, it is very good, compared to the other three algorithms.

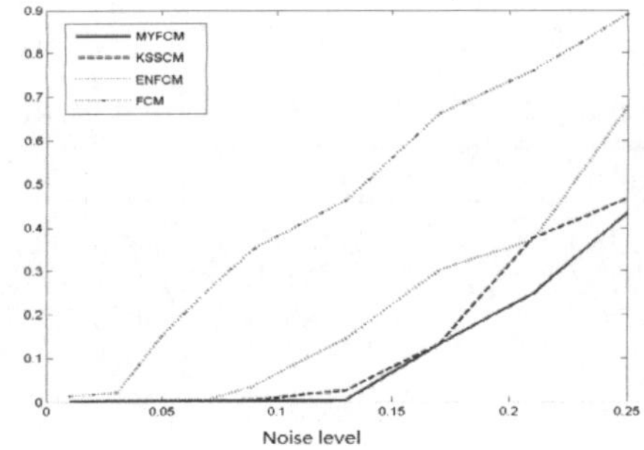

Figure6. The ME distribution map of different methods. The horizontal axis represents the noise level, and the vertical axis represents the ME value.

We add a parameter λ to the fuzzy distance measurement formula to improve the multi-objective optimization. The parameter λ can adjust the weights of the pixel local information. The improved multi-objective optimization mathematical model is applied to the FCM algorithm, so that the new FCM algorithm (MYFCM) has a good processing effect, which can be obtained from Tables. In order to further illustrate the influence of image noise on the algorithm, we do anti-noise test to prove the effect of the algorithm in noisy image processing.

Acknowledgments
This work was supported by Natural Science Fund for Colleges and Universities of Anhui Province of China (KJ2016A666).

References

[1] Carvalho,F.Fuaay c-means clustering methods for symbolic interval data.Pattern Recgnition Letters28(4),2007.pp:1-18.
[2] Krishnapuram R,Keller J M.A possibilistic approach to clustering[J].IEEE Trans on Fuzzy System,1993,1(2),pp:98-110.
[3] J.H.Xue,A.Pizurica,W.Philips,E.kerre,R.Van de Walle and I.Lemahieu.An Integrated Method of Adaptive Enhancement for Unsupervised Segmentation of MRI Brain Images[J].Pattern Recognition Letters,Vol.24,No.15,2003,pp:2549-2560s.
[4] A.W.Liew,S.H.Leung and W.H.lau.Fuaay Image Clustering Incorpoting Spatial Continuity[J].IEEE Proceedings of Vision,Image and Signal processing,Vol.147,No.2,2000,pp:452-455.
[5] D.Dham and J.Prine.Adaptive Fuzzy segmentation of Megnetic Resonance Images[J].IEEE Transations on Medical Imageing.VOL18.NO 19,1999.pp:737-752.
[6] Szilágyi,S.M.Szilágyi and I Benyó. A Modifien FCM Algorithm for Fast Segmentation of Brain MR Images[J].ICIARLNCS.Vol.4633,2007.pp:866-877.
[7] Long.Chen,Jing Zou and C.L.Philip CHEn.Kernel spatial Shadowed C-Means for Image Segmentation[J].International Journal of Fuzzy Systems,Vol.16,No.1,2014,pp:46-56.
[8] Saha I, Maulik U, Plewczynski D. A new multi-objective technique for differential fuzzy clustering[J]. Appl Soft Comput.2011.11,pp:2765–2776.
[9] D.Dham and J.Prine.Adaptive Fuzzy segmentation of Megnetic Resonance Images[J].IEEE Transations on Medical Imageing.VOL18.NO 19,1999.pp:737-752.
[10] Szilágyi,S.M.Szilágyi and I Benyó. A Modifien FCM Algorithm for Fast Segmentation of Brain MR Images[J].ICIARLNCS.Vol.4633,2007.pp:866-877.
[11] Sadi-Nezhad,A Noroozi-Yadak,A Makui.Fuzzy distance of triangular fuzzy numbers[J].Journal of Intelligent & Fuzzy Systems,2013,25(4),pp:845-852.
[12] Szilagyi L, Benyo Z, Szilagyi SM et al (2003) MR brain image segmentation using an enhanced fuzzy c-means algorithm. In: 25thAnnual international conference of IEEE engineering in medicineand biology, ancun, Mexico, pp 17–21
[13] M.A.Yasnoff,J.K.Mui,J.W.bacus. Error measures for scene segmentation[J].Pattern Recognition.1997,pp:217-231.

IOP Conf. Series: Journal of Physics: Conf. Series **1004** (2018) 012036 doi:10.1088/1742-6596/1004/1/012036

People-oriented Information Visualization Design

Zhiyong Chen[1] **and Bolun Zhang**[1,*]

[1]Hebei Normal University Shijiazhuang, Hebei, China

*327479685@qq.com

Abstract. In the 21st century with rapid development, in the wake of the continuous progress of science and technology, human society enters the information era and the era of big data, and the lifestyle and aesthetic system also change accordingly, so the emerging field of information visualization is increasingly popular. Information visualization design is the process of visualizing all kinds of tedious information data, so as to quickly accept information and save time-cost. Along with the development of the process of information visualization, information design, also becomes hotter and hotter, and emotional design, people-oriented design is an indispensable part of in the design of information. This paper probes information visualization design through emotional analysis of information design based on the social context of people-oriented experience from the perspective of art design. Based on the three levels of emotional information design: instinct level, behavior level and reflective level research, to explore and discuss information visualization design.

1. Introduction

In the context of informatization in today's society, information visualization as a new discipline becomes more closely related to the public's life. It plays a vital role in the process of social development, especially in Internet environment under the big data context. Information visualization is an indispensable development trend. In short, information visualization makes the complicated information data into images, clearly conveying information to the users. This is an interdisciplinary integrating a variety of categories, involving computer graphics, computer-aided design, human-computer interaction, and art and so on. In the long process of history, the data information presented in a graphical way can be traced back to the time before text existing, when ancient human convey information by using stones to draw graphics in the caves they lived in. After analysis, human's acceptance of the figures is much higher than the acceptance degree of the characters. Meanwhile, figures can also convey information across the nation and language, so in today's society with the frequent exchange of information, information visualization can reduce communication difficulties and save the time of communicating and understanding. Information and interaction design is the presentation method of information visualization, which attracts increasing attention from the present society. At the same time, under the background of experiential social economy, the service focus of information visualization design is shifted from "task-centered" to "customer-center". [1] Therefore, the design mode focusing on users' feelings and experiences and taking users' personal experience as the prime premise becomes the most popular one in contemporary art design field as well as the design with topicality, so people-oriented design has become the focus of current art design circle.

2. Emotional analysis of information visualization design

Information visualization design involves a wide range of fields. Its purpose is to present the information data graphically to customers by designer's summarizing, organizing, and extracting of the information data, so that the audience can have more intuitive and convenient understanding to find the hidden characteristics, relationships and patterns of data information. With the rapid development of the Internet, the customers of information design become more popular and generalized, so in the design organizing, we should take into account the users' personal experience and emotions.[2] When doing the visual design, we should integrate aesthetic concept in addition to the efficient transmission of information, so that the audience can feel the aesthetic experience and immersion of information exchange. Therefore, we should do enough research on the audience, and deeply mine the information. Besides, we need to analyze the emotional factors to produce a good enough visual information design. The book “Emotional Design” writes that “the human brain has different levels --visceral level, behavioral level, and reflective level. The visceral level responses fastest, which can judge the good or bad and safe or dangerous quickly; the behavioral level is people's behavior; and the reflective level is the highest level, which means it only monitor and reflect the behaviors and try to make the behavioral level have a preference. “[3]Hence, the author makes emotional analysis of information design from the visceral level, behavioral level, and reflective level.

2.1. Visceral Level

“Design at visceral level is the natural law. “[4]The reflection of the visceral level is the characteristics and nature survived in the choice and reinforcement of survival of the fittest generated from the evolution of mankind and the nature. For example, the males have beautiful and gorgeous feathers to attract the females and get the mating right; plants bloom colorful flowers and exude seductive smell to attract bees and insects to spread pollen and so on. Generally, it is the reaction at visceral level of user to judge whether a design work is “beautiful”, which is from the most instinct and straightforward level that users experienced. Therefore, the aesthetic visual display including chart plate, color matching and using of text of information design can influence the users' emotional experience in visceral level most directly. “Design principle at visceral level is innate, regardless of race and culture. If you follow these principles, you can make a very attractive design even with very simple things. “A good information design work should not simply convey the data information, but should explore from the intuitive visual aspects to design the work that people can't help looking at it once see it and have immersion experience.

There are many examples about raising the sales through adjustment of appearance. Apple launched iPhone 4 mobile phone in 2011 as “Figure 1”, which turned out at the time when keyboard phones are the majority. This is an epoch-making products, with the extremely exquisite appearance, simple and delicate icons, interactive design that is easy to operate making sales of Apple mobile phone increase steadily, and receive the users' praise. The automobile industry also has such an example. Volkswagen studied the preferences of Chinese users before replacing a series of products and designing the appearance as shown in “Figure 2”, resulting in the monthly sales of a number of models rise straightly and rank the first in China. Therefore, in the social context of such a high intensity, a good design can seize the users at the visceral level because it is the level with most rapid response. The design with a sense of beauty can attract the attention from more users in shortest time, so that the user's emotions can enter a higher emotion level “from bottom up”.

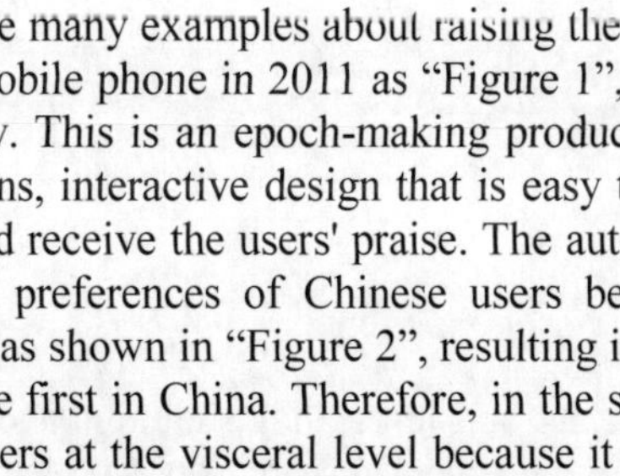

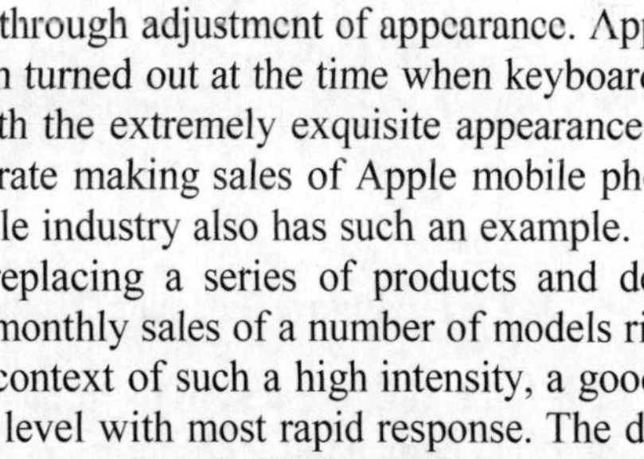

Figure 1. iPhone 4. **Figure 2.** Volkswagen.

2.2. Behavioral Level

"The behavioral level is related to the use and experience of the product. The experience itself contains a lot of aspects: function, performance and usability. "When a product break through the level of "beautiful or not" at the visceral level, it will meet the question of "useful or not" at the behavioral level, which is also the value of this product. If the work only has fine appearance while it has defects in function and use or has insufficient attraction, it will not produce any value, which may lead the users feel irritable and depressed during using process and thus deny the love to the appearance at visceral level.

In the information visualization design, behavioral level refers to the interactive design of information, of which the quality will directly affect the user's operation and emotional experience. The interaction design mainly includes two parts: one is the interaction between data and themselves in the screen; the other is the interaction between users and data, also known as human-computer interaction, and emotional analysis at behavioral level mainly reflects in human-computer interaction.[5] A good interactive system needs to contain two important features--user experience and usability, of which usability is the core, so good user experience is the ultimate goal to achieve as shown in "Figure 3". "The first step of a good behavioral level is to understand how users use the product." [6] Therefore, designers need to deeply dig and organize the data to achieve this goal in the visualization of information, and understand the needs of users correctly as well as the operation at subconscious level. In the design phase, designer should apply the people-oriented design principle to get the best results, and focus on understanding and meeting the real users of the product.

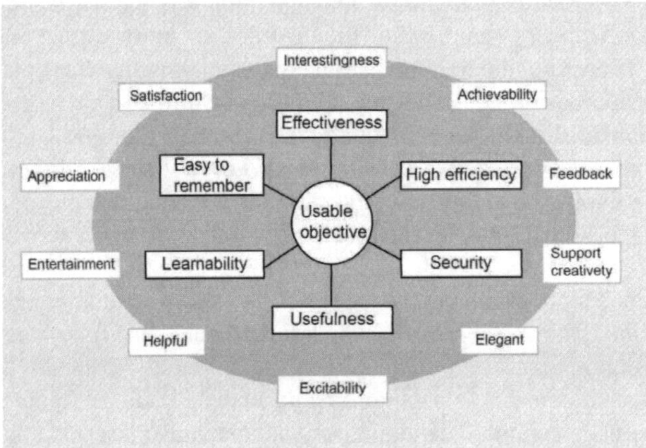

Figure 3. Goal of usability and user experience.

Therefore, when the designer design the product at behavioral level, they should abandon the self-centered design ideas, combining with the users' feedback and more research on product design, and continuously modify and improve the product with the users' comments to ultimately produce outstanding design work. In the information design, the engineers and designers need to work together in order to better complete the information visual design. Engineers transform the data information and develop the algorithm and system; designers research and care the user's inner needs and beautify and design the visual information from the angle of art to make the work achieve excellent level in function, understandability, usability and feel. At the beginning of design, they should let potential users participate in it, and involve in the design process with the team through starting from the initial model and sketch. When it is completed finally, the product also passed a variety of using tests, and can come onto the market after the final test and adjustment. This kind of work that involves users' participation and goes through repeated modification combining with the actual user experience is the real interactive design taking user experience as the core and criteria.

2.3. Reflective Level

"Design at reflective level covers many areas, closely related to the information, culture and the meaning and use of the product. Only at the reflective level, there is the sense of consciousness and higher level of feeling, emotion and perception; only at this level can we experience the full integration of thoughts and emotions. "Reflective level is the top level, which is the unique human emotions. It is also the embodiment of personal feelings that People have different experiences and understanding based on different education, culture, race, and age and so on. The reflective level is constantly changing since as time goes on, the individual's environment and the status constantly change and emotions at reflective level will gradually change. For example, the product that we once liked has not changed at visceral level and behavioral level, but we no longer like it at reflective level, so we reject the appearance and functionality of the product, and vice versa.

Reflective level is not only related to the individual differences, but also able to bring satisfaction to users, which other levels cannot complete. For instance, watches have the same functionality at the behavioral level, but at the reflective level, watches of luxury brand will bring the users satisfaction and the feeling of personal pride, which cannot be brought by ordinary watches. Except the level of material and workmanship, added value of culture connotation of the brand directly impact on the users' reflective level. In the aspect of information exchange, that of Apple's product is excellent enough in the industry. In addition to the attraction of representation of the product, the convenience of the interactive system, easy operability and user-friendly operation are better than mobile phones of other brands, which attracts users to choose at the reflective level. Meanwhile, the price higher than that of other mobile phones also gives users the satisfaction at reflective level. Therefore, the information design product that is outstanding at reflective level can improve users' overall impression, and good user experience can even eliminate some initial negative emotions.

3. Discussion on people-oriented from the perspective of design

Information visualization includes a wide range of disciplines, and people-oriented information visual design requires cooperation of multiple disciplines. This chapter explores people oriented from the perspective of art design. From the denotation, people oriented means the work should take full account of the user's identification, aesthetic and emotional experience of design work, and the design should accord with the user-centered creation idea, namely user-friendly design. And enhance the user's emotional identity through the application experience and interactive feelings.

A good information design product must provide the user with the amazing application experience, which is the approval of design product at behavioral level, namely the applicability of information design.[7] At the beginning of the users experience of the product, we must first improve the work at visceral level to make users have the feeling of "I want" from the aspects of color and graphic design, and then they will develop the psychological feelings of "what it can do" at behavioral level. [8] Therefore, we need to deeply explore when collecting data and implementing user research. Another core of information design is interaction. Good interactive design allows users to have a sense of immersion and take participate in it initiatively, and thus approve the design works, which is also the reflection of user-centered design. Therefore, the process of interaction and experience is the key point that needs the designers' focus. They should make complex product simple and easy to use, which is influenced by different culture, race and experience of participating users. Complicated and poorly designed interactive operation will affect the user's experience emotion, since it may bring frustration to the users, making the users feel depressed, and even generate negative emotions to affect the feelings of interaction, resulting in the failure of the work Hence, to enhance the user's interactive experience, interactive design must strengthen the interestingness, so that users will be immersed in them, and their emotional identity will be enhanced, to stimulate the reflective level.

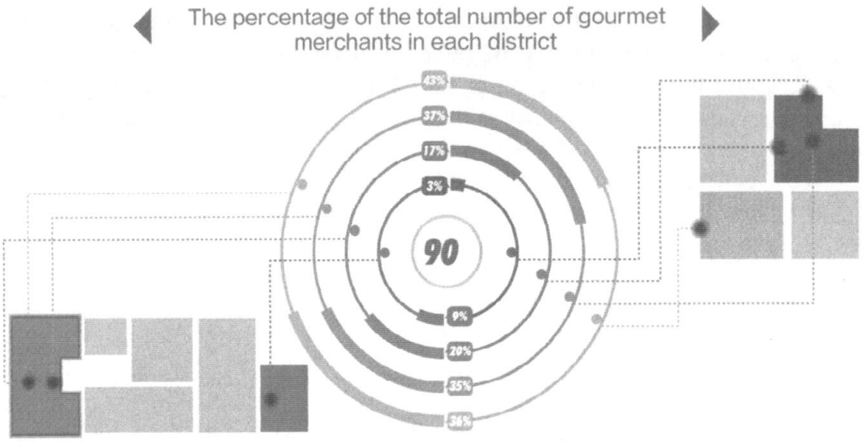

Figure 4. Information design of food in a normal university.

When entering a university, the freshmen will pay more attention to the food and entertainment projects, so we deeply explore the food information of a normal university based on users' need, and obtain the data information by integrating our own experience and network evaluation for information visualization. Age of the freshmen is about 18 years old and the number of boys in normal university is much smaller than that of girls who have higher recognition of the color, so we need to choose saturated and contrasting colors, and use cold and warm colors to distinguish eastern area and western area in school. Besides, we can supplement the new graphics and text words that are more easily for freshmen to understand, and add the map guide to complete the information design of food in the normal university as shown in "Figure 4". First of all, from the perspective of visceral level, "beautiful" color schemes and graphic design is more able to seize the users' eyes compared to other old school information maps. The information design form and color form of food in Normal University is clear and attractive, we can carry out clear and understandable visual design aiming at the flat design style that can be accepted by young people through the redesign plan of the map supplemented by circular data charts. In regard to the color scheme, distinguishing the information of eastern area and western area by the two color tones of cold and warm can let students know the location of shops through a glance, which can help users to build a clear understanding in their own minds. We select the relatively advanced color scheme in the cold and warm tones—orange on behalf of the warm fire and green tones on behalf of health, so that users can feel pleasant when reading the information and this color scheme is more likely to be accepted and loved by young college students, which is the manifestation of deep research and personal experience of designers. Secondly, if we skip the visual image of visceral level to analyze from the functionality and usability at behavioral level, this information visualization design arranges the data through the deep research of users group initially. Displaying diagram design by circle can make users no longer feel confused and puzzled in the using process, and comparison and display of all the information is simple and clear with strong sense of design. Particularly the campus map of flat design through the induction and reorganization, and teaching building and school buildings through the modular and geometric process is easy to understand during using. At the same time, extracting the information and constructing the scene of the overall layout and planning of the school in the brain will not bring strangeness and confusion, and can provide the users with a good interactive experience as well as enjoyment during the use. Finally, when analyzing from the reflective level, the data arrangement and visual design of good campus map information and information of food that is inseparable from life can bring a strong sense of honor and the sense of belonging to student groups, narrow the distance between the users and the school. Especially for the freshmen who have just entered school, we can eliminate the tension caused by just entering an unfamiliar environment, so that students will love and accept the school, which is worthy of approval for this design from all aspects. Therefore, from the above analysis of visceral level, behavioral level, and reflective level, this kind of design based on users emotions, centered on users, and considering for users time for the user all the time, is the people-oriented design.

4. Conclusion

From the above analysis and discussion, "user experience" has become a hot topic in today's society, and the development and progress of all walks of life has been optimized and reformed centering on this word. With the rapid development of mobile data end and the gradual realization and complete of various functions, users will be eager for better interactive experience. With the advantage of multiple disciplinary, information visualization design has become the design discipline of greatest concern in the 21st century, which not only prospers in the user experience-oriented social economy, but also plays a guiding role in the future direction of other design disciplines. User-centered design ideas will also become the mainstream of the future design as well as the center soul of information design. People-oriented design principle is also guideline that information visual design must practice. Information visualization concept has developed rapidly for nearly three decades since it was proposed, and the design disciplines tend to converge. In the future, it is difficult for the design of single direction to meet the various users' needs, so only when it integrates into a variety of cross-disciplines can it keep up with the progress of the times, to provide better services for the community and bring enjoyable emotional experience to users. Therefore, development prospect of information visualization design in the future is very broad, more designers will be devoted in it, and they will cooperate with other researchers of non-design disciplines to produce better products.

References

[1] (America) Jesse James Garrett, tr. Fan Xiaoyan: User Experience Element: User-Centered Product Design [M]. Beijing: China Machine Press, 2011.6.
[2] Liu Pai. "The sight stirs up one's feelings" emotional information design centering on users[J]. Journal of Nanjing Arts Institute Fine Arts & Design, 2009. 03.
[3] (America) Norman, tr. xiaoke. Design Psychology 1: Daily design [M]. Beijing: China citic press, 2015.5.
[4] (America) Norman, tr. He Xiaomei, Ou Qiuxing. Design Psychology 3: Emotional Design [M]. Beijing: China citic press, 2015.6.
[5] Lu Xiaobo, Interactive Design Method in Information Design [J]. Industry leader, 2007.13.
[6] Qin Jingyan, Analysis of Interactive Design Method in Information Visualization [J], decoration, 2007.03.
[7] Xiao Yong, Zhang Youliang, Tu Ya. Information Design [M]. Wuhan: Hubei Fine Arts Publishing House, 2010.
[8] Song Meimei, Research on the Relationship between Picture Design and Audience Perception in Information Design [J]. Packaging Engineering, 2015.8.

IOP Conf. Series: Journal of Physics: Conf. Series **1004** (2018) 012037 doi:10.1088/1742-6596/1004/1/012037

MLESAC Based Localization of Needle Insertion Using 2D Ultrasound Images

Fei Xu[1], Dedong Gao[1,*], Shan Wang[1] and A Zhanwen[1]

[1]School of Mechanical Engineering, Qinghai University, Xining 810016, China

*Corresponding author: gaodd@qhu.edu.cn

Abstract. In the 2D ultrasound image of ultrasound-guided percutaneous needle insertions, it is difficult to determine the positions of needle axis and tip because of the existence of artifacts and other noises. In this work the speckle is regarded as the noise of an ultrasound image, and a novel algorithm is presented to detect the needle in a 2D ultrasound image. Firstly, the wavelet soft thresholding technique based on BayesShrink rule is used to denoise the speckle of ultrasound image. Secondly, we add Otsu's thresholding method and morphologic operations to pre-process the ultrasound image. Finally, the localization of the needle is identified and positioned in the 2D ultrasound image based on the maximum likelihood estimation sample consensus (MLESAC) algorithm. The experimental results show that it is valid for estimating the position of needle axis and tip in the ultrasound images with the proposed algorithm. The research work is hopeful to be used in the path planning and robot-assisted needle insertion procedures.

1. Introduction

Percutaneous needle insertions have been widely used in a variety of clinical contexts in recent years, including biopsy [1] ,brachytherapy [2], anesthesiology [3], ablation [4] and fetal surgery [5]. As image-guided robotic assisted medical technology, ultrasound navigation technology have been widely used in the percutaneous surgery because of its noninvasive, low cost and real-time transmission compared with CT, MRI and nuclear medicine imaging. In practice, because of the existence of sensitive tissues or organs near the target, the detection and segmentation of the needle and the tracking of its trajectory are particularly prominent. However it is difficult to accurately detect and segment the needle from ultrasound image due to the speckle gives ultrasound image characteristic granular appearance and it inherently exists in coherent imaging [6], which affects the resolution and quality of image and increases the difficulty of diagnosis and treatment of doctors. At the same time, due to attenuation, diffraction, diffusion and artifact of ultrasound imaging [7], localization and tracking of 2D ultrasound image in the process of ultrasound-guided percutaneous needle insertion is a crucial factor in image degradation.

Needle detection and segmentation algorithms are based on the line feature detection methods. There existed lots of literatures about the line detection, and a summary of the Hough Transform (HT) was given by Leavers [8], image processing with HT have been widely used for line detection in different areas. Ding et al. used the Hough transformation technique to implement the segmentation of a biopsy needle in real time. Compared to conventional manual segmentation techniques, their approach reduced the computational time by an order of magnitude [9]. Okazawa et al. applied the HT to the medical image, extracted 3 parameters of the slope, intercept and curvature of the line through the coordinate transformation of the ultrasound image, and used the polynomial least squares regression method to fit the line of ultrasound image [10]. Vrooijink et al. used the HT to locate the centroid of the needle from the radial cross-sectional view of the needle, which is affected by the comet tail artifact (CTA) known as reverberation in 2D ultrasound image [11, 12]. In 3D environment, Aboofazeli et al. used anisotropic diffusion filter deal with speckle reduction in 3D ultrasound image, then used HT copy with the needle detection and segmentation in projected 2D image obtained using a ray casting process, and used the polynomial of the midpoint and endpoint coordinates of the segment to fit the needle trajectory with curvature [13].

Neubach et al. proposed tissue stiffness classification to distinguish soft and stiff tissue in the virtual spring model [14]. Their proposed method used the block-matching algorithm in the elastography to calculate the displacement of region of interest (ROI) pixels, and they estimated the displacement of the local motion of tissue, and detected and tracked the position of the needle tip. The speckle tracking algorithm was used to validate the block-matching algorithm using ultrasound image in the "field Ⅱ" [15]. Kaya et al. created new needle coordinate system in 2D ultrasound image, and used Gabor filter to denoise the image, in order to strengthen the outline of the needle while inhibiting the other tissue structure. In the frame difference image, the principal component analysis algorithm was used to estimate the needle insertion angle, and then the position and trajectory of the needle were estimated with the RANSAC line estimator [16, 17].

Waine et al. proposed image contrast stretching in ultrasound image to improve the pixel between the needle and soft tissue, and after morphological processing and thresholding, a set of needle point candidates were obtained in the image. At the same time, the position of needle and the shape estimation of needle in 3D space ultrasonic image were obtained by using the M-Estimator Sample Consensus (i.e. MSAC) in 2D ultrasonic image, and the shape modeling of 2D needle was established on the basis of the Euler-Bernoulli beam after Kalman filtering. They realized the needle localization, needle deformation detection and the manipulation of the needle tip, hence the semi-automatic needle steering system is implemented [18, 19]. Uhercik et al. developed a thin surgical tool in 3D ultrasound imaging, such as a biopsy needle or microelectrode, and used the random sample consensus (RANSAC) to fit the model to determine the position of the needle shaft. The position of the needle tip was determined by an intensity drop along the axis [20]. As an extension of [20, 21], Zhao et al. added Kalman filter, RANSAC algorithm and speckle tracking method to simulate 2D ultrasound image in the "field Ⅱ", and it improved the stability of RANSAC algorithm in dynamic situation [22, 23]. Neshat et al. and Novotny et al. employed Radon transform to realize real-time detection needle in 3D ultrasound image on the GPU [24, 25].

These methods rarely consider the effect of noise which is caused by image acquisition or transmission on needle detection and segmentation. In this paper, we propose new method to detect biopsy needle and locate the position of biopsy needle axis and tip. This is helpful for ultrasound-guided medical assisted robot to detect the position of needle axis and tip. The rest of the paper is organized as follows. In Section 2, the wavelet soft thresholding processing based on BayesShrink rule for the 2D transverse ultrasound image of needle insertion is used to despeckling, and then the image binarization is executed. In Section 3, the maximum likelihood estimation sample and consensus (MLESAC) algorithm is applied to estimate the needle axis and tip position in 2D ultrasonic image. In Section 4, experiments and results are described. Discussion and conclusion are summarized in Section 5.

2. Image preprocessing

In this section, image preprocessing is comprised of the wavelet transform and image binarization. The wavelet transform is used to decompose the original image, then the soft thresholding and BayesShrink rule technique is dealt with the subband in order to despeckling, image binarization is employed Otsu's thresholding method and morphologic operations to deal with ultrasound image.

2.1. Wavelet transform

The speckle, which is multiplicative noise, is caused by the limitation of imaging environment and coherent superposition during ultrasonic imaging [26], the speckle decrease the resolution of the ultrasound imaging system, which seriously affected the image quality. The effect of additive noise of ultrasound imaging is small, which is generally negligible. Previous researchers have identified that the despeckling of the wavelet transform outperforms the other standard speckle filters, such as Wiener filter, Lee filter, Kuan filter, enhanced frost filter, Gamma filter and SRAD filter [27], hence we apply the wavelet transform in order to remove the speckle in ultrasound image. In order to separate speckle from the original ultrasound image, the speckle multiplicative noise is transformed into speckle additive noise by logarithmic transform [28], and the ultrasound image with speckle is decomposed by wavelet transform, the steps involved in the wavelet transform are given in the Algorithm 1.

Algorithm 1: Despeckling of ultrasound image using the wavelet transform
Input: Original ultrasound image.
Output: Despecked ultrasound image.

Procedure:
(1) Apply logarithmic transform of input image;
(2) Perform wavelet transform in multi-level of subband decomposition on the logarithmic transformed image;
(3) Execute thresholding of transformed image of the step (2);
(4) Implement inverse transform on the thresholded image of the step (3);
(5) Despeckled ultrasound image can be obtained by exponential transform on the inverse transformed image of the step (4).

2.1.1. Soft thresholding. After the discrete wavelet transform, the detail coefficients of the wavelet function corresponding to the high frequency components and the approximate coefficients of the scaling function corresponding to the low frequency components are obtained from the ultrasound image. The magnitude and number of the signal's wavelet function coefficients are large and less, the magnitude and number of the speckle's wavelet function coefficients are small and more. Hence the denoising method of wavelet domain can select threshold on a different levels, and deal with the wavelet function coefficients directly by using threshold or weight. The function of wavelet thresholding is aimed at removing the small wavelet function coefficients and keeping the large wavelet function coefficients. In this paper, the soft thresholding function is chosen to deal with speckle in ultrasound image [29]. The soft thresholding scheme shown in Eq.(1).

$$\mathrm{W}_w = \begin{cases} sign\{X_w\}\left(|X_w| - T\right), & \text{Where } |X_w| \geq T \\ 0 & |X_w| < T \end{cases} \tag{1}$$

Where X_w denotes input wavelet transform coefficients, W_w denotes output wavelet transform coefficients after thresholding, T denotes the threshold limit.

2.1.2. BayesShrink rule. Another crux for the desnoising method of wavelet domain is the selection threshold. In wavelet domain denoising, Generalized Gaussian Distribution (GDD) is empirical used as the distribution model of wavelet coefficients in a subband of a natural image [30], Chang et al. [31] proposed the BayesShrink rule based on the assumption. Compared with the VisuShrink rule and SureShring rule, the BayesShrink rule performed the best for ultrasound image with speckle [32].
In BayesShrink rule, the formulation for the threshold T on a given subband is given by

$$T = \hat{\sigma}_n^{\,2} / \hat{\sigma}_x, \tag{2}$$

Where

IOP Conf. Series: Journal of Physics: Conf. Series **1004** (2018) 012037 doi:10.1088/1742-6596/1004/1/012037

$$\hat{\sigma}_n = \frac{median\left(\{|W_k| \in HH_1\}\right)}{0.6745} \tag{3}$$

$$\hat{\sigma}_x = \sqrt{\max(\hat{\sigma}_y^2 - \hat{\sigma}_n^2, 0)} \tag{4}$$

Where $\hat{\sigma}_n^{\ 2}$ denotes the estimated noise variance, $\hat{\sigma}_x$ denotes the estimated standard deviation for the subband coefficients and HH_1 denotes the diagonal subband after multiscale wavelet decomposition.

$$\hat{\sigma}_y^2 = \frac{1}{N_j}\sum_{k=1}^{N_j} W_k^2 \tag{5}$$

Where N_j denotes the number of transform coefficients W_k in the subband, $\hat{\sigma}_y^2$ denotes the estimated variance of the original image observation.

In [33, 34], the experimental results have shown that wavelet filter bior 6.8 with level 3 yields better performance than other speckle filters in reconstructed image. Hence, the wavelet filter bior 6.8 with level 3 is used in this study to reduce the speckle. An example of the 18 gauge needle insertion with the soft thresholding and BayesShrink rule technique for ultrasound frame sequences is shown in Fig 1.

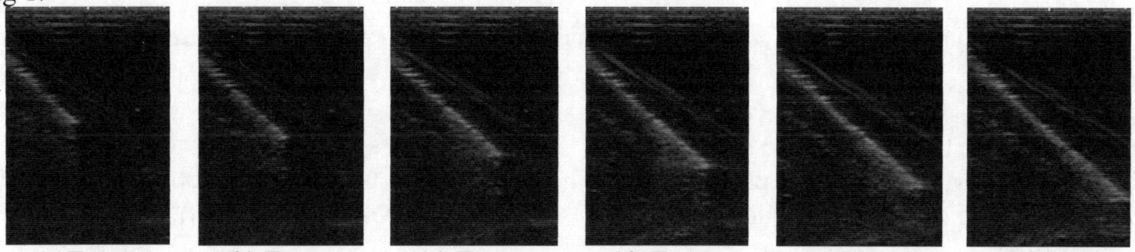

(a)Frame number = 17 (b) Frame number = 22 (c) Frame number = 30 (d) Frame number = 35 (e) Frame number = 42 (f) Frame number = 48

Figure 1. Wavelet thresholding

2.2. Image binarization

After the soft thresholding of wavelet domain, the ultrasound image binarization is achieved in three steps, including smoothing, Otsu's thresholding method and morphologic operations.

It is found in the trials that the image is smoothen and reduce noise by using the disk type structuring element which its radius is 5 pixels, and it is helpful to filter out the structure with no relation with the needle. Then the optimum global thresholding using Otsu's thresholding method is disposed of the ultrasound image in order to separate needle from the background. Finally, we apply the disk type structuring element which its radius is 5 to morphologic erosion and dilation operations, which is conducive to removal of residual speckle, and clear the spur pixel and the isolate foreground point in the binary image. An example of the 18 gauge needle insertion with the Otsu's thresholding method for ultrasound frame sequences is shown in Fig 2.

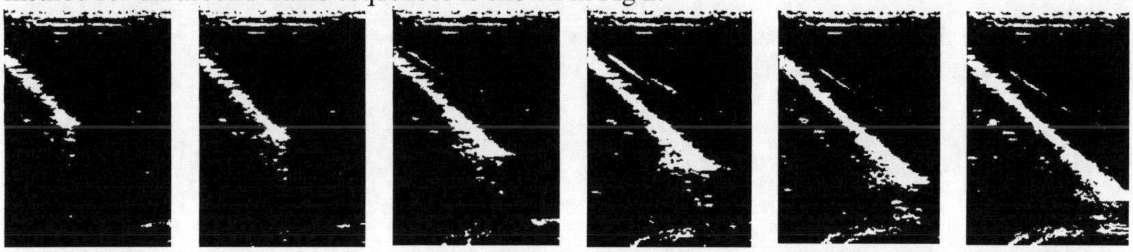

(a) Frame number = 17 (b) Frame number = 22 (c) Frame number = 30 (d) Frame number = 35 (e) Frame number = 42 (f) Frame number = 48

Figure 2. Otsu's thresholding method

3. Localization of needle axis and tip

3.1. Location estimation of needle axis

The location of the needle is estimated by the maximum likelihood estimation sample consensus (MLESAC) [35] and the MLESAC originates from the RANSAC [36] which is a robust line estimator. Instead of using as numerous data as possible to estimate the model, RANSAC uses as few data as possible to estimate the model. Now that RANSAC has developed into a RANSAC family, it can be seen from their cost function that it is better than the traditional RANSAC and m-estimator sample consensus (MSAC) [37]. Compared with the latter two methods, MLESAC has considered the quality of inlier more, we apply the MLESAC.

There are a host of outliers near the needle shape of the ultrasound image. The shape and location of the needle axis can't be discerned from the naked eye. Often times the error distributions for the inliers of the image is modeled with the Gaussian distribution, the error statistics for the outliers is described by the uniform distribution, then the estimation model of the inlier of the needle under ultrasound guidance is:

$$p(e|M)=\gamma\left(\frac{1}{\sqrt{2\pi}\sigma}\right)^{d}\exp\left(-\frac{e^2}{2\sigma^2}\right)+(1-\gamma)\frac{1}{v} \tag{6}$$

Where γ is mixing coefficient and v is a constant diameter of the search window of mixture model.

The two pixels by random selecting in ultrasound binary image are used as the cardinality d of the needle model M, that is d = 2, the parameter is estimated with the selected two pixels as the minimal sample sets (MMS), and the negative log likelihood of the cost function is obtained. We select minimum of sum of the cost function $\text{Loss}(e|M)$ as the largest consensus set (CS), CS is maximum estimation $\hat{M}$ of the needle model *M*:

$$\text{Loss}(e)=-\log p(e|M) \tag{7}$$

$$\hat{\text{M}}=\arg\min_{M}\left\{\sum_{i=1}^{n}\text{Loss}(e|M)\right\}. \tag{8}$$

Initial estimate of γ is 0.5, η_i is the index for the *i*th sampling, where the η_i =1 denotes the inlier of corresponding data, η_i =0 denotes the outlier of corresponding data. In [34], it offered the method to tune γ through Expectation Maximization (EM) approach:

$$\gamma=\frac{1}{n}\sum_{i=1}^{n}P(\eta_i=1|\gamma)=\frac{1}{n}\sum_{i=1}^{n}\frac{p_i}{p_i+p_o} \tag{9}$$

Where n is the total number of samples, p_i is the likelihood of a datum given that is the inlier:

$$p_i=\gamma\left(\frac{1}{\sqrt{2\pi}\sigma}\right)^{d}\exp\left(-\frac{e^2}{2\sigma^2}\right) \tag{10}$$

And p_o is the likelihood of a datum given that is the outlier:

$$p_o=(1-\gamma)\frac{1}{v}. \tag{11}$$

The position of the needle axis can be estimated by the iteration of the algorithm, as shown in Algorithm 2.

Algorithm 2: Needle location based on MLESAC algorithm
Input: Binary despeckled ultrasound image.
Output: Needle axis of the input image.

Parameters:
d: Cardinality of MMS;

σ: Standard deviation of Gaussian noise, default is 1;
n: Number of iteration.

Procedure:

(1) Randomly selecting two pixels as MMS from data of binary image;
(2) Use existing RANSAC methods and selecting two pixels to estimate needle model M;
(3) Calculate error e;
(4) Estimate mixing coefficient γ using EM approach;
(5) Compute the cost function $\mathrm{Loss}(e\,|\,M)$ and select minimum of sum of $\mathrm{Loss}(e\,|\,M)$ as the largest CS, i.e. maximum estimation $\hat{M}$ of the needle model M;
(6) Repeat from step (1) to step (5) n times.

3.2. Location estimation of needle tip

As a consequence of the needle tip pixels is similar to the background in the original image, after the image binarization, the position coordinate of the needle tip is obtained by the background difference method and morphologic operations. Then the coordinate is substituted in the needle model $\hat{M}$ in order to obtain the position of the needle tip, and the position of the actual tip is estimated. MLESAC algorithm for estimating the axis and tip of a 18 gauge needle frame sequences image is shown in Fig 3.

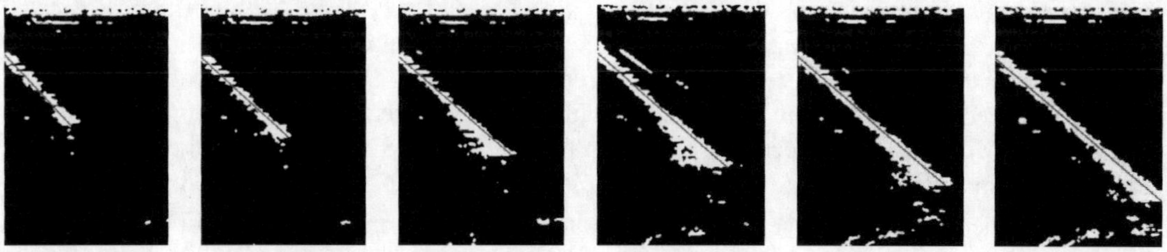

(a) Frame number = 17 (b) Frame number = 22 (c) Frame number = 30 (d) Frame number =35 (e) Frame number =42 (f) Frame number = 48

Figure 3. Location needle axis and tip

4. Experiments and results

4.1. Experimental Setup

(1) The ultrasound image are collected using a SPC-2000CIII B-mode ultrasound imaging system (Shengpu Medical Instrument Technical and Equipment Co. Ltd., China) and a 7.5 MHz linear array 2D ultrasound transducer with 128 elements, the lateral resolution and axial resolution of probe is approximately 1mm. The acquired image sizes are 768×576 pixels.
(2) The transparent tissue-mimicking phantom is constructed of a 50mm thick polyvinyl alcohol (PVA) [38] sheet 100mm wide and 150mm long, the composition ratio of PVA mass (g), distilled water volume (ml) and dimethyl sulfoxide volume (ml) is 8:40:60, 18 gauge, 20 gauge, 21gauge needles are used in the trials.
(3) All algorithms are implemented in MATLAB on a standard PC and run on a 64-bit Window 7 workstation, which has an Inter (R) Core (TM) i5 3470 CPU running at 3.2GHz and 4GB of RAM.

4.2. Experimental Results

In order to prove the effectiveness of the wavelet filter, the signal noise ratio (SNR) is computed using the following equation,

$$SNR = 10\log_{10}\left(\frac{meanI(x,y)}{stdI(x,y)}\right) \tag{12}$$

Where $I(x,y)$ is the input image. The result of the SNR value is described in Fig 4. It is found that SNR value of the original image is smaller than the after processing the image, which improves quality of the image and indicates that the wavelet soft thresholding technique based on BayesShrink rule is effective in suppressing speckle.

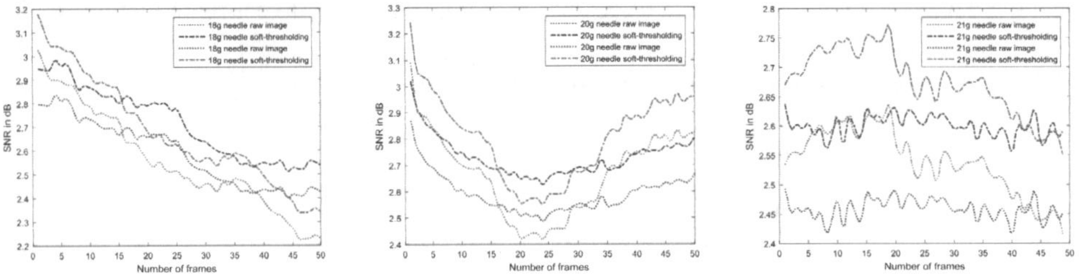

Figure 4. SNR values for different type needles

Nevertheless, there are still artifacts in the ultrasound image, including reverberations, side lobes, and comet tail artifacts, while the presence of air bubbles in the phantom due to the absence of sufficient cooling of the phantom during the preparation, all of which affect the estimation of the needle axis and tip of the MLESAC algorithm.

Through the 18 gauge, 20 gauge and 21gauge needle experiment, 300 single ultrasound images and 6 frame sequences are obtained, and the feasibility and validity of the MLESAC algorithm for needle detection under the 2D ultrasound image are validated. It is found that the pixel of the needle is less in the first ten frames ultrasound image, the estimation error is larger. As the needle insertion movement, the pixel of the needle outline in the 2D ultrasound image is increased and the phantom is deformed, the robustness of MLESAC can effectively overcome these changes. Six iterations of the EM approach can ensure the convergence of the mixing coefficient γ and reduce the time complexity of the algorithm experimentally. When the algorithm iterates 200 times, the execution time of proposed MLESAC line filter method for the single group including 50 frame sequences is 5.5178±0.1581 seconds (mean and variance). The result of proposed method which is dealt with the frame number = 35 is demonstrated in Fig 5 by using 18 gauge, 20 gauge, 21gauge, respectively.

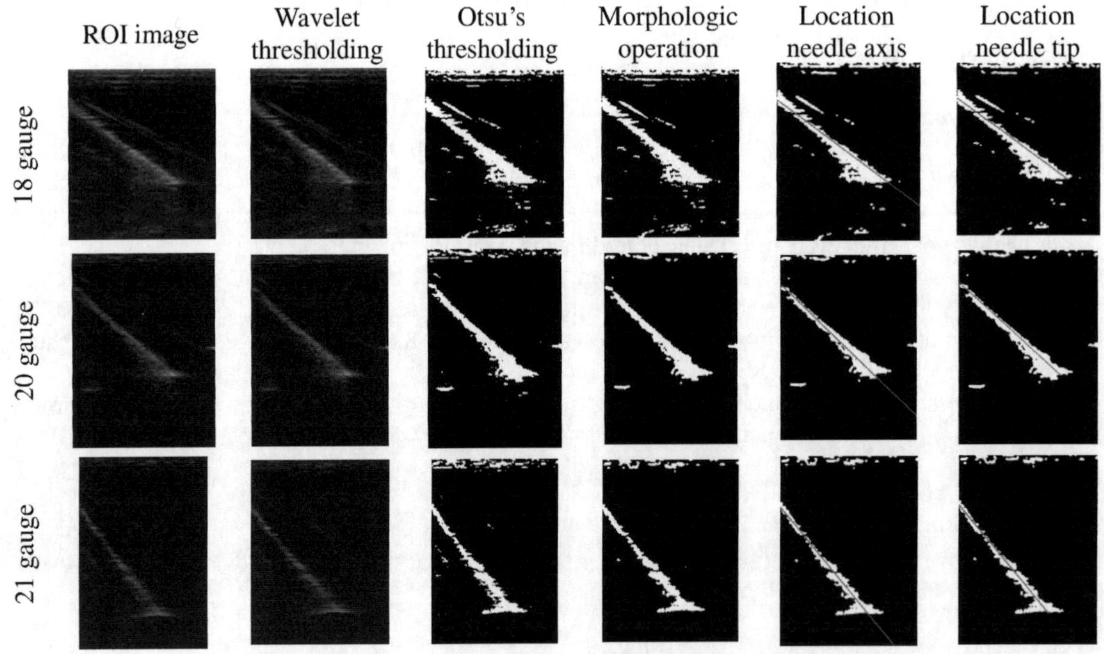

Figure 5. Results of the proposed method in 35 frames

5. Discussion and Conclusion

In this paper, a novel MLESAC-based algorithm is proposed to localize the needle axis and tip in a 2D ultrasound guided. Although the speckle is reduced by the wavelet soft thresholding method based on the BayesShrink rule, the profile of the needle structure may still be intermittent because of the influence of the artifacts in the ultrasound image. After denoising the ultrasound image getting by the probe, the position of needle axis is identified and segmented with the MLESAC algorithm iteration, and the position of needle tip can be tracked through the processing of background difference sequence. Through the percutaneous needle procedures experiments of 18 gauge, 20 gauge and 21gauge, it shows that MLESAC is effective and robust in the 2D ultrasound image, which solves the localization of needle under ultrasound-guided medical assisted robot in medical diagnostics and therapeutics.

In the future work, we will use the camera to take the needle insertion experiment, compute needle displacement error in the estimation of proposed method and the reality, evaluate the accuracy of the MLESAC algorithm, analysis the validity and robustness of this method to color ultrasound image such as Doppler ultrasound image. Further extensions include localizing position of 3D ultrasound needle and combining these techniques with the cantilever beam model in order to path planning, facilitate target hitting and obstacle avoidance.

Acknowledgements

This work was sponsored in part by the National Natural Science Foundation of China (Grant no. 51665049), Natural Science Foundation of Qinghai Province (Grant no. 2015-ZJ-906), and the Youth Research Fund of Qinghai University (Grant no. 2016-QGY-3).

References

[1] V. G. Mallapragada, N. Sarkar and T. K. Podder, 2009, Robot-assisted real-time tumor manipulation for breast biopsy. *IEEE Transactions on Robotics*. **25**(2): p. 316-24.

[2] G. Wan et al., 2005, Brachytherapy needle deflection evaluation and correction. *Medical physics*. **32** (4): p. 902-09.

[3] K. J. Chin, A. Perlas and R. Brull, 2008, Needle visualization in ultrasound-guided regional anesthesia: Challenges and solutions. *Regional Anesthesia and Pain Medicine*. **33**(6): p. 532-44.

[4] R. Wen et al., 2013, Projection-based visual guidance for robot-aided rf needle insertion. *International Journal of Computer Assisted Radiology & Surgery*. **8**(6): p. 1015-25.

[5] S. P. Emery et al., 2007, Computer-assisted navigation applied to fetal cardiac intervention. *International Journal of Medical Robotics & Computer Assisted Surgery*. **3**: p. 187-98.

[6] A. Noble and D. Boukerroui, 2006, Ultrasound image segmentation: A survey. *IEEE Transactions on Medical Imaging*. **25**(8): p. 987-1010.

[7] P. Suetens, *Foundationals of medical imaging*. 2009, Cambridge University Press: New York.

[8] V. F. Leavers, 1993, Which hough transform? *Computer, Vision Graphics and Image Processing*. **58**(2): p. 250-64.

[9] M. Ding and A. Fenster, 2003, A real-time biopsy needle segmentation technique using hough transform. *Medical physics*. **30**(8): p. 2222-33.

[10] S. H. Okazawa et al., 2006, Methods for segmenting curved needles in ultrasound images. *Medical image analysis*. **10**(3): p. 330-42.

[11] G. J. Vrooijink et al., 2014, Needle path planning and steering in a three-dimensional non-static environment using two-dimensional ultrasound images. *International Journal of Robotics Research*. **33**(10): p. 1361-74.

[12] G. J. Vrooijink, M. Abayazid and S. Misra, 2013, Real-time three-dimensional flexible needle tracking using two-dimensional ultrasound. *IEEE International Conference on Robotics and Automation*,(2): p. 1668-73.

[13] M. Aboofazeli et al., 2009, A new scheme for curved needle segmentation in three-dimensional ultrasound images. *Proc IEEE Int Symp Biomed Imaging*: p. 1067-70.

[14] D. Glozman and M. Shoham, 2004, Flexible needle steering and optimal trajectory planning for percutaneous therapies. *Lecture Notes in Computer Science*. **3217**(1): p. 137-44.

IOP Conf. Series: Journal of Physics: Conf. Series **1004** (2018) 012037 doi:10.1088/1742-6596/1004/1/012037

[15] Z. Neubach and M. Shoham, 2010, Ultrasound-guided robot for flexible needle steering. *IEEE transactions on bio-medical engineering*. **57**(4): p. 799-805.
[16] M. Kaya and O. Bebek, 2014, Gabor filter based localization of needles in ultrasound guided robotic interventions. *IEEE International Conference on Imaging Systems and Techniques*: p. 112-17.
[17] M. Kaya and O. Bebek, 2014, Needle localization using gabor filtering in 2d ultrasound images. *IEEE International Conference on Robotics & Automation*: p. 4881-86.
[18] M. Waine et al., 2015, 3d shape visualization of curved needles in tissue from 2d ultrasound images using ransac. *IEEE International Conference on Robotics and Automation*: p. 4723-28.
[19] M. Waine et al., 2016, Needle tracking and deflection prediction for robot-assisted needle insertion using 2d ultrasound images. *Journal of Medical Robotics Research*. **1**(1): p. 1640001.1-01.11.
[20] M. Uherık et al., 2010, Model fitting using ransac for surgical tool localization in 3-d ultrasound images. *IEEE transactions on bio-medical engineering*. **57**(8): p. 1907-16.
[21] M. Barva et al., 2008, Parallel-integral-projection-transform straight electrode localization in 3-d ultrasound images. *Ultrasonics Ferroelectrics & Frequency Control IEEE Transactions on*. **55**(7): p. 1559-69.
[22] Y. Zhao, H. Liebgott and C. Cachard, 2012, Tracking biopsy needle using kalman filter and ransac algorithm with 3d ultrasound. *Acoustic*: p. 231-36.
[23] Y. Zhao, H. Liebgott and C. Cachard, 2012, Tracking micro tool in adynamic 3d ultrasound situation using kalman filter and ransacalgorithm. *IEEE International Symposium on Biomedical Imaging*. **88**(1): p. 1076-79.
[24] H. R. S. Neshat and R. V. Patel, 2008, Real-time parametric curved needle segmentation in 3d ultrasound images. *IEEE/RAS-EMBS International Conference on Biomedical Robotics & Biomechatronics*: p. 670-75.
[25] P. M. Novotny et al., 2007, Gpu based real-time instrument tracking with three-dimensional ultrasound. *Medical image analysis*. **11**(5): p. 458-64.
[26] J. W. Goodman, 1976, Some fundamental properties of speckle. *Journal of the Optical Society of America*. **66**(11): p. 1305-10.
[27] J. Kaur, J. Kaur and M. Kaur, 2011, Survey of despeckling techniques for medical ultrasound images. *International Journal of Computer Technology & Applications*. **2**(4): p. 1003-07.
[28] H. H. Arsenault and G. April, 1976, Properties of speckle integrated with a finite aperture and logarithmically transformed. *Optical Society of America*. **66**(11): p. 1160-63.
[29] D. L. Donoho, 1995, De-noising by soft-thresholding. *IEEE Transactions on Information Theory*. **41**(3): p. 613-27.
[30] K. Z. Abd-Elmoniem, A.-B. M. Youssef and Y. M. Kadah, 2002, Real-time speckle reduction and coherence enhancement in ultrasound imaging via nonlinear anisotropic diffusion. *IEEE Transactions on Biomedical Engineering*. **49**(2): p. 997-1014.
[31] S. G. Chang, B. Yu and M. Vetterli, 2000, Adaptive wavelet thresholding for image denoising and compression. *IEEE Transactions on Image Processing* **9**(9): p. 1532-46.
[32] S. Sudha, G. R. Suresh and R. Sukanesh, 2009, Comparative study on speckle noise suppression techniques for ultrasound images. *International Journal of Engineering and Technology*. **1**(1): p. 57-62.
[33] P. S. Hiremath, P. T. Akkasaligar and S. Badiger, 2010, Visual enhancement of digital ultrasound images using multiscale wavelet domain. *Pattern Recognition and Image Analysis*. **20**(3): p. 303-15.
[34] P. S. Hiremath, P. T. Akkasaligar and S. Badiger, 2011, Performance comparison of wavelet transform and contourlet transform based methods for despeckling medical ultrasound images. *International Journal of Computer Applications*. **26**(9): p. 34-40.
[35] P. H. S. Torr and A. Zisserman, 2000, Mlesac: A new robust estimator with application to estimating image geometry. *Computer Vision and Image Understanding*. **78**(1): p. 138-56.
[36] M. A. Eishler and R. C. Bolles, 1981, Random sample consenus: A paradigm for model fitting with applications to image analysis and automated cartograph. *Commun. ACM*. **24**(6): p. 381-95.
[37] S. Choi, T. Kim and W. Yu, 2009, Performance evaluation of ransac family. *Proceedings of the British Machine Vision Conference*: p. 81.1-81.12.
[38] M. O. Culjat et al., 2010, A review of tissue substitutes for ultrasound imaging. *Ultrasound in Med. & Biol.* **36**(6): p. 861-73.

IOP Conf. Series: Journal of Physics: Conf. Series **1004** (2018) 012038 doi:10.1088/1742-6596/1004/1/012038

Research on Automatic Ticketing Interface Design of Tianjin South Station under the Background of Aging

Zhao Zhenghui[1,*]

[1]Tianjin University of Technology, Tianjin 300384

*765369170@qq.com

Abstract. Based on the context of increasingly serious aging problem in China, the psychological characteristics of elders in using public self-service facilities and the development status and the future trend of public self-service ticketing service. The approach is analysing physiological and psychological characteristics, education level of the elderly and studying its characteristics of consumer psychology and regional cultural characteristics profoundly before conducting comprehensive analysis and research in combination with the interface features of public self-service ticketing machine. The interface design will be more personalized, intelligent, regional and international. Strategies of caring for the elderly in the regional public self-service facility interface design innovation develops the concept of taking care of the elderly in the entire region as an indispensable people-benefiting optimization system in the modern social services.

1. Introduction

The continuous use of digital interfaces in public self-service devices is facilitating our daily life constantly with great adaptability to the fast-paced social life, while some social problems have also arisen subsequently, namely, problems caused by the friction between the increasingly serious ageing and constantly developing digitalization. In recent years, how to coordinate the relationship between public self-service facility interface and aging has become a hot issue in the society as a whole. Based on the research on the Automatic Ticketing Interface of Tianjin South Station, this paper carries out in-depth research and innovative design in combination with the physiological, psychological and lifestyle habits of elders in the vicinity of the South Railway Station.

2. Research background

2.1. Background of Aging Society

The increasing population of elderly people in China with the increasing social aging problem will occupy a higher proportion of population, as how to take the aging problem into account on public self-service facilities in the society and how to enable elders at all levels of society to use public self-service facilities conveniently has become a prominent issue for development of public self-service nowadays. TVM at railway stations is a typical case, as multiple physiological and psychological states reflected in the elder's use process can reveal problems of enjoying the convenience brought by TVM better. The most representative problems of interface design are identified via investigation and analysis, and comprehensive design research based on the living habits of elderly at Tianjin South Station is conducted to get access to better innovative design methods [1].

IOP Conf. Series: Journal of Physics: Conf. Series **1004** (2018) 012038 doi:10.1088/1742-6596/1004/1/012038

2.2. Status Quo of Elderly in Using TVM

First of all, it is undeniable that TVM renders the daily life of the elderly with further convenience, but a series of problems of TVM for the elderly are also identified via field investigation and research on the use of TVM operation interface:

2.2.1. The problem of colour matching of TVM interface. Many elderly people may suffer from failure of eyesight and even a series of eye diseases over time, such as cataracts and glaucoma. However, colour in design of TVM at the South Railway Station is rather gorgeous and bold in obvious contrast for consideration of young people with fast-paced life, as the elderly people will suffer from visual fatigue in the ticket purchase process, and even feel discomfort.

2.2.2. Fear of science and technology. With the rapid development of science and technology, new high-tech things are emerging constantly, as young people enjoy them a lot for the great curiosity. However, not all the new things are acceptable for elders, as many of them have more or less a sense of fear of newlywed high-tech products, including distrust of their own abilities.

2.2.3. Aging of sensation and perception. It is also caused by physiological problems of aging people, as all the functions of the body are constantly aging over time, so they have to face various problems in operation.

3. Development at home and abroad

3.1 Research status at abroad

Foreign self-service interface is rather mature with particularly high requirement on the user interface, as change of people's usage, especially rendering the elderly with a convenient form of use has become a new trend of future interface design. (Figure 1) Cartoon elements are integrated into the interface design of subway TVM of Seoul, South Korea, with an intensive affinity; the form of TVM interface design in Japan is not only rather vivid and lively, but also easier for users to identify. (Figure 2)

Figure 1. The subway self-service ticket machine interface in Seoul, South Korea

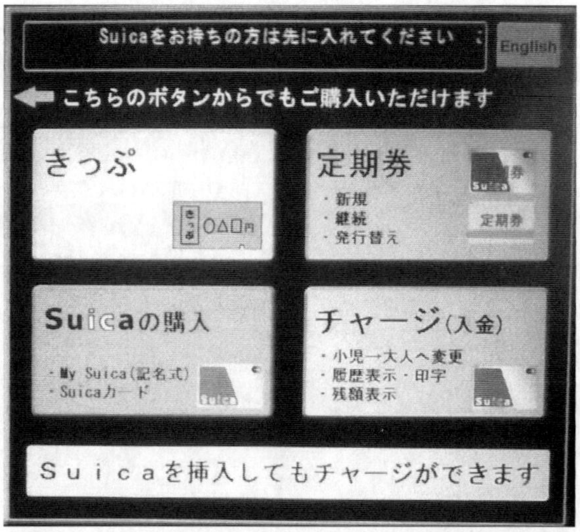

Figure 2. Japan's self-service ticket machine interface

There has been a dramatic shift in technology from its early days to today, such as Heliodisplay technology introduced by the latest technology development company (Figure 3). As an airborne projection imaging and interactive technology, it utilizes electronics and thermodynamics, as the features of air will be changed after entering the machine. In this way, image will be projected from the bottom to the top after re-projection and suspended in the air, so that people can not only see the image in different angles, but also achieve interaction with users' finger movements tracked by laser. This design can provide the elderly with a vivid reminder to facilitate their next operation, with a lot of time spared. Meanwhile, it can also encourage them to use TVM actively from the side.

Figure 3. The latest man-machine communication - air projection and imaging

3.2. Research status at home

Since 1970, the skyrocketing development of metro rail transportation has led to the rapid development of TVM terminal. Especially in the information age, development of computer field and network communication technology have enabled subway TVM, a high-tech product that only existed in imagination of people in early years, to enter the daily life of people at tremendous speed.

IOP Conf. Series: Journal of Physics: Conf. Series **1004** (2018) 012038 doi:10.1088/1742-6596/1004/1/012038

Before 1990s, due to the monopoly of key technologies for the production of self-service equipment by some major foreign manufacturers, almost all the self-service devices of China were imported. Up to the late 1990s, all walks of life in China began to recognize the market value and service value of application system of currency smart identification, with imports replaced by domestic products gradually. However, countries with advanced development of science and technology are still at the forefront of the world. (Figure 4) Domestic subway TVM interface.

Figure 4. Beijing Metro TVM Interface

Compared to foreign countries, TVM of China still has a great deal of drawbacks, as it is obvious in the contrast of pictures that iconic marks on domestic TVM interface are not so eye-catching and vivid. It is a major drawback for the elderly of poor eyesight in use. In addition to a series of aging problems such as slow behaviour, the elder people are rather vulnerable for congestion of ticket purchase, as they may have a psychological shadow in the next use of TVM under the hit of this time [2]-[3].

4. Principles of interface design for TVM for the elderly

4.1. The principle of consistency

Consistency of interface operation usually includes the following aspects:

4.1.1Style: Consistency of colour matching, system terminology, icons and symbols, etc.;

4.1.2 Space: Consistency of classification & combination of space according to the operating habits and interaction requirements;

4.1.3 Operation: For those important or commonly used functions, consistency of operating mode shall be kept with attention paid to conformity with use habits of users.

4.2 The principle of rationality

For the overall layout of interface, the elderly shall be oriented for interface design of elder TVM to provide the elderly with a "simple, quick and easy to operate" man-machine interface. Proper layout of screen can not only relieve the memory burden on elders and improve comfort in operation, but also make it easier for the elderly to control the system. It is mainly reflected in the following aspects:
4.2.1 Whether content on the screen is too much;

4.2.2 Whether significant information is prominent;
4.2.3 Whether different types of information can be distinguished clearly;
4.2.4 Whether regional distribution of the screen and the link structure are reasonable, and so on.

4.3 The principle of coordination
As different colours have different wavelengths, excessive colour may also cause a dazzling feeling, which will affect the user's emotion, mind and psychological activity, so uniform and coordinative colours shall be used. Screen effect can be enriched by adjusting the overall transparency, saturation and using small areas of contrast colour.

5. Research on universal design of TVM interface for the elderly

5.1 Research on universal design of railway TVM

5.1.1 Universal design concept of public facilities and status analysis. The principle of equality, fairness and indifference of human beings is the core idea of universal design. It emphasizes that design should treat everyone equally without discrimination, rather than regard physically weak people or other vulnerable groups as people who need special care to avoid discrimination. A mature universal design system is yet to be formed in China in the face of such a severe trend of aging, so designers shall pay attention to the principle of fairness of "human beings" and carry out design activities from the perspective of humanity on the basis of being "people-oriented".

5.1.2 Universal design of railway TVM facilities. TVM facilities emphasize an undifferentiated design that adopts an undifferentiated approach to satisfy all users and avoids differential treatment to certain groups (such as people with disabilities and the elderly). TVM facilities shall be user-friendly, as embodied as convenience and effort-saving of equipment on the material level. TVM shall have good adaptability in the purpose of meeting requirements of different people; on the spirit level, TVM facilities shall be easy to learn and grasp in line with social and cultural background of users, avoiding increasing learning and memory burden of users; simple and obvious removal of unnecessary complex design is advocated to guide users in operation with simple and clear design semantics.

The consequences of accidents or mistakes that may be caused in the use of TVM should be minimized. Warning design can be applied in the self-service ticketing process to remind users of dangerous operation with timely and effective feedback provided for users to check easily. Recoverable design shall be preferred as users can remedy mistakes or errors. On the one hand, users shall be provided with enough prompts in the operation process, with multiple forms such as voice, image and text adopted to ensure the integrity and effectiveness of information delivery. On the other hand, functions such as returning, cancelling and resetting shall be equipped to reduce the negative consequences of mistakes.

5.2 Research on elderly-centred TVM facility interface

5.2.1 Segmentation of elder user groups. It is noteworthy that the criteria for dividing the elderly are not merely linear. In addition to measuring from the age dimension, perspective of demography, such as gender, education level, place of residence and original occupation, shall be taken for detailed comparison. In short, multi-dimensional criteria are determined by the difference inside the elderly group in subdividing elder users(Table 1)[4].

Table 1. Physiological & psychological characteristics at different stages of the elderly

Stage	Age Range	Physical	Psychological

Elder-to-be	45-59	Conversion period of higher morbidity	Adjustment period, being vulnerable to negative and pessimistic emotion
Young elder	60-74	Decline stage, decline slowly and subtly	Dependence, being self-centred
Old elder	75-90	Decline stage, decline rapidly and obviously	Dependence, with stereotypy increased

5.2.2 Design strategies for TVM aiming at features of the elderly. Gradual weakening of visual ability, slowing down process of visual information and reduction of visual attention are inevitable for the elderly over time. For features of visual ability of the elderly, four strategies in interface design are given:

5.2.2.1 Increasing the size of font and icon. The basic visual ability of the elderly is lower with blurred vision. Readability shall be taken into account for interface text designed for the elderly, including font size, word-spacing and contrast with the background, where the font on the screen shall be over 24pt and not less than 12pt. What's more, the size of button and icon shall be enlarged with target content distinguished from the background significantly to enhance efficiency of reading.

5.2.2.2 Improving colour contrast. Due to the diminished ability in visualizing colour and recognizing colour, the elderly prefers mid-key colour of apparent contrast. In interface design, colour contrast shall be improved via several approaches such as adjusting saturation, brightness and hue, avoiding the use of colour group of low contrast. Compared with red and yellow, blue and green are even harder for the elderly to recognize. In addition to the impact of psychological factors such as possible sense of loss and loneliness after retirement, the elderly prefers warm tone subconsciously, as red and yellow of mid-key shall be considered in interface design[5].

5.2.2.3 Providing a clear visual hierarchy. The elderly is more vulnerable to irrelevant stimuli due to the reduced visual attention ability. The principle of being easy to distinguish shall be adhered to in interface design with the interface being concise and easy to operate; it is particularly significant to provide clear visual hierarchy. A clear visual logic enables elder users to identify the content more relevant to their mission objectives easily when checking information, as both text and graphic symbols shall be used with easy and simple symbols. Graphic symbols should be able to express the functions they present clearly without making users confused.

5.2.2.4 Making full use of multi-sensory information. Vision is the major channel for the elderly to receive information, but the efficiency of information reception can be affected by the decline of basic vision, as it can be compensated for by multi-sensory approach such as increasing auditory and touch. For example, shortage of visual ability can be made up by the voice broadcast function: information can be understood and operation can be completed interactively via voice prompt.

5.3 For characteristics of auditory ability of the elderly, the following four strategies are given in interface design: When using two or more auditory signals, there should be obvious differences between the signals with same information meaning represented all the time for a certain signal to improve the speed of human auditory reaction.

The frequency of sound device can avoid noise masking, so that people can hear the accurate signal despite the noise

In the case of simultaneous use of several TVMs in the booking hall of railway station, sound signal shall be concise and accurate with intensity controlled within a suitable range to avoid mutual interference. Other transmission devices shall be used in conjunction.

IOP Conf. Series: Journal of Physics: Conf. Series **1004** (2018) 012038 doi:10.1088/1742-6596/1004/1/012038

6. Conclusion

This subject mainly studies the design of TVM facilities in the background of aging. On the basis of theories such as universal design and interaction design, user interface of railway TVM is improved with innovative design in orientation of users. In this way, frequency efficiency of such products can be improved with a more pleasant experience provided for users. The memory burden on users is reduced and the usability of product is improved by simplifying the interface layout. In terms of software design, function architecture and system operation flow are optimized, with system functions improved and interactive experience enhanced according to user needs.

Reference

[1] REN Luo-sheng D 2008 *Agedin China Talent Development Resources* (Changsha: National Defense University) P7

[2] LIU Lu M 2013 *Study on the Interface Design of Subway Ticket Machine* (Xi'an: Packaging Engineering) P4

[3] YU Hai M 2012 *Study on Human-computer Interaction Design of Subway Automatic Ticket Vending Machine* (Nanjing: Packaging Engineering) P6

[4] CHEN Hong M 2011 *SubwayAutomatic Ticketing System Interaction Design Research* (Shanghai: Packaging Engineering) P46

[5] XU Wen-juan J 2008 *On the Software Interface Design Using Color* (Computer Knowledge and Technology) P891

IOP Conf. Series: Journal of Physics: Conf. Series **1004** (2018) 012039 doi:10.1088/1742-6596/1004/1/012039

Research on the Perforating Algorithm Based on STL Files

Han Yuchuan[1,*], Zhu Xianfeng[1], Bai Yunrui[1] and Wu Zhiwen[1]

[1]College of Precision Instrument and Opto-electronics Engineering, TianJin University, TianJin 300072

*hanyuchuan2012@163.com

Abstract. In the process of making medical personalized external fixation brace, the 3D data file should be perforated to increase the air permeability and reduce the weight. In this paper, a perforating algorithm for 3D STL file is proposed, which can perforate holes, hollow characters and engrave decorative patterns on STL files. The perforating process is composed of three steps. Firstly, make the imaginary space surface intersect with the STL model, and reconstruct triangles at the intersection. Secondly, delete the triangular facets inside the space surface and make a hole on the STL model. Thirdly, triangulate the inner surface of the hole, and thus realize the perforating. Choose the simple space equations such as cylindrical and rectangular prism equations as perforating equations can perforate round holes and rectangular holes. Through the combination of different holes, lettering, perforating decorative patterns and other perforated results can be accomplished. At last, an external fixation brace and an individual pen container were perforated holes using the algorithm, and the expected results were reached, which proved the algorithm is feasible.

1. Introduction

STL（Stereo lithographic）is an interface file of the three-dimensional solid modelling system proposed by 3D Systems Company in 1987, which has become the most commonly used 3D data format and the actual interface standard in the field of rapid prototyping presently [1]. The STL file approximates the real 3D model with small triangular facets, and has the characteristics of controllable precision, format conversion conveniently and segmentation easily. The STL file has been widely used in rapid prototyping, displaying of image real sense, finite element analysis, and reverse engineering [2].

Because the STL file format is simple and universal, it is supported by almost all platforms, and most of the three-dimensional modelling systems' interface data is saved as STL file format, such as 3D scanning data and 3D models produced by modelling software. However, due to the lack of topological information for the small triangular facets and the loss of precision of real models [3], the STL files are only used for displaying and translating, which means it is hard to be edited [4-6]. In order to meet the need of ventilation, individuation, saving material and reducing weight in some special scenes, this paper proposes a perforating algorithm based on STL files. The algorithm can help perforating round holes, elliptical holes, square holes and triangular holes, and through the combination of a number of holes, the complex structure of holes can be perforated, including character holes, pattern holes, etc.

2. The STL file's format and its specification

The STL file uses a lot of triangular facets to represent the 3D object, and each triangular facet includes two parts: vertex coordinates and the normal vector [7]. The process of representing a cube as the STL file is shown in figure 1.

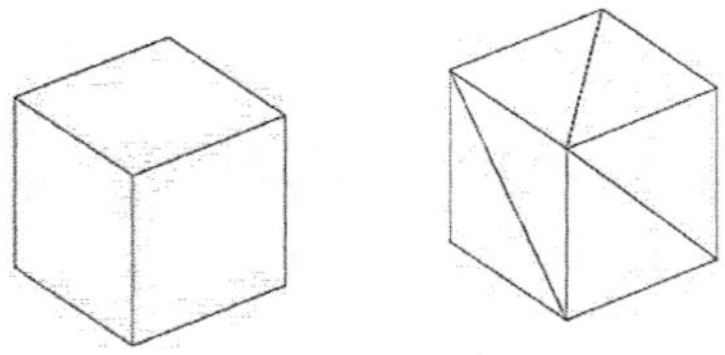

Figure.1 Triangulation of a cube

The STL file has two formats: ASCII code and binary. ASCII code file is readable, while binary file takes up less memory, which is easy to read and store. The perforating process of STL files needs to do various operations on a large number of triangular facets, and speed is a very important factor, so this paper selected binary file as the research object. Its format is shown in table 1[8]:

Table 1. The binary format of the STL file

	File header	The number of triangle facets	The triangle facets(50 bytes)					…
Bytes	80	4	3×4	3×4	3×4	3×4	2	…
Data type	Unsigned char	Unsigned long integer	Float	Float	Float	Float	Unsigned short integer	…
Storage information	Text information	The number of triangle facets	Normal vector	Vertex 1	Vertex 2	Vertex 3	Attribute byte	…

The STL file is a collection of triangular facets, and each triangular facet is represented by one normal vector and three coordinates of the three vertices. The STL file must follow the following rules [9]: (1) Common-vertex rule. Each triangle must share two vertices with adjacent triangles, which means the vertices of a triangular facet cannot lie in the edges of adjacent triangles. (2) Orientation rule. The normal vector of each triangular facet and the three edges of the triangle satisfy the right-hand rule, and the normal vector direction points to the outside of the model. (3) Full rule. Triangular facets must cover all the outer surfaces of the model, without any omission. (4) Value rule. The coordinates of each vertex of the STL file must be nonnegative, i.e., the model must be in the first quadrant of the coordinate system.

3. Basic thought

Use closed space surfaces (such as hollowed cylindrical surface) to intersect with the STL file, delete the STL files inside the closed space surface, reserve the STL files outside the closed space surface, reconstruct triangles on the space closed surface to form the inner surface of the hole, and then complete perforating.

Therefore, the perforating processes are as follows:

1).Construct the imaginary closed space surface mathematical equation whose image in the Descartes coordinate system is a closed space surface, which intersects with the STL file.

2).Calculate the crossover points of the closed space surface and the STL file, reconstruct triangles by the crossover points and the vertices of the triangular facets at the intersection, delete the original triangular facets so that all the triangular facets are totally inside and outside the closed space surface.

3).Delete the triangular facets inside the closed space and form a hole on the STL file.

4).Using the triangular facets to fill the inner surface of the hole, and the perforating process is realized.

4. Perforate round holes

4.1. The crossover points of a hollowed circular cylinder and a sphere

For the convenience of research, this paper chose a spherical STL file as the research object, and the perforating hollowed cylindrical equation is given by

$$\left(z-\bar{z}\right)^2+\left(x-\bar{x}\right)^2=r^2 \tag{1}$$

This equation describes a y-infinity cylindrical surface, where $\bar{z}$ is the mean value in the z-direction of the points on the cylindrical surface, $\bar{x}$ is the mean value in the x-direction of the points on the cylindrical surface, r is the radius of the cylinder [10]. The intersection of the hollowed circular cylindrical and the sphere is shown in figure.2.

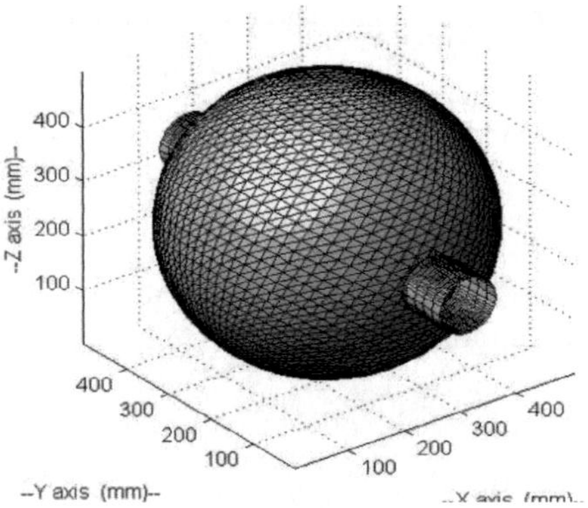

Figure.2 Intersection of one hollowed circular cylinder and a sphere

The STL file consists of numerous triangular facets, each of which has three vertices, and each vertex corresponds to a space coordinate (x_i,y_i,z_i) [11]. The position relation between the triangular facet and the hollowed circular cylinder can be found by judging the position relation between the three vertexes' coordinates and the hollowed circular cylinder. By searching for the number of vertices of each triangular facet in the outer and inner surface of the hollowed circular cylinder, the intersecting edges of the triangle can be found, seeing table.2.

Table.2 Intersection of triangular facets and hollowed circular cylinders

	inner	outer	surface	intersecting lines	Intersecting situations
case1	1	2	0	2	inner-outer1 inner-outer2
case2	2	1	0	2	inner1-outer inner2-outer
case3	1	1	1	1	inner-outer
others	m	n	p	0	none

After knowing the intersecting lines, the crossover points can be obtained by the equation of the line intersecting with the surface.

The parameter equation of the intersecting line is [12]:

$$\begin{aligned} x &= x_0 + t*l \\ y &= y_0 + t*m \\ z &= z_0 + t*n \end{aligned} \tag{2}$$

IOP Conf. Series: Journal of Physics: Conf. Series **1004** (2018) 012039 doi:10.1088/1742-6596/1004/1/012039

Where (x_0, y_0, z_0) is a known point of the intersecting line, and (l, m, n) is the direction vector of the line. By taking (2) into (1), we can get the t, and then get the crossover point.

4.2. Reconstructing triangles at intersecting lines

Reconstruct triangles with crossover points and the corresponding triangles' vertexes to form a circular contour on the surface of the sphere, to help to perforate a circular hole. There are five kinds of intersections between the surface and the triangles [13]. 1. Two edges intersect with the surface. 2. One edge intersects with the surface as well as one point is located on the surface. 3. One edge is located on the surface. 4. One point is located on the surface and the other two points are located outside the surface. 5. The three vertexes are located outside the surface. See figure.3.

When reconstruct a triangle, there is no need to deal with the case 3, 4, and 5. For case 1, the two crossover points construct a triangle with a single vertex on one side of the surface, and the two crossover points construct two triangles with the other two vertices. For case 2, the crossover point, the interior point, and the upper and lower separate vertices construct two triangles. After the constructions are finished, completed triangles can be formed on both sides of the surfaces, seeing figure 4.

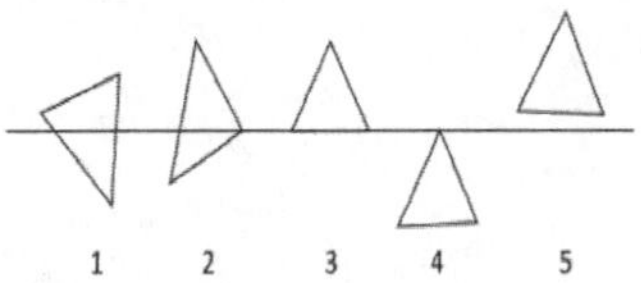

Figure 3. The five situations of triangles and a surface

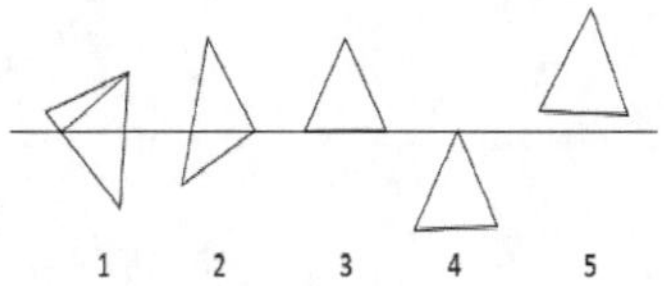

Figure 4. Reconstruct triangles

The result of reconstructing triangles at the intersection of the sphere and the hollowed circular cylinder mentioned above is shown in figure 5.

With the hollowed circular cylinder as the boundary, delete the triangular facets inside the hollowed circle cylinder, and a sphere with a circular hole is formed, as is shown in figure 6.

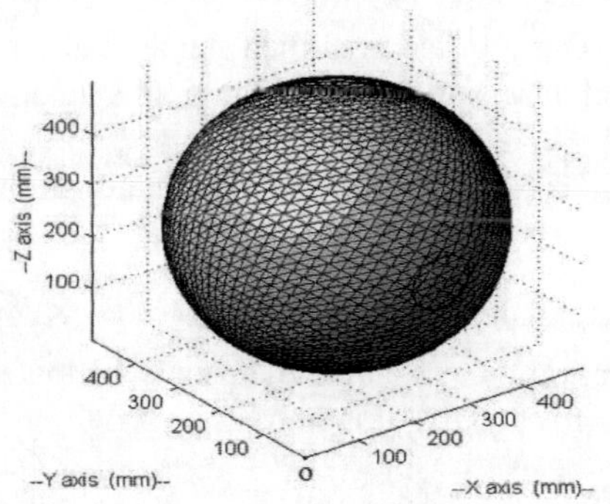

Figure 5. Reconstructing triangles at the intersection

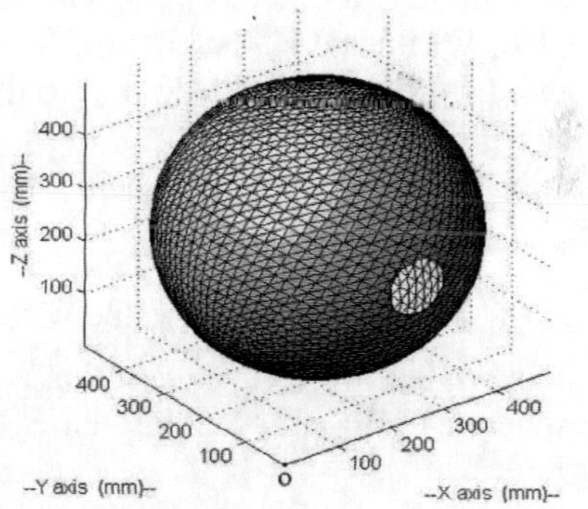

Figure.6 The spherical with a circular hole

4.3. Clockwise sorting of boundary points

According to the common-vertex rule and full rule of STL files, each edge of the triangle has only two triangles sharing [14]. The STL file shown in figure.6 is incomplete, which has to be filled with triangular facets inner the hole. If an edge has only one triangular sharing, the edge is the boundary. According to this feature, we can find the boundary edges, and then find the boundary points [15, 16].

The boundary points are unordered, and if using the boundary points of the two ends constructs triangles in the inner surface directly, intersections and holes will occur. To construct the complete

IOP Conf. Series: Journal of Physics: Conf. Series **1004** (2018) 012039 doi:10.1088/1742-6596/1004/1/012039

inner surface, we need to sort the boundary points, and then can we use the top and bottom boundary points to construct the triangles sequentially [17].

Clockwise sorting method is used here to sort the boundary points based on its tangent value. Project the boundary points to the x-z plane for reducing dimension, and then sort the boundary points according to their tangent value on the x-z plane. First, calculate the centre of the boundary points. Second, reconstruct an x-z coordinate axis with the centre point as the origin, and divide the boundary points into the left and right sides of the z-axis. Third, sort the boundary points of each side by the tangent value. Forth, connect the two-part points to achieve a clockwise sorting.

For the point on the z-axis, if it is on the top, divide it into the right boundary point ring, and if it is at the bottom, divide it into the left boundary point ring. Thus avoid the denominator of zero, and then achieve the sorting, seeing figure.7.

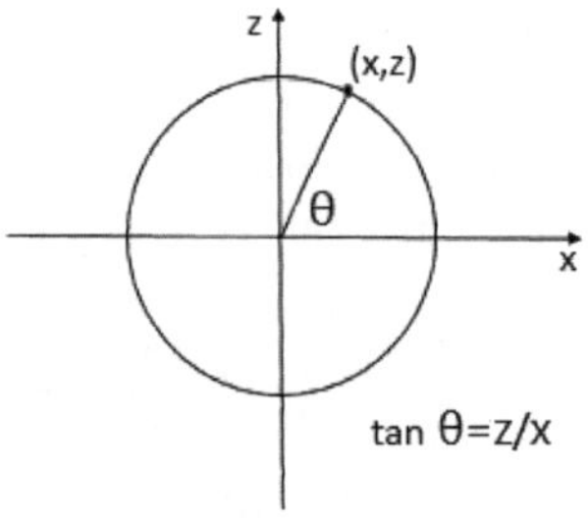

Figure 7. Tangent sorting

4.4. Inserting points in the way of interpolation

If the number of points in the top and bottom boundary is the same, construct triangles with the top and bottom points directly to form the inner surface of the hole. Otherwise, it is necessary to insert points in the fewer points' boundary, or the inner surface will not be smooth. When inserting points, first calculate the difference of the number of the top and bottom points, and set it as P, which means we need to insert P points to the fewer points' boundary. Set the number of its boundary points as W, insert the first point between the first and second boundary points, and insert the second point between the second and third boundary points... If P is not bigger than W, insert P points in this way. If P is bigger than W, update P's value as P-W after traversing a circle, and then iterate back until P points are inserted. Add each inserted point to the back of the W, and then sort the points again after all the insertions are completed.

Assume that the first point is being inserted (set it as point P). Set the coordinates of the first point and the second point on the boundary points as (x_1, y_1, z_1) and (x_2, y_2, z_2). Set the inserted point's x-value as $(x_1 + x_2)/2$, and the value of z_p can be calculated by $(z-\bar{z})^2 + (x-\bar{x})^2 = r^2$. Knowing the x-value and z-value, we can find out the point's projection on x-z plane fallen in which triangular facet of the sphere STL file. If point P satisfies the formula (3) and (4), it is in triangle ABC.

$$BP = u \times BA + v \times BC \tag{3}$$

$$u >= 0, v >= 0, u + v <= 1 \tag{4}$$

The u and v are the coefficients, and the schematic is as figure 8.

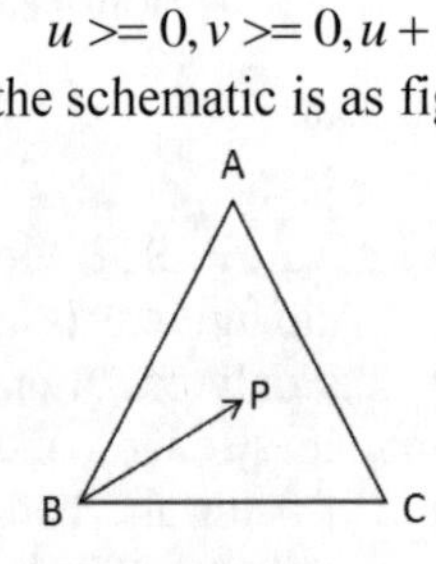

Figure 8. The judgment of whether one point is in a triangle

This triangle is the projection of the space triangle on the x-z plane. By traversing the sphere STL file in a certain range, we can find the triangular facet satisfied with the formula (3) and (4). The triangle facet is corresponding to the point P. Then obtain the crossover point of the triangular facet with the straight line parallel to the y-axis as well as goes through the point P. The equation of the triangular facet is:

$$AA \times (x - A(1,1)) + BB \times (y - A(1,2)) + CC \times (z - A(1,3)) = 0 \tag{5}$$

AA, BB, CC are the three components of the triangle ABC's normal vector. $A(1,1)$, $A(1,2)$, $A(1,3)$ are the three components of point A. By bringing $x = x_p$ and $z = z_p$ into the equation (1), the value of y can be obtained as y_p, and the (x_p, y_p, z_p) is just the coordinate of the point P.

When the number of the top and bottom boundary points is the same, the inner surface of the cylinder can be constructed using the top and bottom boundary points in order. See figure 9.

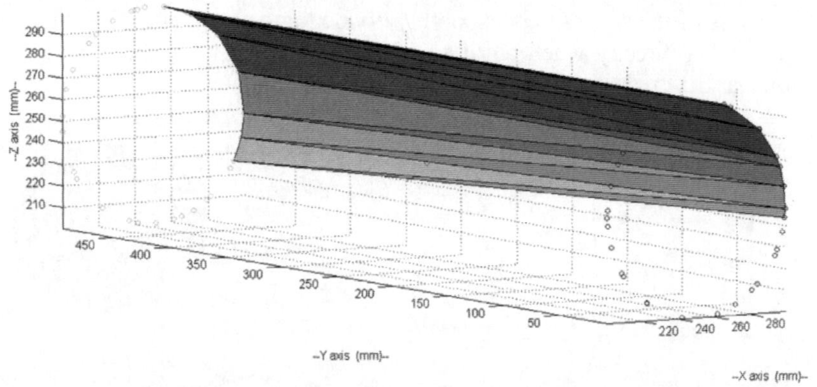

Figure 9. The process of constructing the inner surface

Combine the inner surface of the cylinder with the outer triangle facets of the cylinder on the sphere can realize the perforating, seeing figure 10.

If the cylindrical equation is changed into elliptic cylinder equation, the elliptic hole can be perforated, as is shown in figure 11.

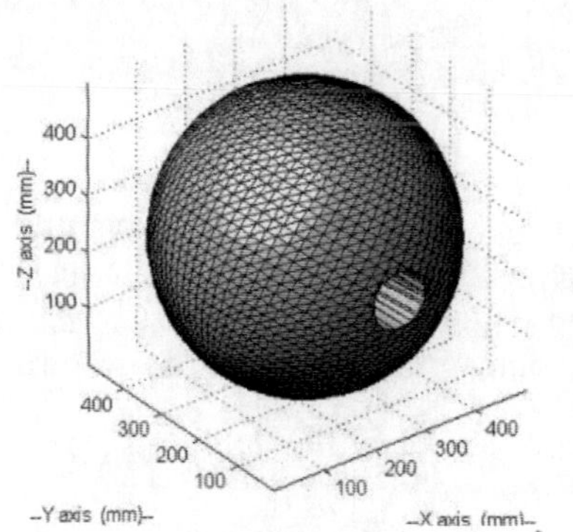

Figure 10. Perforate a round hole

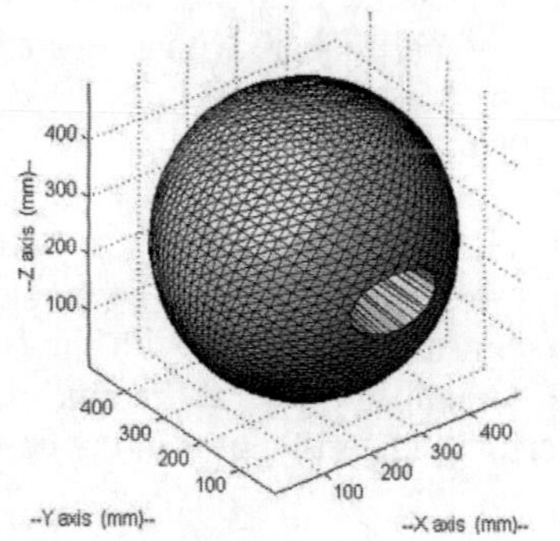

Figure 11. Perforate an elliptic hole

5. Perforate polygon holes

For a space polygon prism that cannot be represented by a single equation, the method proposed above cannot be used for perforating polygon holes. At this time, the space polygon equation can be decomposed into several surface equations, and each surface equation can form a surface in the Descartes coordinate system, which can be used as one of the inner surfaces of the hole. Combine the inner surface of the hole with the outer triangle facets of the polygon prism on the sphere can realize the perforating.

5.1. Set vertices
For the convenience of research, this paper chose the z-infinity square prism as the perforating model, and selected the sphere mentioned above as the perforated object. First of all, set the four vertices as the projection of the cuboid's prism on the x-y plane, which were (200,200), (200,300), (300,200), (300,300) here.

Obtain the crossover points of the square prism's edges with the triangular facets of the STL file. In order to find the triangular facets intersected with the edges, it is necessary to judge each of the projected points falls into which triangle. After the triangle is found, the crossover point of the edge and the triangle can be obtained. Use the crossover point and the three vertices of the triangle reconstruct three triangles and delete the original triangle, as shown in figure 12. This allows the triangles at the vertices of the square to be smoothly divided into small, complete triangles. Thus, STL files can be manipulated directly when cutting with a square prism without having to deal with the complex situation that the prism's edge is located in a triangle.

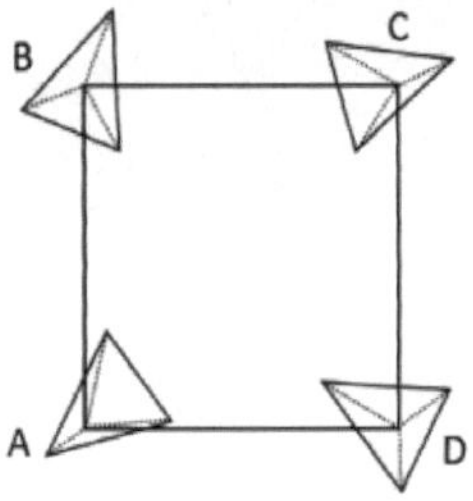

Figure 12. Reconstruction of triangles at the square's prisms

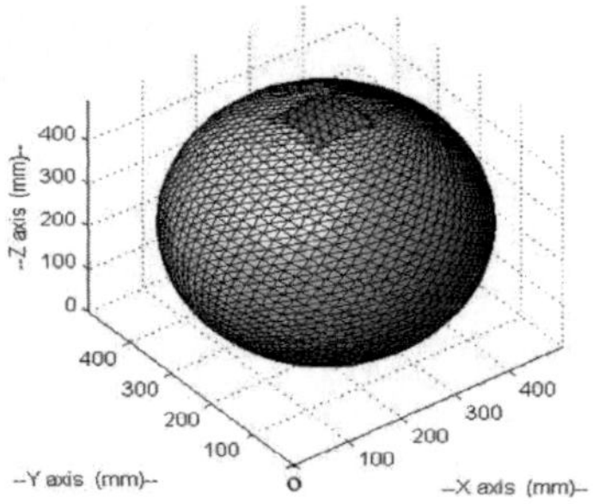

Figure 13. The sphere with a square hole

5.2. Intersections of planes and the sphere
Using line segments AB, BC, CD, and DA to divide the triangular facets intersected with them separately. Delete the triangles facets inside the square prism, and the rest of the triangle facets of the STL file can form a sphere with a square hole, see figure 13.

5.3. Triangulate inner surfaces
Find the top and bottom boundary points of the STL file shown in Figure 13. Each side's boundary points are distributed on four edges. Sort the points in each edge. If the number of points in the two boundaries is the same, construct triangles directly, otherwise, we need to insert points to make them the same. Construct triangles with the corresponding top and bottom boundary points, which means complete the inner surface of the hole, and together with the outer STL files of the square prism can realize perforating, as is shown in the figure 14.

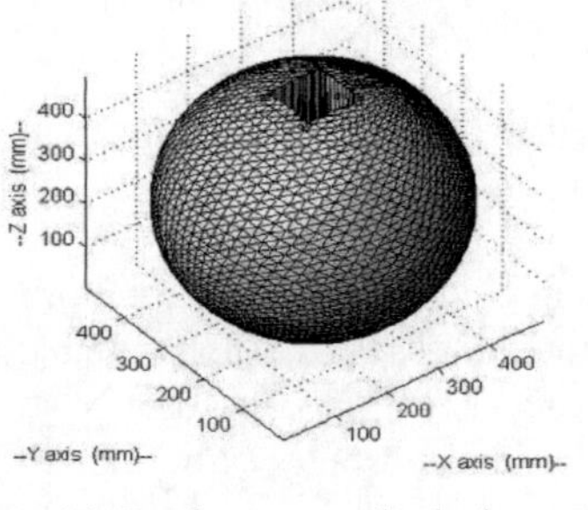

Figure 14. Perforate a cube hole

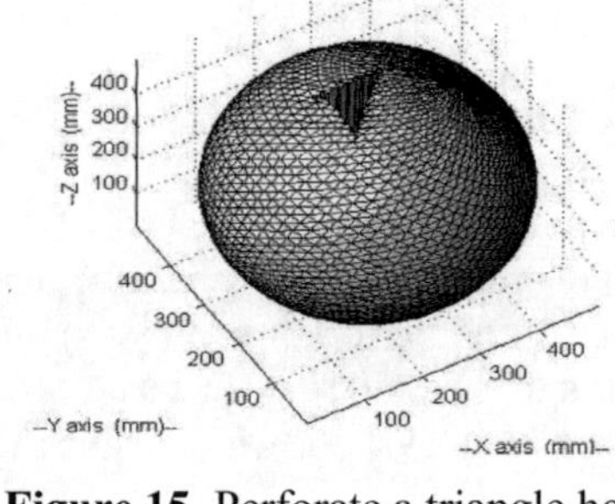

Figure 15. Perforate a triangle hole

When the square prism equation is changed into a triangular prism equation, the triangular hole can be perforated, as is shown in figure 15.

6. Examples and analysis

Through coordinate translation and matrix transformation, it is possible to perforate holes of multiple directions with different shapes on the STL file. The rotation matrices in the x-direction, y-direction, and z-direction are as follows [18]:

$$R_x(\alpha)=\begin{bmatrix}1 & 0 & 0\\ 0 & \cos\alpha & \sin\alpha\\ 0 & -\sin\alpha & \cos\alpha\end{bmatrix}\quad(6)\qquad R_y(\beta)=\begin{bmatrix}\cos\beta & 0 & \sin\beta\\ 0 & 1 & 0\\ -\sin\beta & 0 & \cos\beta\end{bmatrix}\quad(7)\qquad R_z(\gamma)=\begin{bmatrix}\cos\gamma & \sin\gamma & 0\\ -\sin\gamma & \cos\gamma & 0\\ 0 & 0 & 1\end{bmatrix}\quad(8)$$

6.1. Individual external fixation brace

Figure.16 shows the STL model of a wrist personalized external fixation brace, which was generated by offsetting the triangle facets of a scanned wrist STL file. In order to make the model breathable and light, it needs to be perforated. By perforating several round holes, the perforating effect can be realized, as is shown in figure 17.

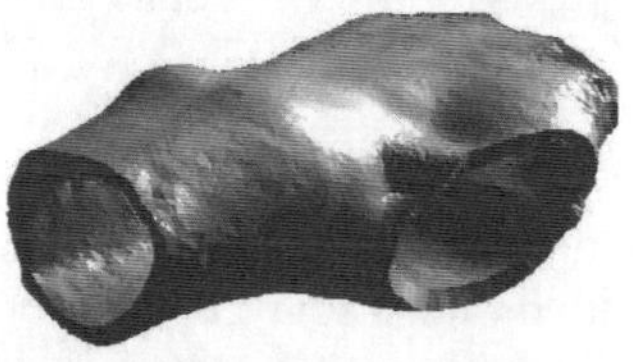

Figure 16. The wrist's personalized external fixation brace

Figure 17. Overall perforating effect

After 3D printing out the individual external fixation brace, it can be worn on the arm, as is shown in figure 18. It can be seen that these holes can make the brace breathable, which can prevent the side effects such as swelter and itching in the treatment of fracture, meanwhile the hole can lighten the whole weight of the brace and reduce the burden for the patient.

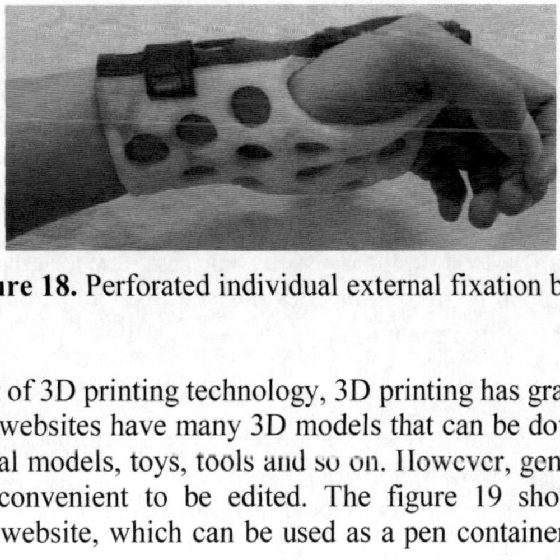

Figure 18. Perforated individual external fixation brace

6.2. Pen Container

With the rapid development of 3D printing technology, 3D printing has gradually entered the homes of ordinary people. 3D model websites have many 3D models that can be downloaded, printed, and used directly, such as architectural models, toys, tools and so on. However, generally the models' format is the STL file, which is inconvenient to be edited. The figure 19 shows a pen container model downloaded from 3D tiger website, which can be used as a pen container as well as a mobile phone holder.

IOP Conf. Series: Journal of Physics: Conf. Series **1004** (2018) 012039 doi:10.1088/1742-6596/1004/1/012039

Figure 19. Pen container

Figure.20 personalized pen container

However, the pen container's shell has no patterns, which is slightly dull, and will consume much material when it is printed. At this time, it can be perforated by the perforating algorithm proposed above. Through perforating different text and patterns can achieve personalized perforating effort so that it becomes one's own personalized pen container. By using different spatial equations such as circle, ellipse, rectangle, and triangle to perforate the STL file, meanwhile control the drilling depth so that it does not run through the whole model, we can achieve the perforating effect as shown in figure 20.

7. Conclusion

Formed 3D data is generally STL file format, the sources of which are 3D scanner, 3D model designing software, etc. This format file is easy to be transmitted and displayed, but not easy to be edited. In this paper, a perforating algorithm based on STL file is proposed, which can be used to perforate the three-dimensional model, to meet the special requirements of ventilation, reducing weight and so on. Aiming at perforating mesh files, this paper proposed a method of spatial surface cutting. The crossover points of the cutting surface and the STL file are sorted by the tangent value, and then the inner surface is triangulated, that is, the perforating is completed. Through the coordinate transformation and matrix transformation, a plurality of different holes can be perforated on an STL file, which means complete the whole perforating. This paper verified the effect of the whole perforating with the personalized external fixation brace and the personalized pencil container, and proved the feasibility and practicability of the algorithm. Nevertheless, this paper's perforating pattern is just basic pattern of the circular, square, and other simple geometric shapes, and if we can establish the complex pattern (such as Chinese traditional auspicious clouds pattern) equation's 3D mathematics, the corresponding perforating operation can also be achieved. This is the future research direction of this topic.

Acknowledgments

At this time I want to thanks to my tutor—Zhu Xianfeng, who gave me many academic and constructive advices in composing this paper. Besides, he also helped me in other aspects such as reading and thinking which benefited me a lot.

References

[1] Wei Wei,Zhou Laishui,Zhang Liyan. Rapid STL File Loading and Visualization. Mechanical Science and Technology, 2006, 8: 935-938

[2] WANG Zeng-bo. Research of 3D Model Visualization Based on STL Files. Journal of Hengyang Normal University, 2008, 6: 21-24

[3] Liu Jinyi, Hou Baoming. Efficient Topological Reconstruction of Solids in STL Format. journal of engineering graphics, 2003, 4: 34-39

[4] Quigley N, Lyne J E. Development of a Three-Dimensional Printed, Liquid-Cooled Nozzle for a Hybrid Rocket Motor. Journal of Propulsion & Power, 2015, 6: 1-2

IOP Conf. Series: Journal of Physics: Conf. Series **1004** (2018) 012039 doi:10.1088/1742-6596/1004/1/012039

[5] Campbell T A, Ivanova O S. 3D Printing of Multifunctional Nanocomposites. Nano Today, 2013, 2: 119-120
[6] Giannatsis J, Dedoussis V. Additive Fabrication Technologies Applied to Medicine and Health Care: A Review. Int J Adv Manuf Tech, 2009, 1-2: 116-127
[7] PAN Hai-peng ZHOU Tian-rui HU Shi-fei et al. Implementation of the visualization for STL model. Journal of Plasticity Engineering, 2006, 6: 94-98
[8] WANG Congjun ZHANG Lichao HUANG Shuhui. A New STL Data Compressed Storage Format. china mechanical engnieering, 2001, 5: 558-560
[9] LI Jiang-feng , ZHONG Yue-xian, LI Dian-sheng. Research on errors identifying and repairing of STL file. Machinery Design & Manufacture, 2002, 2: 40-42
[10] Hong Taoqing. A Discussion on the Uniform Equation of the Cylindrical Section, Journal of Lishui University, 2008, 05: 72-74
[11] XIAO Renxian ZHANG Junjian FENG Hao et al. STL Files Reading, Displaying and Manipulating. Journal of Ceramics, 2011, 02: 285-288
[12] DING Dian-kun, BIAN Ping-yong. The Parameter Equation of Space Curve's Projection on the Plane. Journal of Luoyang Teachers College, 2005, 5: 23-24
[13] Li Busong, Zhou Gang, Wang Congjun et al. Realization of arithmetic for dividing solid model based on STL file format. Journal of Huazhong University of Science and Technology(Nature Science), 2002, 3: 40-42
[14] Xie Cunxi, Li Zhongyang, Cheng Xiaoyang. Research on the Establishment of adjacency relation of STL file and its slicing algorithm. Journal of South China University of Technology, 2000, 3: 33-38
[15] Cline A K, Renka R J. A Storage-Efficient Method for Construction of a Thiessen Triangulation. University of Texas at Austin, 1982. 119-140
[16] Fang T P, Piegl L A. Algorithm for Delaunay Triangulation and Convex-Hull Computation Using a Sparse Matrix. Comput Aided Design, 1992, 8: 425-436
[17] Zhou Xiaoyun, He Dazeng. The Algorithm of Triangulation of Scattered Data Points on the Plane. Journal of Computer-Aided Design and Computer Graphics, 1994, 4: 256-259
[18] Zhao Dongmei, Zhang Jialei. Complex multiplication and rotation matrix. Journal of Chifeng Unimersity, 2011, 6: 22-23